CAMBRIDGE LIBRARY COLLECTION

Books of enduring scholarly value

Life Sciences

Until the nineteenth century, the various subjects now known as the life sciences were regarded either as arcane studies which had little impact on ordinary daily life, or as a genteel hobby for the leisured classes. The increasing academic rigour and systematisation brought to the study of botany, zoology and other disciplines, and their adoption in university curricula, are reflected in the books reissued in this series.

Tutira

In 1880, William Herbert Guthrie-Smith (1862–1940) emigrated from Scotland to New Zealand, where he learned the basics of sheep farming and acquired Tutira, a disused sheep station of 20,000 acres in the Hawke's Bay region of the North Island. *Tutira*, published in 1921, describes every aspect of Guthrie-Smith's enterprise, including the redevelopment of the land and comprehensive advice on sheep farming. The book also covers the history of the local Maori and of European settlement, and provides an extensive account of the farm's natural history including its geological configuration, meteorological patterns, the formation of lakes and waterways, and the native plant and bird species Guthrie-Smith discovered on his land. It also draws attention to the impact of introduced, 'alien' plants and animals. *Tutira* is one of the great classics of New World environmental consciousness; it was reprinted in 1926, and a posthumous revised edition appeared in 1953.

Cambridge University Press has long been a pioneer in the reissuing of out-of-print titles from its own backlist, producing digital reprints of books that are still sought after by scholars and students but could not be reprinted economically using traditional technology. The Cambridge Library Collection extends this activity to a wider range of books which are still of importance to researchers and professionals, either for the source material they contain, or as landmarks in the history of their academic discipline.

Drawing from the world-renowned collections in the Cambridge University Library, and guided by the advice of experts in each subject area, Cambridge University Press is using state-of-the-art scanning machines in its own Printing House to capture the content of each book selected for inclusion. The files are processed to give a consistently clear, crisp image, and the books finished to the high quality standard for which the Press is recognised around the world. The latest print-on-demand technology ensures that the books will remain available indefinitely, and that orders for single or multiple copies can quickly be supplied.

The Cambridge Library Collection will bring back to life books of enduring scholarly value (including out-of-copyright works originally issued by other publishers) across a wide range of disciplines in the humanities and social sciences and in science and technology.

Tutira

The Story of a New Zealand Sheep Station

H. Guthrie-Smith

CAMBRIDGE UNIVERSITY PRESS

Cambridge, New York, Melbourne, Madrid, Cape Town,
Singapore, São Paolo, Delhi, Tokyo, Mexico City

Published in the United States of America by Cambridge University Press, New York

www.cambridge.org
Information on this title: www.cambridge.org/9781108040013

This edition first published 1921
This digitally printed version 2011

ISBN 978-1-108-04001-3 Paperback

The original edition of this book contains a number of colour plates, which have been reproduced in black and white. Colour versions of these images can be found online at www.cambridge.org/9781108040013

TUTIRA

The Story of a New Zealand Sheep Station

Lakes Waikopiro and Tutira.

TUTIRA

The Story of a New Zealand Sheep Station

BY

H. GUTHRIE-SMITH

"Pinea rawatia ki Tutira ra;
Ki te ue pata, ki te kai rakau.
A ehara e hine i te roto hou;
He roto tawhito tonu no matou ko o nui.
Ina tonu te raro i potau atu ai e hine."

William Blackwood and Sons
Edinburgh and London
1921

TO

MAJOR-GENERAL WILLIAM SMITH,

LATE R.A.

PREFACE.

So vast and so rapid have been the alterations which have occurred in New Zealand during the past forty years, that even those who, like myself, have noted them day by day, find it difficult to connect past and present—the pleasant past so completely obliterated, the changeful present so full of possibility. These alterations are not traceable merely in the fauna, avifauna, and flora of the Dominion, nor are they only to be noted on the physical surface of the countryside: more profound, they permeate the whole outlook in regard to agriculture, stock-raising, and land tenure.

The story of Tutira is the record of such change noted on one sheep-station in one province. Should its pages be found to contain matter of any permanent interest, it will be owing to the fact that the life portrayed has for ever vanished, the conditions sketched passed away beyond recall. A virgin countryside cannot be restocked; the vicissitudes of its pioneers cannot be re-enacted; its invasion by alien plants, animals, and birds cannot be repeated; its ancient vegetation cannot be resuscitated,—the words "terra incognita" have been expunged from the map of little New Zealand.

In regard to the construction of the volume, when the writer first found himself in the family way—as authors wish to be who love their books—his intention was only to have attempted the natural history of the run. As, however, he proceeded, chapters on physiography, native life, pioneer work, and surface alterations have been added, and the volume thus increased to its present bulk. Every subject treated has been treated deliberately from a local point of view. Pererration beyond the marches of the run has never willingly been indulged; the writer lays no claim to other than local knowledge on any one of the subjects treated. No apology, therefore, is offered for the microscopic size of the canvas. Nor again is apology offered for apparent egotism in chapters devoted to the stocking of the run with man. The early failure of *homo sapiens* on Tutira, his ultimate acclimatisation, has been noted, as far as may be, in terms of the weasel or rabbit; he has been treated without fear or favour as a beast of the field. First and last, then, 'Tutira' is a record of minute alterations noted on one patch of land: for the author's purpose, indeed, New Zealand is bounded on the west by the Mohaka river, on the east by the Arapawanui run, on the south by the Waikoau, on the north by the Waikari.

Every man has his idiosyncrasy: it has been that of the writer for half a lifetime to note small things; it has interested him. Perhaps, therefore, there may be found, if not a hundred, then haply ten righteous men to share that interest—to read, mark, learn, and inwardly to digest the subcutaneous erosion of a countryside, the ancient way of the Maori, the fortunes of pioneer man and beast, the acclimatisation of an alien flora and fauna, the disappearance of the squatter, the rise of the bold yeoman in his stead.

THANKS.

MY thanks are due to Miss Beatrix Dobie for her physiographical sketches, and for her careful and accurate restorations of the old-time *pas* of the station. I consider myself most fortunate in having secured her services. I should like here also to acknowledge indebtedness to Mr T. F. Cheeseman, of the Auckland Museum, and to Mr W. W. Smith, of the New Plymouth Botanical Gardens, for assistance in the nomenclature of plants. I have also to thank Mr Percy Smith, the erudite editor of the 'Polynesian Journal.' Although an exceedingly busy man, he has found time to transcribe and correct papers written from time to time for me by native friends. Lastly, as a sheep-farmer in search of truth, my case has been commiserated, my presence condoned, in many museums and libraries of the Old World. I take this opportunity of reiterating thanks to many learned men for gifts of valuable time.

CONTENTS.

CHAPTER I.

TUTIRA—ITS PROMINENT PHYSICAL FEATURES.

CHAPTER II.

ROCK CONSTITUENTS OF THE RUN.

CHAPTER III.

THE LAKES.

CHAPTER IV.

THE SOILS OF TUTIRA—PAST AND PRESENT.

CHAPTER V.

SUBCUTANEOUS EROSION.

CHAPTER VI.

SURFACE SLIPS.

CHAPTER VII.

THE FOREST OF THE PAST.

CHAPTER VIII.

TWO PERIODS OF MAORI LIFE.

CHAPTER IX.

TRAILS FROM THE COAST TO TUTIRA.

CHAPTER X.

TRAILS ROUND TUTIRA LAKE.

CHAPTER XI.

THE TRAIL TO THE RANGES.

CHAPTER XII.

VEGETATION OF THE STATION PRIOR TO SETTLEMENT.

CHAPTER XIII.

THE FERNS OF TUTIRA.

CHAPTER XIV.

THE AVIFAUNA OF THE STATION PRIOR TO SETTLEMENT.

CHAPTER XV.

IN THE BEGINNING.

CHAPTER XVI.

THE LURE OF IMPROVEMENTS.

CHAPTER XVII.

HARD TIMES.

CHAPTER XVIII.

THE RISE AND FALL OF H. G.-S. AND A. M. C.

CHAPTER XIX.

FERN-CRUSHING.

CHAPTER XXVI.

GARDEN ESCAPES.

CHAPTER XXVII.

CHILDREN OF THE CHURCH.

CHAPTER XXVIII.

BURDENS OF SIN.

CHAPTER XXIX.

FIRE AND FLOOD WEEDS.

CHAPTER XXX.

PEDESTRIANS.

CHAPTER XXXI.

THE STOCKING OF TUTIRA BY ALIEN ANIMALS.

CHAPTER XXXII.

OTHER ALIENS ON TUTIRA PRIOR TO 1882.

CHAPTER XXXIII.

ACCLIMATISATION CENTRES AND MIGRATION ROUTES.

CHAPTER XXXIV.

THE INVASION FROM THE SOUTH.

CHAPTER XXXV

THE INVASION FROM THE NORTH.

CHAPTER XXXVI.

DOMESTIC ANIMALS "WILD."

CHAPTER XXXVII.

RECONSIDERATIONS.

CHAPTER XXXVIII.

VICISSITUDES

LIST OF ILLUSTRATIONS.

FULL-PAGE ILLUSTRATIONS.

ILLUSTRATIONS IN TEXT.

MAPS.

TUTIRA.

CHAPTER I.

TUTIRA—ITS PROMINENT PHYSICAL FEATURES.

TUTIRA STATION is situated in the Hawke's Bay province of the North Island of New Zealand. The homestead itself lies a few miles inland midway between the ports of Napier and Wairoa. Tutira proper extends over 20,000 acres—about one-third of the size of the lands to be described; lands which have at one time or another been occupied by the writer. The larger area is bounded by three considerable rivers. The largest, rising in the interior of the North Island, flows along the base of the Maungaharuru range, eventually reaching the sea twenty miles north of the run. There may be here and there a crossing to this deep, swift, and dangerous river. I know of none. Another river, running from source to sea between cliffs, is impassable except where crossings have been constructed in modern times. The third, rising on the high lands of inland Tutira, is crossable at the old pack-horse ford, where in the 'nineties a "cage" on wires was slung for the convenience of degenerate modern wayfarers, and where in still more recent times, for still more degenerate travellers, a bridge has been thrown across the river. There is another ford nearer the sea where the ancient Maori foot-trail passed inland, otherwise this stream also was practically uncrossable until beyond the Tutira boundary.

To account for the external configuration of the run and for the material of which it is built, vast general changes over the whole east coast of the North Island—over the whole of New Zealand indeed—would have to be considered. For such a review the writer lacks both knowledge and space. Certain main facts, however, can be accepted on authority. The first is, that the bay of the province extending now between Cape Kidnappers and the Mahia Peninsula has been, within a comparatively recent geological period, high and dry; the second, that this vast "half-moon, this monstrous cantle" of land, has sunk, and that simultaneously with its subsidence there has been a general fall of the coastal area towards the east—towards the ocean.

In regard to local geology there has been no special inducement for detailed study of the Hawke's Bay district. It is the land of the Golden Fleece, rich only in flocks and herds. There exist in the province neither oil, coal, iron, nor gold to stimulate minute research.

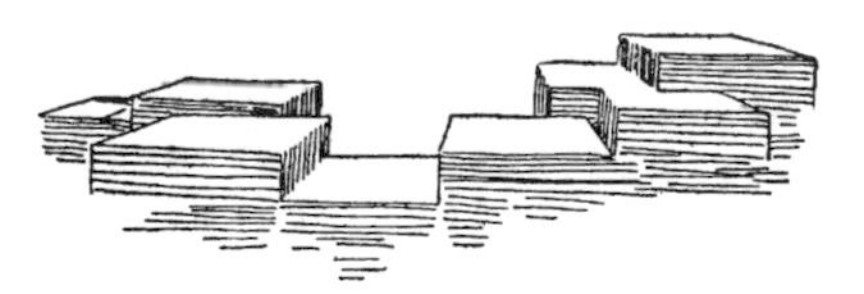

Sections of original plateaux.

Tutira and the adjoining lands have, I imagine, risen from the ocean as plateaux of different heights. It is probable that during the subsidence of what is now the bay of the province this formation was altered. At its termination sections of the original plateaux lay on their edges inclining to the east: the countryside had changed from an agglomeration of elevated plains to a series of tilted terraces. Consentaneously the hill chains of the run must have been created. In sympathy with the tilting process there must have taken place an increase in the height of each of them; as the one edge sank, the other rose, until the run assumed approximately its present outlines.

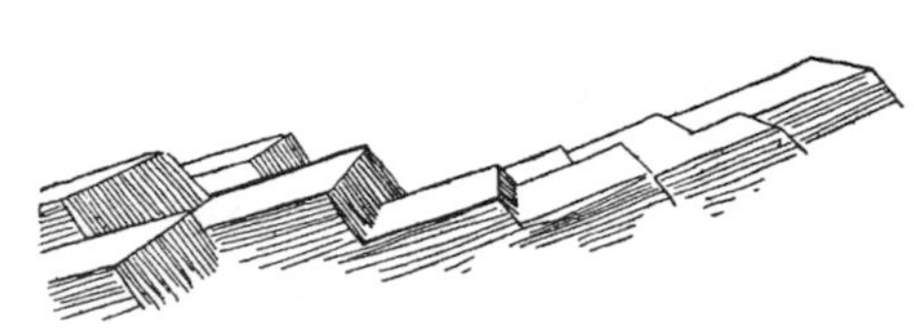

Plateaux tilted.

Situated between two parallel chains of hills, and containing a centre of low-lying lands, the station may in shape be compared to an elongated trough. The western edge of this trough is the Maungaharuru range, reaching the height of 3200 feet, and containing minor eminences of over 2000 feet; the eastern edge, the Newton range, is considerably lower in elevation, its highest top not rising above 1400 or 1500 feet. The physical appearance of Tutira is in the main that of the adjacent regions north and south, but the general geological features noticeable on them are on Tutira marked in a peculiarly definite manner. One pattern only, sometimes sharp and sometimes

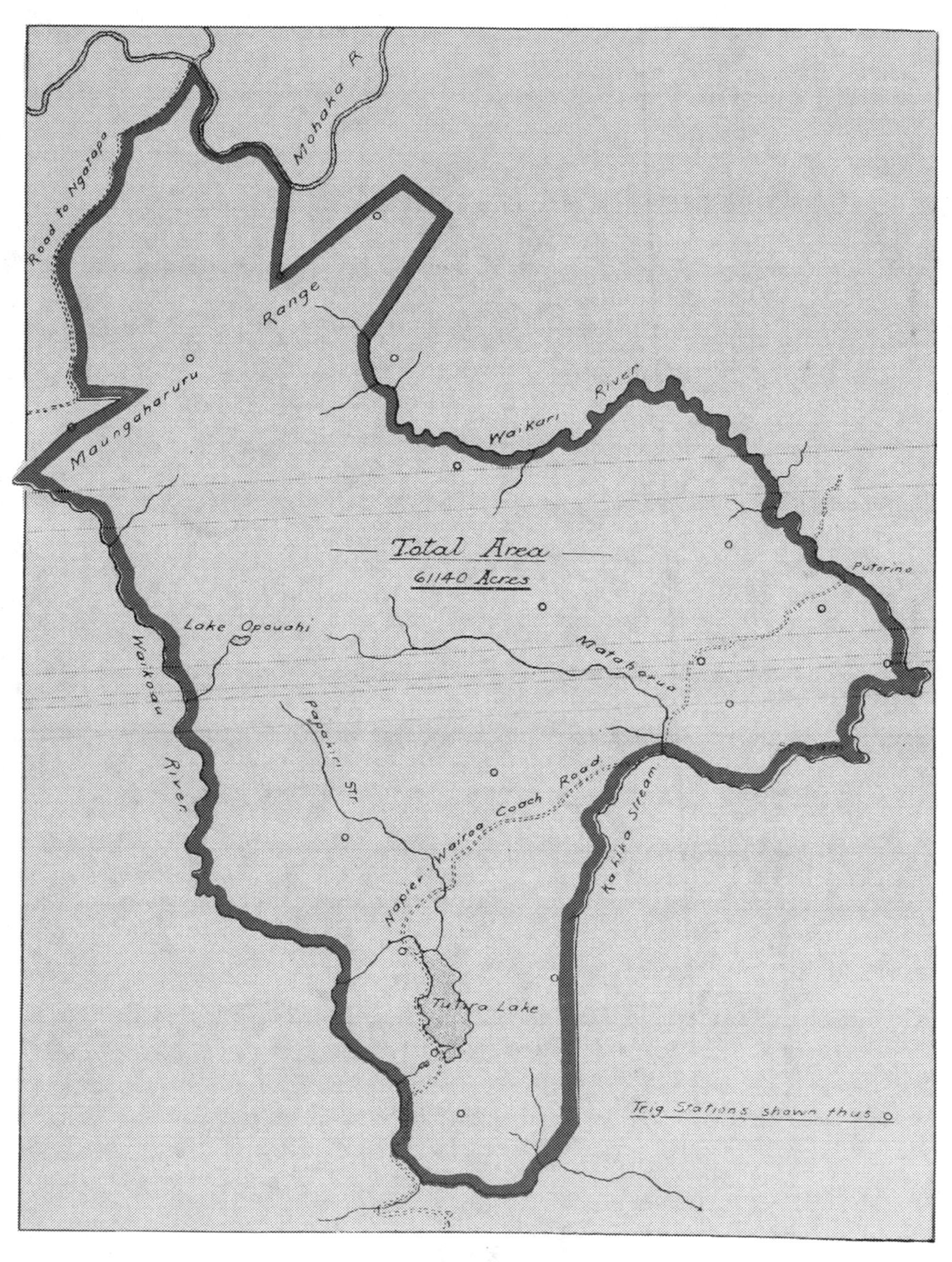

TOTAL ACREAGE LEASEHOLD AND FREEHOLD FARMED BY WRITER.

blurred, obtains, or has obtained, over the whole of the run. It is very simple: every range, or portion of range, runs north and south, the western face of every range is precipitous, the eastern face of every range falls away gradually, the eastern face of every range is split by fissures, the sides of every fissure are perpendicular.

The native name of one of these ranges, Heru-o-Tureia, the comb of Tureia,[1] admirably illustrates the general geological pattern of the run—the unbroken line of top, the western cliff, the fissured eastern slope. It is typical on a great scale of every hill and mountain chain on Tutira; indeed, if the signification of the name be firmly grasped, the reader will hold in his mind an easy key to the physical outlines of the run. The long, even, unbroken ridge itself is the "back" of the titanic comb, the spurs running at right angles from it the "teeth," the cracks which never penetrate the solid summit or "back," and which, therefore, never completely bisect the range, the "interstices" between the teeth. The likeness of these geological formations to vast combs is still further heightened by the even, perpendicular edges of the "teeth."

These are the features of the comb system broadly outlined to let the reader visualise its strange cleavage pattern. A modification must now, however, be noted; it is this, that although the fissures start at right angles to the main range, their sides do not remain parallel. These gaps, their shape at base more or less that of an inverted horse-shoe, widen as their distances from the back of the comb increase. Another minor modification of the parallel hill chain system is the presence here and there of narrow linking spurs that jut forth east and west as if wedding the ranges to one another. Running at right angles to the general north and south trend of the ranges, though infrequent, they are well-marked features in the landscape.

The breadth of the run can be traversed and its surface viewed if, in imagination, the reader will take up his position on the western-most rim of the trough to which the whole station has been compared, and proceed thence eastwards towards the ocean. The Heru-o-Tureia range marks the limit of limestone and divides the sandstones, conglomerates, and marls of the coastward belt from the more ancient slates and ryolites of the interior. Moving from it eastwards we shall

[1] Tureia was sixth in descent from Tamatea, who reached New Zealand in the Takitimu, one of the fastest of the canoes of the great heke or migration from Hawaiki.

pass over successive lines of hills till we arrive at the Newton range—the easternmost edge of the trough. It too illustrates the prevailing feature, the continuity of top and upright rock rampart facing west, the "back" of the "comb," its cloven spurs sloping towards the east, the "teeth."

Everywhere, therefore, on Tutira we discover one pattern, one principle, one type of formation dominant; we find furthermore throughout the run narrow "tooth" valleys enclosed by perpendicular walls—valleys which may deepen but which never can expand, and out of which over the bulk of the run no water whatsoever visibly flows. The precipices containing them form what may be called, for convenience sake, the dry cliff system of the run.

Blue Duck on Waikoau river.
(This and other sketches from photographs taken by H. G.-S.

Strongly contrasting with it exists another which may be equally well termed the wet cliff system. Unlike the former, its sculpturing offers no difficulty to the imagination. It has been cut out by processes which are still at work. Its streams still chiselling out their beds flow far beneath the surface. Its cliffs, saturated with moisture percolating through the pervious soils above, are from top to bottom feathered with ferns and delicate greenery. There are in fact two conspicuously distinct series of cliff—the rock walls of the one dry, bare, and prominent; the rock walls of the other damp, densely overgrown, and, until closely approached, invisible.

Besides the boundary rivers named, there are within the compass of the station several streams of lesser volume, also flowing between narrow perpendicular walls; the only stream, indeed, not imprisoned by precipices during its whole length is the Papakiri, which ends its career in Tutira lake. Every upland lake is a sea to the rivers that feed it; to the Papakiri the lake is the ocean of its extinction. For the same reason also that the Waikoau curbs the speed of its current and deposits its silt upon approach to the Pacific, the Papakiri drops its burden of soil as it nears Tutira lake. Save for a mile or so in the course of this little river, and the equally brief run of a few brooks on the uplands of Opouahi, the drainage system of the station has to be searched for. It lies beneath the level. Barring the two or three miles that march with Arapawanui, the boundaries of the run are wet cliff; except on the Newton range, the paddocks are enclosed by wet cliffs. Within each paddock are wet cliffs; within many of them are miles of wet cliff. There are in addition miles of dry cliff in almost every one of these natural enclosures. The reader will not grasp the coming story of Tutira if he fails to understand that there are, wet and dry, several hundred miles of precipice on the run, varying in height from 20 to 150 feet.

Other prominent natural features of the station are its water surfaces. Of these the largest is Tutira lake, next in size is Waikopiro—the two, conjoined in wet weather, covering some 500 acres. Within a couple of chains distance from the last-named, and at a lower level, is situated Orakai, five or six acres in extent. There is a lakelet, Opouahi, of about similar size, on the uplands of the west; a deep clear lakelet, Te Maru, on Putorino, and a couple of tarns on Heru-o-Tureia. Tutira lake, about two miles long, resting at the foot of the Newton range, is drained by a meandering serpentine creek of the same name, which, after crossing the old Maori foot-trail, breaks into a series of overfalls, and finally leaps at Te Rere-a-Tahumata into a magnificent chasm of 157 feet in depth.

As in the shaping of the run water has played so prominent a part, it will be well in this initial chapter to devote a few lines to the rainfall. The heaviest deluges are blown up from the north-east, east, south-east, south, and south-west. During some three or four days' duration, not infrequently one foot and over, and on one occasion nearly two feet, have been registered. Except in the form of showers, rain seldom reaches Tutira in appreciable quantity from the north and west. Thunderstorms, which cling to the coast and the ranges, the station almost entirely escapes. Snow falls but rarely—only thrice in my time has it lain for more than a few hours; during one of these blizzards, however, it certainly fell in the same whole-hearted manner as have done the greatest of the rain-storms. Everywhere on the low lands two feet deep, it lay still thicker on the Newton range, completely blotting out the sheep for a couple of days.

The rainfall of eastern Tutira is different in character from that of

RECORD OF RAINFALL FOR 1917 AT TUTIRA.

Height above Mean Sea-level, 500 *feet. Hour of Observation,* 9 *a.m.*

Date.	Jan.	Feb.	Mar.	April.	May.	June.	July.	Aug.	Sept.	Oct.	Nov.	Dec.
1	·83	...	...	·28	...	...	...	...	·07	...	·54	·20
2	2·70	...	·32	...	...	...	...	...	·15	...	·05	...
3	3·94	·57	·10	...	·01	...	...	...	·35	...	·02	...
4	...	·09	...	...	1·65	...	...	·26	...	·52	...	...
5	...	·03	·03	...	...	...	...	3·40	...	...	...	...
6	...	...	...	...	·08	...	·11	·36	...	...	...	...
7	...	...	·05	...	...	...	·05	·24	·44	...	...	...
8	·18	...	·08	...	...	...	...	1·91	·08	·16	·15	...
9	2·23	...	·55	...	...	...	...	1·67	·04	...	...	...
10	·02	...	·27	·03	...	1·70*	·01	·01	·08	·35	·04	...
11	...	...	·04	·98	...	8·40	·02	...	2·92	·14	...	...
12	...	·12	·09	·02	2·95	8·40	·02	...	2·44	...	...	...
13	...	·30	·07	...	6·80	1·61	...	...	·02	...	...	...
14	...	·39	·16	...	·47	...	...	1·63	...	...	·36	...
15	...	...	...	...	...	...	·14	·13	...	...	...	...
16	...	...	...	·86	...	...	·16	...	...	...	·14	...
17	...	...	...	·08	...	...	...	...	...	...	·01	...
18	...	1·68	...	...	...	...	·11	·10	...	·27	...	·85
19	...	·61	...	...	...	...	...	·22	...	...	...	·60
20	...	·73	...	...	...	...	·57	...	...	...	1·14	...
21	·10	1·80	...	...	...	...	...	...	...	...	1·88	...
22	·66	·03	...	...	...	...	·11	·10	...	...	·93	·17
23	...	·55	...	...	...	·24	·01	·12	...	...	·16	...
24	...	...	...	·01	...	·29	...	...	...	·01	...	...
25	...	...	...	...	·25	...	...	·01	...	·46	...	...
26	...	...	...	·27	...	·15	·02	...	·01	...	...	...
27	·12	...	·11	...	·01	...	·27	...	...	...	·06	·33
28	...	...	...	...	...	1·30	·09	...	·04	...	...	...
29	...	...	·08	·02	...	...	·33	...	...	...	...	·08
30	...	...	...	·02	...	...	...	...	...	...	...	...
31	·06	...	...	...	...	...	...	...	...	·01	...	...
Total.	10·84	6·90	1·95	2·57	12·22	22·09	2·02	10·16	6·64	1·92	5·48	2·23
Number of days.	10	12	13	10	8	8	15	14	12	8	13	6
	Jan.	Feb.	Mar.	April.	May.	June.	July.	Aug.	Sept.	Oct.	Nov.	Dec.

Total for Year, 85·02 *inches.*

The reader will note the great variations in rainfall of these two years—years, doubtless, exceeded both in minimum and maximum by others whose

* The storm of 10th, 11th, 12th, and 13th of June registered just over 20 inches. What weight of water may have fallen in addition we cannot tell, for on the 11th and 12th the rain-gauge was found to be filled and overflowing.

RECORD OF RAINFALL FOR 1919 AT TUTIRA.

Height above Mean Sea-level, 500 *feet. Hour of Observation,* 9 *a.m.*

Date.	Jan.	Feb.	Mar.	April.	May.	June.	July.	Aug.	Sept.	Oct.	Nov.	Dec.
1	...	...	...	...	...	...	...	·15	...	...	...	...
2	...	...	...	...	...	...	·05	...	...	...	·08	·30
3	...	·25	·01	1·20	·02	...	·03	...	·15	·05	...	...
4	...	·03	·02	...	...	...	...	...	...	1·80	·01	...
5	...	...	...	...	...	...	...	·37	...	1·32	·03	...
6	...	...	...	...	...	...	·06	·06	...	·17	·16	...
7	...	...	·43	...	...	·18	...	·14	·04	...	...	...
8	...	...	...	...	...	...	...	·40	...	...	...	...
9	...	...	...	...	·06	...	...	·12	...	...	...	·16
10	...	...	·06	...	...	...	...	...	1·75	...	...	...
11	...	...	...	...	...	...	...	...	1·19	...	...	...
12	...	·50	...	...	...	...	·18	·01	·40	...	...	...
13	...	4·55	...	...	...	...	...	...	·11	...	·16	·26
14	...	...	...	·12	...	...	...	...	...	...	...	·02
15	...	...	...	...	...	·01	·04	...	...	...	...	...
16	...	...	·17	·14	...	...	·23	·28	...	...	·55	...
17	...	...	...	...	...	·01	...	·72	...	...	...	...
18	...	...	...	...	2·15	...	...	...	...	...	...	...
19	...	·01	...	...	·21	·05	...	...	...	...	...	·12
20	...	...	·20	...	·22	...	...	...	...	...	...	·02
21	...	...	...	...	...	·22	...	...	...	...	...	...
22	...	·20	...	...	...	...	...	...	·02	...	...	...
23	...	·27	...	...	...	·09	1·40	...	...	...	...	...
24	...	...	...	·07	·05	·19	·07	...	...	...	...	...
25	...	...	...	...	...	1·78	·48	...	·01	...	...	...
26	...	...	...	...	·11	·53	·42	...	...	...	...	·11
27	...	...	...	...	·20	...	·32	...	·01	...	...	...
28	...	...	...	·01	...	...	·62	1·26	...	...	...	...
29	·76	...	...	...	·02	...	1·10	·87	...	·07	·47	...
30	·07	...	...	...	·25	...	1·95	·35	...	...	...	...
31	...	...	...	...	...	...	2·25	...	...	...	...	...
Total.	·83	5·81	·89	1·54	3·29	3·06	9·20	4·73	3·68	3·41	1·46	·99
Number of days.	2	7	6	5	10	9	15	12	9	5	7	7
	Jan.	Feb.	Mar.	April.	May.	June.	July.	Aug.	Sept.	Oct.	Nov.	Dec.

Total for Year, 38·89 *inches.*

records are not available. In the one, 85·02 fell in 129 days; in the other, 38·89 fell in 94 days. As, however, showers of under 10 are immediately dried up in a climate like that of Hawke's Bay, practically there were but 84 wet days in 1917, and 60 wet days in 1919.

the interior; deluges that break on the coastal hills do not, or at any rate do not always, reach inland. When on one occasion over seventeen inches in two days were measured on the homestead lawn, there was not on the track below the "Image" hill, distant some three miles from the stance of the rain-gauge, enough rain to wash away the dust from the trampled stock route.

Rain on the western rim of the run falls in the form of frequent showers, or when a coastal deluge does reach the ranges of the interior, it falls with an attenuated precipitation. Generally speaking, not only is the rainfall of Tutira about double that of Napier, only ten or twelve miles distant as the crow flies, but there is a sapidity in the atmosphere, perhaps owing to the enormous quantity of deep gorges, increasingly noticeable inland. When southern Hawke's Bay is brown, the hills of Tutira are often green as leeks. Never in forty years have I known grass slopes fronting south or east fit to carry a fire; not more than half a dozen times have I seen hillsides facing north and west burnt brown.

These details of rainfall have been given not merely as meteorological data of an impersonal sort; the climate of Tutira has deeply affected the fortunes of the station. As the reader will later be shown, excessive rainfall has been the bane of the place, retarding its development by years.[1]

To recapitulate: The original shape of Tutira has been either a single elevated plain, or more probably a series of plateaux. At a later period, owing to extraneous subsidence, this terrain has tilted towards the east, leaving the station approximately in its present form—a series of eastward-facing slopes, with a precipitous back to each of them. Consentaneously the many hill-chains of the run have come into being, not through upheaval from beneath, but by the one edge of the plateau rising as the other dipped. The outstanding physical features of Tutira, its lakes, its waterfall, its hidden rivers, its double cliff system—the dry, remarkable for its far-seen alternating bands of strata, the wet, for its hanging curtains of fern—have been described. Lastly, the reader's attention has been drawn to the rainfall of the run.

[1] Not all records are published. One observer whose case I recall was requested by neighbours to cease to forward his returns. "Science may be right enough, perhaps, in its proper place," they declared, "but he was ruining the district and hampering settlement with his blessed rainfalls."

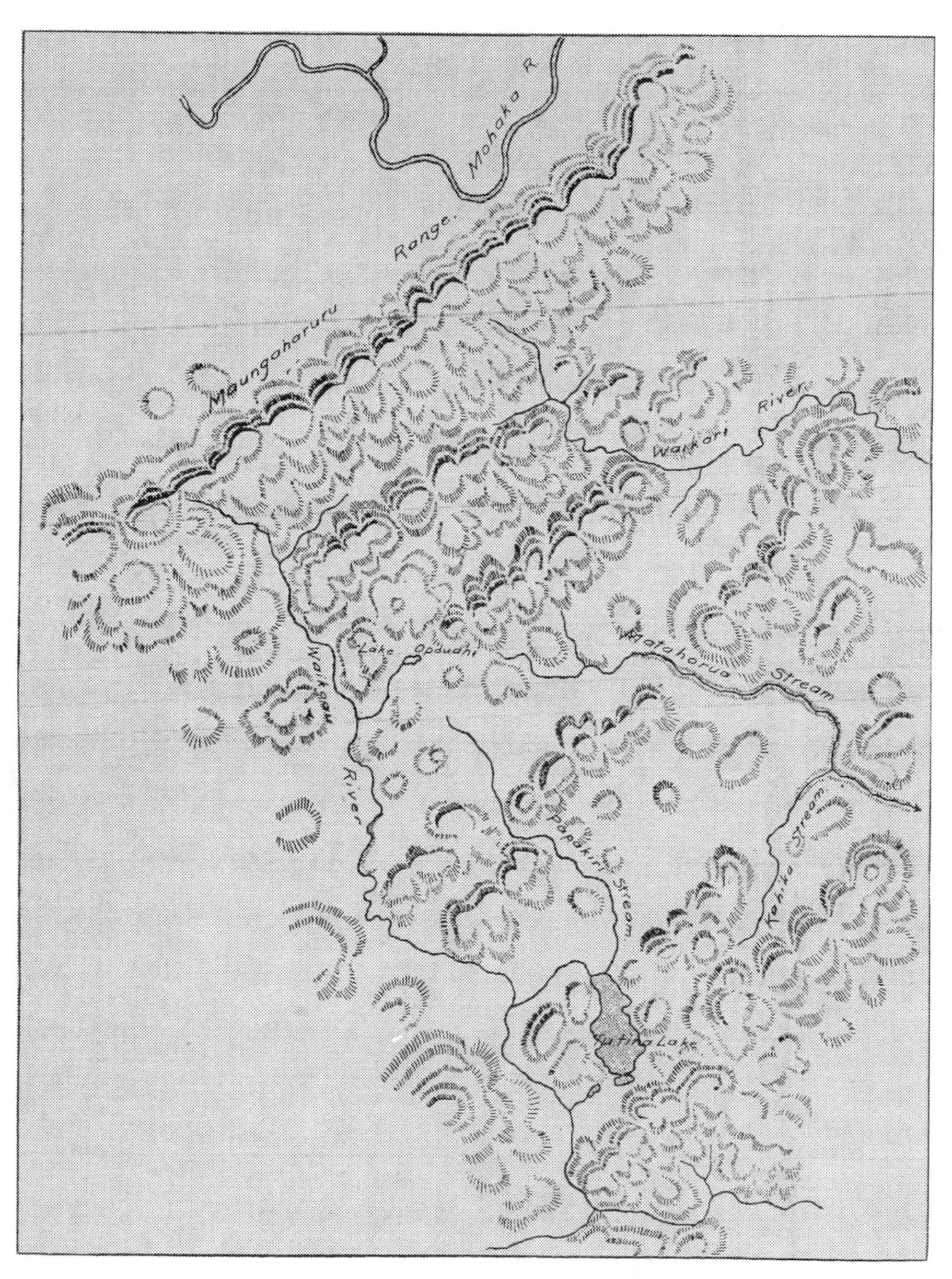

"COMBS" OF WEST, CENTRE, AND EAST TUTIRA.

CHAPTER II.

ROCK CONSTITUENTS OF THE RUN.

THE materials of which the station is formed are marl or "papa," sandstone, sandy marl, limestone, and conglomerate. Except the last, they are obviously of marine origin. Marl is the foundation on which the others lie,—it is the bed-rock of Tutira. Whatever may have been its origin, the remains of an ancient southern continent or not, its constituents seem to have been carried by ocean currents or tidal action. Deposition has been, at any rate, intermittent, not constant. Undulatory lines of sand-grit can be traced on cliffs where flaking is constant and where consequently exposures are clean.[1] Although so faint as to be only decipherable in certain lights, like patterns in watered silk, they nevertheless mark brief periods of quiescence as surely as the grosser pelagic accumulations outcropping elsewhere on the run. The immensity of these deluges of mud can be gauged by sand lines sometimes yards apart. Their close recurrence can be inferred by interpolations of sand so thin as to be practically invisible.

The marls of Tutira vary in fertility and mode of weathering, the least fertile being the most homogeneous and compact, the most fertile those that disintegrate in cubes or exfoliate in peelings.[2]

We can now consider rock formations superposed on this base of marl. To do so it will be convenient that in imagination the reader should as before take his stand on the Maungaharuru range and again traverse the run from west to east.

The great dry bastion on the west is built up of alternate layers

[1] Such lines are particularly noticeable on the beach cliffs between Waikari and Mohaka.

[2] Through the kindness of Dr Lauder, of the Edinburgh and East of Scotland College of Agriculture, a station sample of marl has been analysed. It "contained 34 per cent of calcium carbonate (chalk), fairly large quantities of the oxides of iron and aluminium; smaller quantities of calcium and magnesium oxides. Phosphates were present, but no potash."

of sandstone and limestone, the result, doubtless, of floodings of sand over shell-strewn ocean beds. Each band is of a different thickness. Each slightly differs also in material, the sandstone sometimes less and sometimes more dry, the limestone bands varying in the nature of their shell fragments. Where, as occasionally happens, the topmost sea-floor has been stripped by gales of its earthy coverings, cleavage is revealed in the form of almost exact squares, in a mosaic of grey blocks of limestone set in packed red sand.[1]

Continuing to traverse the run, moving eastwards, we reach central Tutira. Hereabouts limestone disappears, sandstone and conglomerate resting on the marl, the sand of the former seemingly ground out of gravel, the pebbles of the latter rolled, worn, and of a generally ovoid form. In these conglomerates and sandstones, fossils are rare. In the first I have found but a single specimen, a short length apparently of some knotless tree stem; in the second, one or two kinds of bivalves. Usually the weathering of the alternate bands of a cliff face proceeds evenly. Sometimes, however, it happens that in a loose type of sandstone erosion by frost and wind is rather more pronounced than on the conglomerates above. When that occurs the cliff assumes a protuberant air, a curious rotund or pot-bellied appearance. Normally the conglomerates of the central run cap the sandstones. Lacking their protection, the softer rock has melted into cones, domes, razorbacks, and peaks.

Proceeding once more from the west to the east we reach low lines of hills, eminences hardly more than hummocks. What is visible of their rock material differs but little from the earlier mentioned conglomerates; their colour is rusty-red instead of grey, they do not appear to be so thoroughly set, they can be worked with a pick, almost with a shovel. There is perhaps rather less variation in the size of their pebbles.

Again advancing on our traverse of the run, crossing the lakes we reach the eastern range of Tutira—the eastern edge of the trough. Once again the rock formations change: the typical western facing cliff is built up here of limestones, sandstones, and marls, irregularly superposed one upon another in bands of different depths. Compared with the sandstones and limestones of the west, the sandstones and lime-

[1] A very perfect example of such a pattern may be seen on the great wind-blow of the Maungaharuru range.

stones of the east show fresher and fuller signs of marine life. Their sand is more truly sea sand. Their limestones contain a larger number of shell-species closely connected, or actually identical with those now existing in the pertingent seas. Above several of the limestone bands are shallow deposits of a highly friable ferruginous matter. On the cliffs of the east, too, there supervenes a new and most important factor. In this region the marl bands interpolated at irregular intervals, rapidly crumbling, sap and undermine the more durable strata; the cliffs are somewhat less sheer, the strata exposed having weathered less evenly. From the sea-floors thus sapped great quadrilaterals—rock cleavage

Sea-floors on Eastern Tutira.

is similar to that of the western limestones—have already dropped out, whilst others projecting into space also appear to be about to break away.

To recapitulate: The base of the whole station is blue marl rock or "papa." Superposed on it rest the ranges of the west formed of sandstone and limestone, the ranges of the centre formed of compressed sandless pebble-rock and slaty sands, the ranges of the east built up of layers of marl, sandstone, and limestone.

Of these masses of material, the sandstones and conglomerates extend over nine-tenths of the station and are of little value. Such elements of fertility as exist are contained in limestone outcrops, in masses of travertine, but especially in elevated streaks of marl.

CHAPTER III.

THE LAKES.

EXAMINATION of the general physical features of the run has shown a series of parallel hill chains running north and south. Traversing the run from its inland boundary, the reader has crossed them one by one and seen the comb formation, though becoming less and less marked in height, persist through the west and throughout the trough of the run. A continuity of these features might have been looked for, especially as after a brief interruption they reappear in eastern Tutira. Instead we come upon a sheet of water still, after the depositions of centuries of alluvium, ninety feet in depth, two miles in length, and half a mile in width. In lieu of the normal narrow gorge there occur a series of immense hollows, one of which, the "big" swamp, is already filled with alluvium; whilst others still exist as lakes Tutira, Waikopiro, and Orakai.

It is at the base of the Newton range, where the conglomerates of the central run cease and the limestones of eastern Tutira begin, that this change in the plan of the run, this interpolation of a new pattern, occurs. Its presence is an anomaly; it is an extraneous feature to the great general scheme of the station; it can, I think, be accounted for only by processes unlike any yet considered.

Hollows where waters lie may be attributed to erosion, to lodgment of water in craters, to the accumulation of material forming barriers or dams, by subsidence of the crust of the earth.

Except the last, that of erosion is the only theory which might at first seem to fit the facts. The great trough passing through the centre of Tutira,—a trough extending scores of miles north and south of the station,—marked throughout by extensive beds of conglomerate and sandstone, has been described. These deposits, sharply separated from

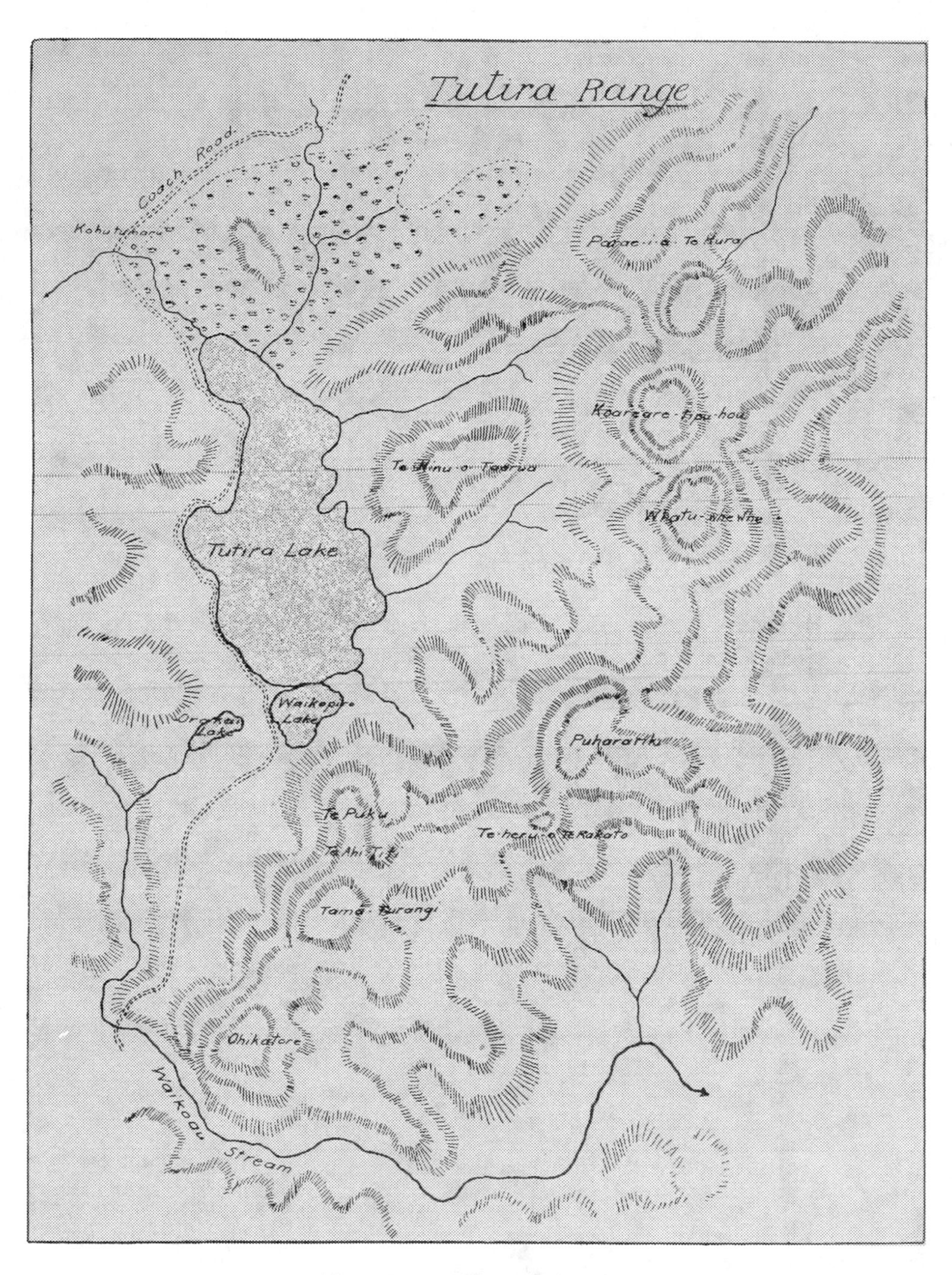

NEWTON OR TUTIRA RANGE.

one another, have presumably been lodged by water action of some sort. In the site of this trough or long hollow a great river might have been imagined to have run at some period, a river which might have left the "Big swamp"—once a sheet of water—and the present lakes Tutira, Waikopiro, Orakai, Opouahi, and Temaru, further to the north Waikaremoana and Waikareiti, further to the south Waipukurau, Te Roto-a-Tara and Wairarapa, as evidences of its former course. It might have been imagined, in fact, that this chain of lakes had been scooped out by some vast old-world river.

There are, however, difficulties in the acceptance of this theory. The sands and conglomerates are not mixed, they are sharply distinct; certainly the latter do not contain that proportion of sand which might be expected in a river-bed. The shape, moreover, of the pebbles suggests neither the grinding of a shelving beach nor the erosion of a running river. In each conglomerate band lie horizontal seams varying in size of stone. Inspection of these seams suggests that their pebbles have been sown in a vast top-dressing, rocked into settlement rather than rushed into position by chance of currents. Perhaps these stones, originally cubes frost-fractured, have been ground by a more remarkable trituration; perhaps their smooth, ovoid form has been acquired by volcanic boilings or tossings. At any rate they are perfectly different from stones of the same material gathered from the top of the Newton range—stones evidently shaped by some little stream that ages ago must have flowed there.

Alternate layers of sand and conglomerate seem in fact to have been laid down very much as on eastern Tutira sands, limestones, and marls have been superposed one on another. Perhaps, indeed, a more profound likeness may be traced—the depths of strata conforming or corresponding to one another on east and west.

There is another difficulty also in regard to the fashioning of the Tutira lake basins by river action: the existence of the coupling spurs already mentioned, spurs here and there linking one range to another. Two of them traverse the breadth of the lakes, one of them at either end of Tutira lake proper, a third barring the way at a slightly higher elevation south of Waikopiro. These coupling spurs, as elsewhere on the run, cannot be reckoned as enormous afterthoughts, as avalanches of rock and soil solidified. They are marl—basic, homogeneous parts of the original scheme of things. Their presence at right angles to the length of the lakes precludes the possibility of river action. Scour sufficiently violent to have scooped out the lake basins must have worn to an equal depth these barriers of solid marl. Their surfaces, now submerged, must moreover have been subject, during a comparatively recent geological period, to superterranean influences such as now obtain elsewhere on the run. They bear evidence, too, that the drainage system ran then as it continues to run. They are mere relics, in fact, of former shelves and terraces, whose material has been worn away when levels were other than they are now. Previous to subsidence they had been sculptured by processes similar to those which have created spurs of kindred shape on the ranges of the west, centre, and east. What, moreover, is true locally of these comparatively small water areas is true of the whole terrain included in the great dip which runs throughout and beyond the length of the Hawke's Bay province. If, indeed, there has been a river flowing at any time north or south along the trough of the run, no signs now remain. It is safe to affirm that since Tutira assumed approximately its modern form the drainage system has been west and east, never north and south.

If, then, the creation of the lake basins is due neither to scour nor to blockage by earthfall of waters once freely escaping, a single possibility remains. Their presence, I believe, is due to subsidence of the crust of the earth — a movement sympathetic with that of the great outside subsidence of what is at present the bay of the province.

Although too much stress need not be placed on local phenomena, it is nevertheless certain that many small facts countenance the belief that Tutira is situated on a line of seismic partiality. During the 'eighties, when the famous Pink and White Terraces were destroyed by the eruption of Tarawera, the waters of Orakai became a dull brownish-green colour, and for many weeks continuously gave forth a strong smell of sulphuretted hydrogen; it has done so on several occasions for shorter periods since that date. The reek of sulphur is distinct, too, in many parts of the gorge of the Waikari. In the Waterfall paddock there is a spring of sulphur water; there is a tepid runnel in the bed of one of the Waikoau tributaries. Earth tremors are frequent, though perhaps not more so than elsewhere in the district. The eastward fall, moreover, of the Newton range seems to be rather less pronounced than elsewhere, as if its base, in close proximity to the lake, had somewhat sunk in local sympathy—that, in fact, the cant of the range had been in some degree readjusted by the subsidence which created the lake basins.

Local weakness in the earth's crust, break in continuity of geological plan, proof in the exposed ocean-floors of many changes of elevation, difference possibly in degree of eastward cant of the range contiguous to the lakes, makes subsidence at any rate a tenable theory in regard to the lakes of Tutira.

Their original depths may be approximately calculated from the present appearance of the ranges on either side and from examination of the cachment area of the Papakiri. The valleys gouged out, the sand and grit borne down, minus that percentage drained off in the form of muddy water, rest at the bottom of one or other of the lake basins. The area covered by Tutira lake has been in past times considerably greater than at present, for the stream that drains away its surplus water has eaten through a bank of sandy marl to the depth of some twenty feet. There remain, nevertheless, no signs whatever of this period—a fact explicable only on the assumption that erosion of the channel of escape must have been at first exceedingly rapid; that it must have been worn through not in years, but weeks, perhaps days.

Had flood-water, containing in suspension mud carried down in landslips from the fertile marls of the Newton range, been deposited on the receptive pumiceous lands which must have been then under water, a distinct vegetation would have arisen. The effect of such submersion would have been apparent still, for after heavy floods the waters of the lake remain wan for months with comminuted clay.[1] Such a top-dressing would, I am confident, have remained distinguishable in its effects to this day. If in no other way, it would have been recognisable by bracken of a deeper green, by manuka of a taller growth.

The conjoined waters of Tutira and Waikopiro at present escape from the north-west corner of the former. Whether they have always done so remains an open question; a 30-foot rise from the present height of the lake would allow its waters to lap over in three directions. These three lines of escape would be—firstly, its present

[1] After the deluge of 1917, when more than 20 inches fell in four sequent days—the rain-gauge twice overflowing—the lake had not regained its usual blue twenty-four months later.

exit; secondly, westwards along the base of the Natural Paddock hill; thirdly, in a southerly direction towards the Race Course flat. A

Tutira lake—as it is.

cant, therefore, quite inconsiderable in a volcanic region would have allowed an escape west or south.

The future of the lakes on present lines of geology is as certain

Tutira lake—as it will be.

as their past is puzzling and obscure. Where their waters now roll, will be steep hillsides and firm ground. Even in my time hundreds of thousands of tons of slips and silt have noticeably filled up the bays Oporae and Kahikanui. The lake is destined ultimately to contract itself into a narrow crooked creek flowing on the west edge of its present formation; for on the west the hill-slopes are less steep, and the slips washed down enormously less in volume. Even this, however, will not be the last change. In imagination we have seen its waters gone, and its basin, through which a narrow streamlet will then flow, completely filled with washings from the hills.

Peering even farther into the future, we shall find not only the lake gone but its very base vanished, and the alluvium, stored for centuries, once more displaced and carried to the sea. Through the

"A long deep valley with arms extending up each of the branch flats, every one of which will have become a gorge."

centre of what once was the lake will then run a long deep valley with arms extending up each of the branch flats, every one of which will have again become a gorge.

At present the lake is drained from its nor'-west corner by the stream Tutira. This stream, after a tortuous course of half a mile through level flax swamp, reaches the old native crossing Maheawha. Immediately below begins a series of overfalls and waterfalls culminating in a leap of over a hundred and fifty feet. This drop is distant some forty chains from the lake, a distance lessened every year by erosion. I imagine that the fall has receded lakewards some two yards since the 'eighties. Exact accuracy is impossible, as the landmarks by which I have tried to gauge wear and tear have themselves moved. There is, however, growing on the stream's edge immediately above the fall, a certain aged kowhai tree whose bole is, I believe, five or six feet farther from the chasm's rim than thirty-seven years ago; the rim has receded that distance. At all events there can be no doubt that the fall is slowly retreating lakewards. Attrition is at present almost imperceptible, yet there are reasons to suppose that under certain circumstances it might become rapid, and that then the alluvial deposits of the lake-basin accumulated during centuries might be washed away in weeks. Because there has been almost no movement for years, it does not follow that such conditions will continue.

Instances of sudden erosion have occurred not infrequently even

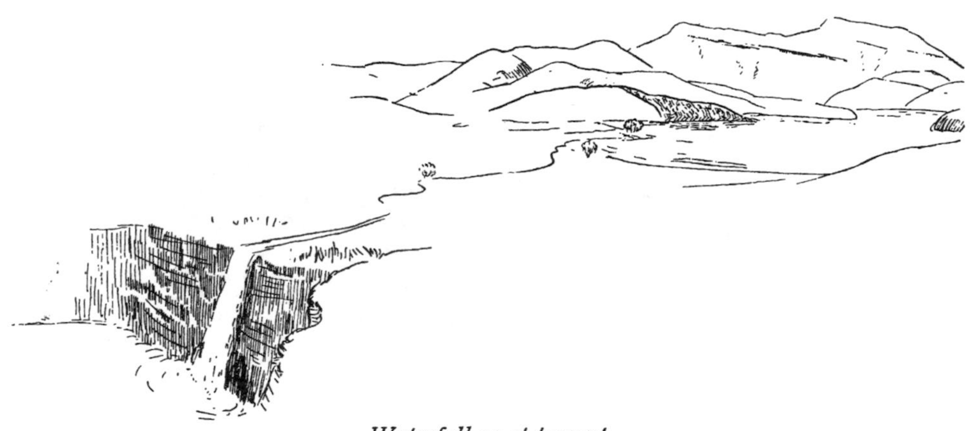

Waterfall as at present.

in my time. After years of quiescence the ditch—three feet deep and two feet across—draining Kahikanui flat, amply wide enough for the normal flow of water, became in a single flood and in a few hours' time a chasm 140 feet wide, 15 feet deep, and 300 feet long. In the three days' deluge of 1917, 600 yards of Tylee's Valley were gutted to the width of a chain and a depth of 20 feet. Flood-water had got into softer strata and gouged out in a few hours these great weights of soil. Some such catastrophe might likewise happen in the far future to the big waterfall. Already there is a cavern extending far beneath the ledge over which its water flows, proving thereby the existence of a softer rock beneath. Should, therefore, the hard upper crust give way or wear out, as must eventually happen, and should the stream's course continue to tap the softer material, the progress lakeward of this deep rift would be relatively rapid. The lake-basin itself would be reached in time and its soft contents quickly washed out. Each little surface rill and brooklet draining the branch flats would deepen into a gorge. The foot-hills resting on these little flats would in their turn begin to move, until in a short time a steep valley, similar to others in east Tutira, would be formed. "The thing that hath been, it is that which shall be; and that which is done is that which shall be done; and there is no new thing under the sun." The lake in fact is no more a permanency than are the great conglomerate cliffs of central Tutira, whose every pebble æons ago has been frost-fractured on the heights of

Waterfall receding towards lake.

old-world hills. Now again they are crumbling into modern river gorges, to be carried down to modern seas and ground to grains of sand.

Ages, however, before this is likely to happen there will occur another change to the waters of the lake. During a portion of each season their escape will be barred. It will be a change consequent not on strictly natural processes, but by reason of the great drain dug in the 'nineties to connect the Papakiri, which used to filter through or overflow the Big Swamp, with Tutira lake. This drain has become in a quarter of a century gouged and gutted into a considerable water-course. Silt which previously had been precipitated on the surface of the swamp is now carried direct to the lake. At the mouth of the drain there has been already created a long sand-spit; the northern bay of the lake is rapidly filling up, and must ultimately become dry land. The stream which now passes into the lake and out again some 60

Northern bay, 1921.

yards distant is destined in the not far distant future to discharge itself directly, without comminglement with the waters of the lake.

Northern bay—say one hundred years hence.

In droughts, or when nor'-westers press its waters south, for a few days or hours every year, Tutira lake will remain landlocked.[1]

To reiterate: Reasons have been given why the lakes of the station

[1] The pressure of wind upon water even on so small a surface as that of Tutira is very noticeable. Upon cessation of violent nor'-west gales I have seen water forced over the low gut separating in calm weather the twin sheets of water, Tutira and Waikopiro, pouring back to the depth of many inches; there has in fact occurred, in a small way, what happens in the Red Sea on an immense scale.

appear to be interpolations in its general geological scheme. It has been shown that the least unlikely theory accounting for their presence is one of land subsidence. Their original depths have been inferred from erosion of the contiguous countryside by floods and landslips. Facts have been adduced making it appear probable that the diminution in size of Tutira lake, otherwise than by deposition of mud and silt, has been accomplished by an extremely rapid process. Lastly, its future has been sketched, and, barring human interference, its eventual annihilation.

CHAPTER IV.

THE SOILS OF TUTIRA—PAST AND PRESENT.

EVERYWHERE on Tutira there has been spread at one period a carpet of dark matted humus. Immediately beneath it has lain a streak of clean grey pumice grit three or four inches in depth; beneath that again, dense, slightly greasy beds of packed red sand. On the eastern run there are traces of pale, valueless clay. Such were the primeval soils and subsoils of ancient Tutira.

The relative poverty of the huge district west and north of Napier is usually but wrongly attributed to the presence of the pumiceous band. Really this shallow layer of grit is of trifling import. The bane of the region, of which Tutira forms a part, are its sheets of packed red sand.[1]

Strata, undisturbed by time and change, are now discoverable only on the tops of tablelands, the backs of the aforementioned combs. No soil from greater heights has ever lodged on them. Owing to poverty and subsequent impossibility of enrichment, also by reason of elevation and exposure, it is likely that scrub and fern only, not forest and bush, have always flourished on their barren wind-swept heights. The original stratification of their soils has remained, therefore, comparatively undisturbed, undislocated, by root growth of any considerable size. The belief that only low growth has always clothed these arid, wretched tops is confirmed by the absence of a certain scrub—manuka (*Leptospermum*

[1] The surface soil of humus, dust, and pumice grit is described in the Edinburgh analysis as "poor, light, sandy material, containing only 7·2 per cent of organic matter. It contains no phosphate." The subsoil of red sand "consisted mainly of silica, the colouring being due to oxide of iron. No phosphates or potash were present."

scoparium)—which at a later date, as we shall see, took possession of the run. These table tops then, almost alone on the station, remained free of the pest. Presumably it had grown on them alternately with bracken for so long that the particular chemical constituents needed for the plants' growth had become exhausted. Corroboration of this theory is to be found in the fact that upon the small proportion of other land on Tutira where manuka had flourished prior to the 'seventies and 'eighties subsequent crops were also scant and poor.

These elevated scraps of plateau or "comb" backs—a few acres altogether—still remain samples of what all Tutira was. To this day on them there can be discovered more clearly than elsewhere the original soils in their original order of deposition—the dark dusty humus, the grey grit, the bed of packed red sand. On their heights not only has

top-dressing by slips been impossible, but they have been less tapped by "under-runners," less torn up and intermixed by uprooted timber, less slipped away by land avalanches than any other portions of the run.

It will be most convenient to consider the lowermost first, and thus take them in the order of red sand, pumice grit, and humus. The red sand deposits are of a firm, impervious, slightly greasy texture. The whole run has been plastered with the worthless stuff. It is common to west, central, and eastern Tutira. It shows up red on the naked "wind-blows." The quadrilateral mosaics of the topmost limestone sea-floors are set in it. It rests in sheets on the conglomerates.

There can be little doubt that these red sand deposits are of volcanic origin. Their substance may prove to be waterlogged pumice-stone fine ground, its greasiness resulting perhaps from some admixture with the scanty pale clays found on the eastern run. Perhaps, too, the coloration of the deposit may be derived from sources similar to those which give the conglomerate beds their rusty ferruginous hue.

Proceeding still from below towards the top, we come upon the band of pumice grit that lies on the red sand. It is about four inches in depth, the grit perfectly loose and dry, its grain very even, in size resembling the granules of a rough brown sugar. It is remarkably unclogged and free from intermixture of foreign matter. Like other bad things, it has reached Tutira from one or other of the inland volcanoes of the North Island. It is, I think, æolian, wind-carried, and has probably descended on a countryside supporting a vegetation thick enough, stiff enough, and high enough to afford immediate shelter and cover. Its light dry substance, falling like rain from above, would thus either percolate direct through this growth or be at once watered on to the ground by wet weather. It would lie evenly on the ground—as in fact we find it in the undisturbed plateau fragments—and during succeeding centuries become covered with root matter and leaf-mould. The vegetation which thus sheltered and harboured this pumice shower must have been then growing on red sand, for it cannot be doubted that the top spit of humus is merely the accumulated result of decayed vegetation. The pumice deposit appears to have been the result of a single volcanic outbreak. Nowhere, at any rate, do I find the slightest trace of alternate layers of grit and humus. The two substances are clearly distinct. Other showers of pumice may have fallen at other periods on naked surfaces and been blown off, but the result of the one particular eruption incorporated in the soil of the station fell, I believe, on a surface protected from wind. Unless we are to suppose different meteorological conditions, no layer of grit could have endured one hour's nor'-west gale on bare baked sand. Even by the less violent action of rain it would have quickly been washed away.

The alternative to the theory of wind-blown pumice is to suppose the grit to have fallen on water and eventually to have sunk. There are many difficulties, however, to be faced. Falling on open sea the light material would have been soon dispersed; even granted that it may have fallen on landlocked waters, the floating masses would have been heaped together by action of wind and wave, the grit would not have been evenly distributed on the bottoms. Moreover, even supposing that the grit had become submerged in a level band, upon the rise of the plateau its surface would have dried and the light top-dressing of pumice would have been dispersed by gales. Lastly, there is nothing in the appearance of freshly-exposed red sand beds to suggest that weighted water-clogged grit has rested upon them. Its grains have not become incorporated in any considerable degree with the lower-lying material. It seems to me that whilst there are insuperable objections in the one case, there is nothing hard of belief in the other.

The volcanic mountains Tongariro, Ngaruhoe, Ruapehu, are all within a hundred miles of Tutira. During the Tarawera eruption of '86 the run would have been thickly dusted with fine grit had the wind happened to have blown towards it. On east Tutira flake pumice is rare; there are banks of it on Maungaharuru varying in size from a crown-piece to a man's palm. The Mohaka river, rising in the chief region of volcanic activity in New Zealand, bears in flood-time from its fountain-head quantities of sponge-shaped blocks; in fact, the size and shape of pumice fragments differ according to distance from source of supply.

The dark, dusty, matted humus of the surface requires little description. There can be no doubt that it is the most modern soil of the run, the outcome of rotted fern fronds, forest debris and fine dust, blown from burnt forests and bracken lands.

From the foregoing description of the soils and subsoils of the run, it is apparent how poor the original surface on every part of the run has been. It is apparent, too, that any surface improvement can only have taken place by the superposition of limestone, travertine, marl, and sand, by an admixture of soils due to overblown forests, and by the deepening of vegetable mould. From a sheep-farmer's point of view, the original state of Tutira must have been worthless: indeed for him New Zealand even now has been discovered and Tutira "taken up" many hundreds of thousands of years too soon.

In another chapter the reader has been, I hope, helped to grasp the general configuration of the station by the metaphor of the "comb." With equal ease he will understand its soils, if another salient fact of another sort be grasped. It is this, that the fertility of the run is in direct proportion to its angle of inclination, that the steepest country is the best, the flattest country—alluvial flats excepted—the worst. From the former the jacket of pumice and red sand has to some extent been overlaid by washings or stripped by slips. The latter remains as it was centuries ago, even its leaf-mould stolen away by subcutaneous erosion.

Kahikanui swamp, lying beneath the highest marl outcrop on the station, and containing washings almost wholly composed of that substance, is the only first-class piece of soil on the run; in miniature it resembles the far-famed flats of Poverty Bay.

Each of these plains—the one of a few dozen acres, the other of many thousands—has been created by alluvium carried down in flood-water from hills of marl. In each case the rougher, larger particles have been precipitated at the apex of the plain, whilst the finest, most highly comminuted silt matter has remained in solution until dammed back—in the one case by Tutira lake, in the other by the waters of the Pacific. The result is that although rich throughout, the physical conditions of the soil at apex and base are widely dissimilar, the apex easily worked and friable, the base heavy and hard set. What Herbert Spencer has termed "the multiplication of effects" can further be traced in different weeds, different grasses, a different permanent pasture.

The flats—alluvial is too sumptuous a term—of the pumiceous trough of the run do not exceed some two or three score acres, the largest single patch perhaps not more than five or six acres in extent. As station assets they are of no great value, it is the means by which they have come into being that deserves notice here. That method is not direct deposition from above, but injection from below of water charged with microscopic quantities of rotted vegetation, the soakage from higher slopes. These pumice alluviums in outward semblance differ but little from neighbouring lands. For long, indeed, they were regarded as equally worthless. It was a belief countenanced by the arid grit of the surface and by the appearance of the vegetation supported thereon—groves of pole manuka, their bark tattered and thin, their harsh leaves brown-green and prickly, a foliage yielding neither shelter from winter storms nor shade from summer heats. On the other hand, covering contiguous slopes, flourished luxuriantly green tutu (*Coriaria ruscifolia*) and koromiko (*Veronica salicifolia*). It was believed that their foliage was creating leaf-mould on the slopes where it fell; really its manurial value soaked through the humus and sandy grit by a process of filtration to be explained later, then, unable on the flat to escape further, rose from below in the form of an alluvium of dirty water no stronger than weak tea.[1]

Sufficient now has been said of the soils of the run and their ancient order of deposition. If the reader has grasped the fact that all steep land is good and that all flat land is bad, he knows everything that need be known.

[1] Riding through some such bit of country with a stock and station agent, he remarked of the deep green tutu-covered slopes, wishing presumably to say something pleasant: "That does not look so very bad, but this——!" Without consideration I replied: "You might not think so, but this is really better." I yet recollect his sour, sick face as he turned in his saddle and looked at me. As I say, he was a stock and station agent inured from childhood to chicanery, yet he was fairly nauseated, not of course at the lie as he necessarily accounted it, but at its ineptitude, inadequacy, futility, the waste and uselessness of such a foolish falsehood.

Shephera's Basket Fungus.

CHAPTER V.

SUBCUTANEOUS EROSION.

THE original appearance of Tutira has been described. We have yet to trace the agencies by which its contours have been modified. It will be remembered that the station was pictured in the beginning as a serious of plateaux afterwards tilted towards the east, these tilted terraces or canted plateaux smooth and unbroken by fissures. The evolution of the comb system, the development of these fissures, has yet to be described. It has, I think, been brought to light by a very remarkable process of subcutaneous erosion, a process akin to the dissolution of a dead beast when first the flesh decays, then the skin shrinks and shrivels, whilst only at the last do the bones protrude. In this transformation water has been the chief agent, but before proceeding to state what water has done it will simplify our task to state what water has not done.

Water has not created the dry cliff system. Water has not scoured out the solid rock of these strange valleys walled in by precipice. Both sense and sight forbid the supposition. There is no lateral soakage whatever into them. In hundreds of instances, except the drops that actually fall on to their surfaces from the sky, no extraneous water has ever reached them. Rock erosion by scour has not occurred, because water has been able to pass away from the contiguous lands on either side through beds of grit and sand. The junction of their double walls of cliff is not lanceolate but horse-shoe shaped. It bears no resemblance to the typical commencement of a water-worn ravine. The difference between the width of the valleys as measured from cliff to cliff at two, twenty, and two hundred yards from their beginnings is out of all proportion to the necessities of scour, supposing such a process had ever been at work. Finally, at the mouths of these valleys there exist no indications of the rock material which must have been deposited had water erosion been responsible for their formation. As regards these chasms, this is, I think, a fair statement of what water has not done.

The task of water has been to remove by infinitesimal quantities the material deposited between the rock walls; to flush, if indeed such a verb can be applied to an enormously tardy process, and scour out material already deposited in the valleys; to render visible what a mightier force had already accomplished. Erosion, in fact, has brought to light, not created, the rock-bound valleys of Tutira. When the plateaux which I have supposed to have been the original shape of the run canted to the east, making the station a series of tilted terraces, these deep interstices had been already moulded.

About the nature of the force that has severed the sea-floors upon which Tutira sheep now run, yet incompletely severed them,—that has severed them, yet not severed them in parallel lines,—any suggestions I can offer are little likely to be of value. There seems, however, to have been a twofold motion—the one cracking them in lines running north and south, the other incompletely parting these strips or oblongs of country into numerous short unparallel asymmetrical gaps east and west. The fissurings extending north and south are attributable probably to the effects of subsidence as segment after segment canted towards the east. The gaps east and west, with their broad horse-shoe beginnings, are less easy to account for. They may be due to shrinkage by evaporation whilst their rocky material was still plastic; their shape forbids the idea of cracking or fissuring.

Although no satisfactory solution can be offered as to their origin, much can be said as to the manner in which they have been brought to view. Each range shows its own slight modification of the general geological plan. In order, therefore, not to confuse the reader and darken counsel with a multiplicity of detail, I propose to work stage by stage from past to present conditions. The reader is invited to contemplate an ideal section of a hill range in the conglomerates of central Tutira before erosion had begun its work.

In the beginning, then, when the main rivers of Tutira ran hundreds of feet above their present levels, the surface of our imaginary section seemed whole, unwrinkled, ungapped, and inclining gently towards the east. It was coated with a matted humus overlying æolian pumice grit, overlying disintegrated red sand. Filtering through the loose surface mould on to the layer of grit and sand, rain-water did not so much penetrate downwards as follow subcutaneously the trend of the slopes. It passed away in an unseen filtration; I imagine it lipping from grain to grain of grit, moving as evenly, as diffusedly, and with as little current, as water spilled on an inclined sheet of blotting-paper. Finally, at the base of our imaginary slope the accumulated soakage would gather against the cliff of the next "comb,"—against, that is, the next western-facing precipice—the next segment of canted plateau. There, unable to press further eastward, it would be temporarily held. Eventually, at the base of our

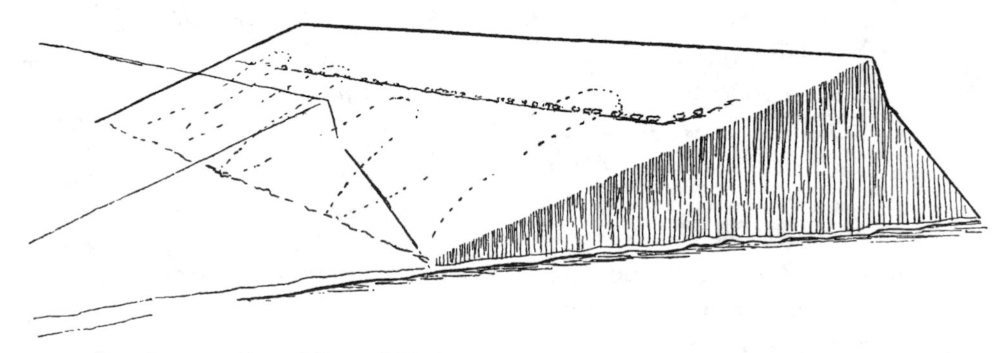

Imaginary section of Central Tutira prior to erosion, the dotted lines showing yet hidden interstices.

block, there would develop as an essential factor in our scheme of erosion, a tiny runnel or rivulet escaping north or south towards one or other of the main rivers, towards the Waikoau, the Matahorua, or the Waikari. Up to this point our imaginary section of hill-slope has seemed a homogeneous whole, but now, with the deepening of the rivulet—a deepening sympathetic with that of the whole drainage system of the station—two different qualities of surface would begin to reveal themselves. The little stream would ooze at the earliest date along the humus; as the deepening process continued, it would trickle alongside the layer of volcanic grit, and later along beds of disintegrated sands. With a prolongation of the process our rivulet, cutting still deeper, would at last begin to skirt alternately two different qualities of material, hard and soft—the "teeth" and "interstices" respectively of the yet hidden "comb." Erosion of the teeth—conglomerate—would be infinitesimal; of the interstices—grit and disintegrated sands—perceptible. The mass of the one would endure; of the other, perish. As this rivulet deepened, the soft material of each interstice tapped would be drawn upon and would sympathetically shrink, whilst the hard would remain unaltered. In process of time this subcutaneous drainage, this subterranean withdrawal of matter from certain portions of our block, would begin to affect the evenness of its surface. The humus skin of the interstices would slightly sag; instead of an unwrinkled soil-sheet over the whole of our imaginary slope, dips and hummocks would alternate. As the sag deepened with the continued drain from beneath of its internal material, the height of the hummocks would seem to rise. As a tooth wears through the gums, the hard edges of the rock would appear to shove through the

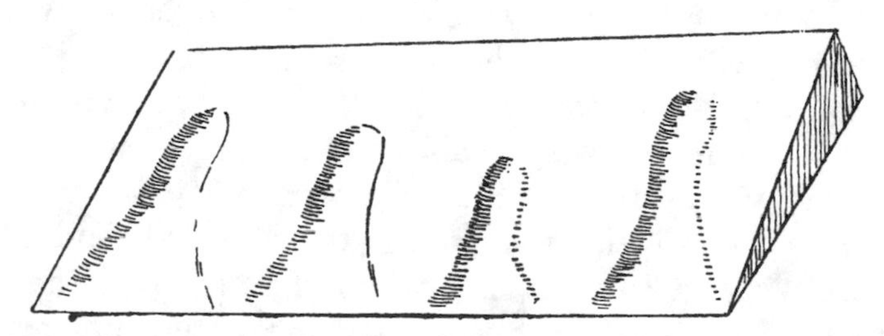

"In proportion to the depth of sag the height of the rock walls would seem to rise."

covering of humus. Unceasing shrinkage of the surface would later reveal low, bare, perpendicular rock walls. Later again, the whole of our once even slope would be an alternation of portions still even and of other portions shrunk and sagged. Uninterrupted subcutaneous erosion of the material of the interstices would year by year continue to deepen the lap, whilst in proportion to the depth of sag the height of the rock walls would seem to rise. At last our imagined block of "comb" country in central Tutira would exist as it does actually exist at this present time,—a slope falling towards the east with gorges of perpendicular cliff, between each of whose walls lies a dry sagging lap or fold still unwrinkled on the surface, showing no sign whatever externally of water action. Enormous rainfalls have never gouged waterways in this porous soil-sheet; no watercourses have furrowed it. Except for pig-rootings and hummocks of overblown trees, there are to this day no inequalities to vary the monotony of its even blanket of dark fibrous humus.

Our first hypothetical block was selected from the conglomerates of central Tutira, where conditions are most simple and where the valleys are invariably dry except for such rain as reaches them from the sky. A further development of the subcutaneous drainage system will again be made easily comprehensible if once more a conjectural block be visualised, taken this time from the limestone ranges of the west. As before, we must imagine the slope towards the east, the skin of matted humus covering grit and pumice sands, the subterranean soakage system, the right-angle rivulet at base, the western precipice; finally, the different stages of sag as exemplified in the history of the conjectural conglomerate block of central Tutira. We can start, in fact, now where we concluded then—that is, with a deep sag between cliffs of 10 or 20 feet. Our western valley, however, to begin with, is many times the length and width of the central valley. Owing to its greater area, there is a quite important rainfall reaching it from the skies. It stands at a higher elevation above sea-level: there is a greater fall for its drainage system. Shrinkage in the sag becomes more and more pronounced, until the rock walls stand up 80 and 100 feet on either side, until there comes a time when the centre of the fold undermined by the soakage system, by the flushing action of springs, which now first come into account, and by rain-water falling within the gorge, changes from a deep lap to an angled incline—from a U, in fact, to a V. The extreme point, the apex of the inverted V of humus skin, is now for the first time in our story directly exposed to water. It is finally worn through by the action of the running water of springs supplemented by soakage of heavy rainfalls; a brook trickles over the lower portion of the sag; a normal valley, in fact, has been formed save for the impossibility of lateral expansion.

Proceeding once more from the less simple to the least simple, the reader is invited to picture a third conjectural section, on this occasion from east Tutira. Once more, then, a block must be imagined showing the typical eastern cant, the even upper covering of humus grit and red sand, the drainage system at right angles to the tilt of the slope. As, however, we find differences between the central and western sections, so again we discover others now. The interstices of the east are wider than elsewhere, the cachment area larger, the rainfall heavier, the presence of clay marked, the number of springs greater, the marls nearer to the surface, in immediate contact moreover with the superposed limestones. These different conditions produce different effects. At first, however, soakage passes down the slope exactly as in the two instances already given. As before, the percolated water banks up at the base of the succeeding western cliff; as before, it soaks and oozes away until an incipient water-channel is established; as before, this

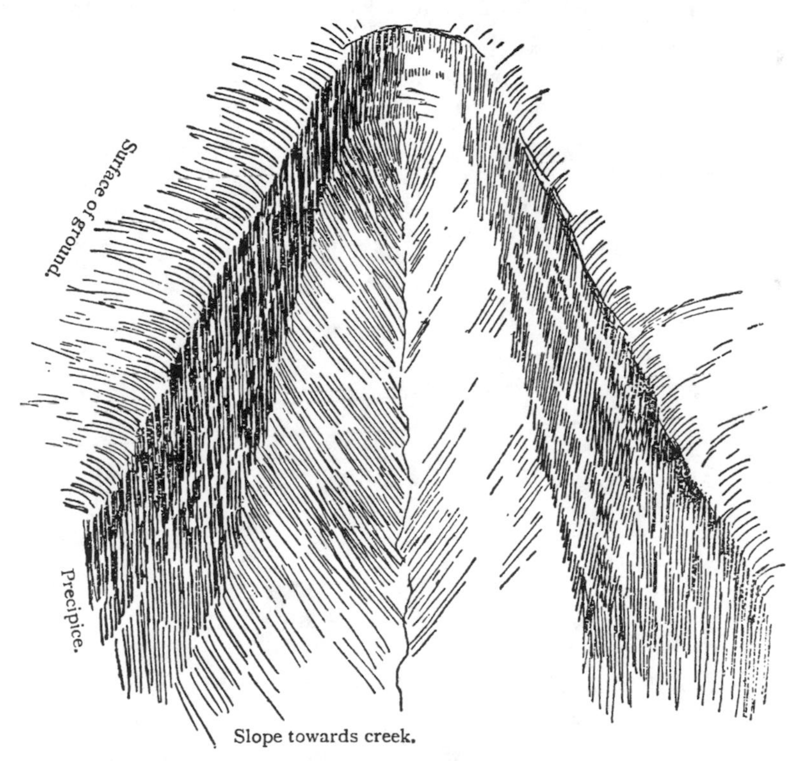

"*Sag deepened from a* U *to a* V*—a normal valley formed save for the impossibility of lateral expansion.*"

water-channel deepens until a scour is created; as before, this scour begins to eat out the topmost constituents of the hidden interstice; as before, a sag which deepens into a lap appears; as before, the lap develops into a shallow U. Now, however, occurs a change. The base of the U rests not on pervious sands, as on our conjectural blocks of the centre and west, but on a stiffer material, marl. Rain passing off between a top spit of fibre and grit and an impervious subsoil, forms a sort of water sandwich, the sands and grit being carried off particle by particle between the floor of marl and the ceiling of humus. In process of time the sharp pumice grit chisels out of the former a minute irregular bed. It deepens into a tiny hidden runnel; at last there is created a subterranean stream, or, in shepherd's phrase, an "under-runner." Its course is at first unseen, then, as through process of time its channel is gouged out, and as the carpet of humus fibre gives way at irregular intervals, great rents and holes betray its presence. The bed continues to deepen, the carpet of turf falls in more and more, until finally the under-runner becomes an open rivulet. Ancillary effects now become

Under-runner, the roof of which has here and there given way.

prominent, along each edge of the rivulet a secondary process of subcutaneous attrition is set up. Subsidiary under-runners become established at right angles to the stream, and in time laterals also to the latter. The stream increases in depth, the angle of inclination of the slope on either side grows more acute. In time this triple process of erosion, still subcutaneous, still veiled by the carpet of rooty humus, saps up to the containing walls of the interstice, and now for the first time on Tutira we see a possibility of a widening as well as of a deepening of the gap. The containing walls are not of conglomerate as in central Tutira, nor are they of limestone and sandstone as in west Tutira. They are composed of alternate layers of limestone and marl. The latter substance crumbles and flakes, the limestone rock cap is undermined and breaks away fragment after fragment; the upper portions of the containing walls, thus exposed to air, frost, and rain, continue no longer perfectly perpendicular—they tend to become exceedingly acute slopes. Shielded by soil only, the unexposed portion of the cliffs, where there is no possibility of wear and tear, remains absolutely perpendicular.

Limestone rock cap undermined.

At length our conjectural section of once even eastward-sloping range is left an open gorge stripped of its original contents, exposing a slope studded with fragments of undermined limestone.

There have now been traced three stages of development in the underground drainage system of Tutira: the first, erosion by percolation of rain-water only, together with a deepening of the valley within upright walls; the second, erosion by percolation of rain-water plus erosion by springs, together with a deepening of the valley within upright walls; the third, erosion by percolation of rain-water plus erosion by springs, and supplemented further by a certain slight slow widening as well as by a deepening of the valley no longer within exactly perpendicular walls.

These conclusions have been reached by working from considerable heights above sea-level downwards. They can, I think, be proved, to use an arithmetical term, by ascent. The reader has but to trace the course of the main streams, to follow up their tributaries, lastly, to mark the sources from which the tributaries themselves are fed. Our conclusions too have been reached by deduction. Corroboration by the dry light of the inductive method is easy. We can drop ideal sections and consider actual conditions.

Shrinkage shows itself in every stage; there are endless modifications, but although details differ, the general principle is unmistakable, the pattern clear. Beginning with instances where the sagging is still in its preliminary stage, a fold can be instanced on the Heru-o-Tureia block parallel with and to the south of the steep horse-trail known as the "Zigzag." Although close to the enormous gorges of the highest range on Tutira, this particular narrow fold remains but a fold; on the other hand, in the "Waterfall" paddock the cliffs are so hummocky as to have remained innominate. The "comb" pattern is hardly recognisable; teeth and interstices alike are so little in evidence that the plough has passed over both. Again, the laps of the "Second Range" have sagged so little that only the outlines of the cliffs show beneath the humus covering; the teeth have not yet broken through the gums.

On the "Sand Hills" the interstices are extraordinarily wide, whilst the back of the "comb" is less emphasised than usual. On the "Tutu Faces," where the "teeth" are set particularly near to one another, folds are to be found varying from those hardly noticeable to others enclosed by cliffs from ten to fifteen feet high. The "Nobbies" range, a duplicate in miniature of the Heru-o-Tureia, is gapped in lines more nearly parallel to one another than elsewhere. Everywhere, however, erosion has taken place subterraneously, subcutaneously; be the sags deep or shallow, wide or narrow, salient or unseen, the ancient original humus still blankets the surface.

Where the "comb" system is distinctly marked on Tutira, there is little more to say of its peculiar system of underground drainage, but there are considerable tracts of country where the subcutaneous processes described are subject to modification, where the different appearance of the countryside itself deserves comment. There is in central Tutira a considerable area of peaked country—the "Dome," the "Conical Hill," the "Razorback," "Mata-te-Rangi," the "Pa Hill," and other solitaries. These hills are of sandstone formation, more or less weathered into points, as their names imply. Each of them is a relic of a small fragment of tilted terrace from which the conglomerate cap has gone, whose uncrowned top has been exposed to the elements; they are the scattered "teeth" of dislocated "comb" systems. Unroofed by the action of rain, and in a minor degree by that of frost and wind,

Fragment of terrace still rock-capped.

Fragment from which cap has slid, melting into a cone.

they have been melted and moulded to cone and dome and razorback. Peak formation in fact represents, on the east coast of the North Island of New Zealand, the intermediate stage between the plateau of the far past and the plain of the remote future.

Subcutaneous erosion has played as curious a part about the bases of these solitaries—these erratics, if I may so call them—as on the slopes of the terrace system. However probable it might have seemed that their dusty weatherings would have been deposited on the surface, no such boon has blessed the land. Everywhere the ocean robs the upland farmer, but nowhere more brazenly than on Tutira. Stuff urgently needed for the amelioration of the surface of the run is borne off by under-runners to the sea—that vast, barren, grassless flat which does not carry a sheep to ten million acres. About the sandstone formations of the run—especially about the softer sandstones—their ramifications are most highly developed. The "Dome" and "Dead Man's Hill"[1] in central Tutira exemplify in an extreme degree this network of tunnellings: their steep slopes are everywhere honeycombed with hollows. Hardly a grain of sand-weathering is deposited on the surface. In rain-storms it is washed directly off the surface of the melting cone into tunnels, whose circular, open, funnel-shaped mouths seem actually to gape for it. As on the marls of the east a water sandwich is formed, so here again similar conditions are re-enacted with the substitution of sandstone rock for marl. Tutira remains unfertilised, constituents that might be supporting grass and sheep are rushed to the hungry ocean, the old original sin of worthless humus persists almost to the rounded sandstone cones. Although the land surrounding these rain-scoured, wind-blown, melting solitaries has sunk scores, even hundreds of feet, yet always the worst soil—the dusty humus—has contrived to remain on top.

[1] So called from the discovery of a human skull and bones scattered by pig, but evidently when first found those of a man but recently dead. We surmise that the poor chap may have at first missed his way on the high tops, may have in an exhausted state seen the lake, and in making for it become trapped in the gorges of the central run. At any rate, a few yards back from the edge of one of these precipices lay the bones. The remains of two other men have been in my day discovered on Tutira. In the one case they were those of a European, in the other those of a Maori. Near the skeleton of the latter lay a fragment of fire-bleached greenstone.

CHAPTER VI.

SURFACE SLIPS.

THE deluges that from time to time pass over Tutira have been mentioned. Readers will have, therefore, no difficulty in picturing their effects on steep marl slopes. Although on the station itself there is only a small proportion of land of this type, yet speaking broadly of the "papa" country of the east coast of the North Island, it is being flattened towards the sea by a mighty melting process, most marked and most discernible in the soft marls of Poverty Bay. As, however, it is the history of Tutira I am writing—limited as is the acreage affected—I shall cull my facts from local sources.

During heavy rainfalls on eastern Tutira the numerous oozes, leakages, and "damps," consequent on alternate bands of marl and limestone, become surcharged with water. The supersaturated subsoils burst with their weight of wet, chasms of many feet in depth are created, the hillsides spew forth mud; under-runners become gulches, or, choked with debris, spill on the hillsides their streams of silt, torn turf, and curious rough-rolled balls of clay.

Eastern Tutira, indeed, after a violent "buster," appears to have been weeping mud. From the edges of all ancient slips the water-sodden fringes drip with clay; new red-raw wounds smear the green slopes, scalp-shaped patches detach themselves, slipping downward in slush and turf. Sometimes a whole hillside will wrinkle and slide like snow melting off a roof, its huge corrugations smothering and smashing the wretched sheep, half or wholly burying them in every posture. Sometimes a slip rushing down a steep incline will temporarily block the creek below, piling itself up until again washed away, and leaving on the opposite slope, yards above the stream, a curious plaster mark of dirt. Gluey streams, hardly moving faster than glaciers, from whose tenacious mud bogged sheep have to be extricated hoof by hoof, make the hillsides a terror to shepherds.[1] After a "southerly buster" or a "black nor'-easter" of three or four days' uninterrupted torrential rain, I have counted on a two-mile stretch of hillside over two hundred slips great and small, new or newly scoured out. Seven or eight times since '82 the grasses and sedges of the valleys around the lake have been overlaid by mud varying in depth from six inches to a couple or three feet. Huge masses of solid hill have slid on to the larger flats. Fencing is buried, roads and bridges washed away, culverts

"*Like snow sliding off a roof.*"

destroyed, stock bogged or caught and buried in the displaced masses of earth.

Besides earth avalanches there remain after every great storm, here and there, fissures on the hillside nine inches or a foot in depth. Sometimes they are mere longitudinal cracks, but more often of an irregularly ovoid shape. They mark areas, often of great extent, where the surface has slid a few inches. They can be detected further by trees slightly out of the natural angle of growth, by the bulging and bellying of fence

[1] Returning directly after the Armistice, I was amused to find that recollections of the flood of the previous year had been adapted to the needs of the nursery. My three nieces had invented the new game of "bogged sheep." There is no necessity to give the exact rules as framed by the little shepherdesses. Suffice it to say that the game can be played as most convenient on carpet or grass, that some of the players are "sheep," others "shepherds," the object of the latter of course being to rescue the entombed animal by dragging it leg by leg out of the mud. Should it bleat piteously during the operation, again fall back into the mire, or, best of all, should its cold cramped legs refuse their office without pastoral support, by so much more is the game quickened.

East Tutira—Landslips after Flood.

lines, by the dry overlapping of turf over turf, by surface wrinklings, by bursting of gate fastenings. Even, however, when thus started on its downward path, the progress of a landcreep is by no means always sustained. Sometimes for years the gaping rents remain unwidened, sometimes they fill with dust and debris. They play, nevertheless, an important part in the promotion of the earth avalanches already described. Water lodging in them penetrates to the marl, greases the base on which the upper soils rest, and expedites the slip.

I believe that even during my brief span on Tutira scarcely a rood of marl in the eastern run has not been affected in some degree by the great rainfall—has not slid seaward, perhaps a few inches, perhaps a few feet.

Landslips and landcreeps may, in fact, be considered complementary to the earlier processes of subcutaneous erosion. In the valley of "Newton" paddock we have an example of surface wear and tear arrested from lack of sufficient fall. There the local stream, blocked and barred with limestone debris, still runs several hundred feet above sea-level. The original upper soils too, torn and patchy, have not yet been completely sloughed.

In the valley of the Maungahinahina, however, where the fall nearly reaches sea-level, and where, moreover, the mouth of the great gap abuts directly on to the Waikoau river, we get an almost perfectly completed bit of water sculpture five or six hundred feet deep and nearly half a mile in width. The fibrous, rooty humus, the pumice grit, the red sands, the clays are gone, the great scoop they used to hide is wholly revealed. Percolation and soakage has developed into the under-runner system, that into the open gulch, the gulch into multitudinous lateral gorges, until the loose heterogeneous mixture of soils that once filled the huge interstice to the brim has been scoured out and the marl basis of the gap exposed. Lastly, unable to cope with and carry off the vast quantity of limestone fragments,—portions of the original rock-cap slid into it from either side,—the little stream has finally left them piled and prominent in a sort of moraine at the mouth of the gap.

Nevertheless, although thus buffeted by deluges and sapped by earthslips, the remaining portion of the rock-cap of eastern Tutira is likely to endure for an almost incalculable period. Attrition is enormously slow.

During my ownership three only of the great grey squares into which the limestone sea-floors split themselves have perceptibly shifted their sites. In 1905 a landslip of some quarter of a mile in length started from the lower part of the Racecourse flat, overwhelmed the road near the Waikoau crossing, swept it out of existence, smashed like matches trees of three and four feet in circumference, finally depositing two great boulders in the Waikoau river. There to this day they stand, monumentally white in their unlichened youth.

Valley of the Maungahinahina.

In 1911 another vast rock moved, not after rain, but after a long spell of particularly dry weather, and on a day so calm as to forbid suspicion of earth tremors. This enormous fragment of limestone cliff broke away from the highest sea-floor of the Racecourse paddock. The sound of the mass moving, the clouds of dust raised, were perceptible half a mile off. Viewed more nearly, it had ploughed a deep chasm into the earthy slope below, parting it as a battleship breasts the water at her bow. The weight of the mass had been so great that notwithstanding its drop—or rather precipitous slide—of ten or fifteen feet, the grasses and flax on its top had perfectly retained their natural angle of growth. Although, however, only three limestone quadrilaterals have thus been detected in motion, I believe that the numberless boulders already broken from the cap and deeply embedded in the hillsides really never cease to move,—that they are being slowly sucked downhill, perhaps an inch or so a year, sapped by the action of innumerable under-runners.

Alterations in the positions of the large-sized river-bed boulders have hardly been more conspicuous. Certain very noticeable fragments have moved a foot or two oceanwards. The ford of a river is more closely scanned than any other portion of its bed: that of the Waikoau has scarcely changed in forty years. We cross now—or did until the bridge was built—within a few yards of where we crossed in '82.

Two minor processes of erosion yet remain unchronicled, the more important I believe responsible for the circular pits found over the relics of the ancient plateau caps of eastern Tutira. Sometimes these funnel-shaped cavities are still skinned over by turf, sometimes the turf has broken through and they are open at bottom. Most of them are of moderate depth. One pit, however, within a short distance of the Tutira boundary, was, until opened up by spade-work, a death-trap for animals. At its base remains of sheep and pigs used often to be visible, —the former presumably tempted over the edge by succulent weeds or trapped by mere bad luck, the latter induced to slide down by the bait of the former and then unable to escape. These pits, great and small, have been probably worn by the action of the carbonic acid of rain-water affecting the limestone rock-cap. Its substance is dissolved and borne away as travertine, masses of which accumulate about the sides of the streams. Consequent on the chemical dissolution of the rock-cap beneath, the unsupported soils slide downwards towards the centre of weakness, thus forming rudely circular pits. Withdrawal of matter from beneath may also be held responsible for the almost perfectly moulded funnels alongside of one of the streams of this part of the station. It runs over a jumble of squares, cubes, and slabs—sections that have broken away from the limestone cap and been carried violently by earth avalanches or mined by under-runners into the valley bottom. Amongst them no doubt are endless cracks and gaps through which water escapes—perhaps to reappear at lower levels of the stream itself, perhaps to reissue elsewhere as fountains. There are also in its bed minute intermittent whirlpools that alternately suck and cease to suck its waters down.

One other form of erosion remains to be described. The sum-total of its effects is so puny that perhaps I should apologise for its inclusion, yet there is a fascination in its strange rapid action. It is an operation readily to be appreciated by those who have attempted to water a steeply-sloping garden bed, dust-dry and in finest tilth. About the bases of bare scarps—the unhealed scars of hillside slips—quantities of the finest dust accumulate in dry seasons. On these miniature skrees of powdered soil fall the first great drops of a western shower. The dust slope can neither retain the drops nor instantaneously absorb them. Striking the slope they gather earth particles in their downward course. While thus in motion, as if by miracle, they change from liquid to solid. Metamorphosed first into ashen-grey and then into brown balls, these earthen pilules, preserving their shape but changing their substance, race madly downhill, bound downhill, no longer clear drops from heaven, but minute circular solid globes of soil. With a faster fall of raindrops the process ends perforce; the dust-heap becomes a mud torrent.

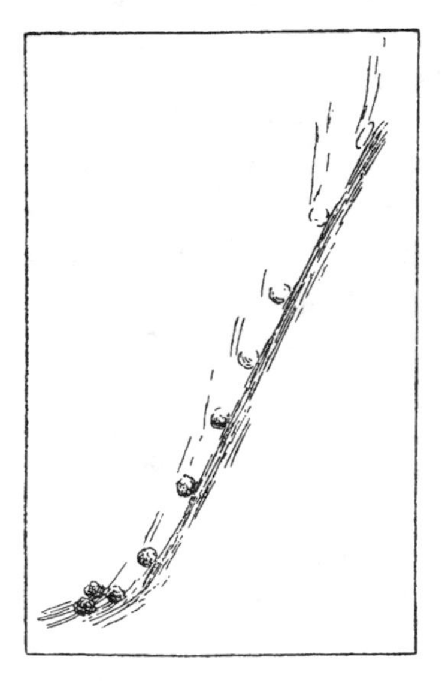

"No longer clear drops from heaven, but minute circular solid globes of soil."

Frost and wind have played but minor parts in the transfiguration of the run. The former on wintry mornings has accelerated the weathering of the cone area by elevation of their sand surfaces on upright spicules of ice, the latter by blowing abroad the desiccated dust. Doubtless also frost has contributed to the disintegration of other surfaces; speaking generally, nevertheless, it is water which has moulded the run to its present shape.

Shrinkage of the station has been compared to the decay of a dead beast: the softer parts dissolve, the skin, shrivelled and sagged, endures. So has it been over the vast bulk of Tutira. Its skin also remains as of yore, shrunk and wrinkled indeed with age, but intact. In spite of torrential rainfalls, the surface of the station remains at this date as it was ages ago. It is now, as it was then, blanketed with a dark, porous, unfertile, rooty humus. So vast a change by processes of internal waste, of subcutaneous dissolution, is perhaps unique in the annals of geology.

CHAPTER VII.

THE FOREST OF THE PAST.

ALTHOUGH Tutira when first taken up as a sheep-run was a wilderness of bracken (*Pteris aquilina*, var. *esculenta*), it had nevertheless been within a very recent period wholly under forest. In the oozy bog runnels of the central run, where the current scarcely stirs the floating weeds or shivers the tall green reeds, timber is plentiful. The swamps, undrained and drained, are full of it. Through the shrunken surfaces of the latter protrude in the drier parts dark peat-preserved boles. In the great drains scoured out by flood-water are to be found the crowns and octopus-like roots of trees. Timber lies in the basin of every lake, lakelet, and tarn on the run. It shows beneath the turf of grassed lands whiter in the morning frosts, browner in summer droughts. Surface timber also, chiefly totara (*Podocarpus Totara*), is, or rather was—for thousands of posts and strainers have been split from it—plentiful. It was most abundant on the most arid parts of the trough of the run. Thereabouts there had been a lesser growth of fern, a lesser accumulation of inflammable material. The fires, which from time to time used to sweep the countryside, had been from lack of combustible matter less fierce and less frequent in these localities. Great lanceolate-shaped spars curiously gouged and chiselled by fire were common, whilst here and there entire boles remained almost intact. Some of these prone trunks were of great girth; one lying in the open gave a diameter of twelve feet.

Totara bole deeply sunk into the soil.

Another deeply sunk into the soil can have been scarcely less than fifteen feet through. Huge boles of less durable species, their shrunken bulk unmarked by the least mound, lie to this day absorbed in the dark gritty soil, unseen and unsuspected until advertised by fire. They appear to have disintegrated into mould, or perhaps more correctly to have been reduced by former fires into a sort of charcoal. It would seem impossible that the material of these rotted boles could once again take form—these dry bones live. They do so nevertheless. After a fire has swept the bracken the long-vanished giants will sometimes rekindle and burn for days in a slow, smokeless smoulder.

Shape of fallen tree rediscovered by fire.

As the invisible image on a photographic plate is revealed by chemicals, so by fire is the entire shape of the fallen tree rediscovered. At first on the dark ground it lies flat, a fragmentary skeleton, the massy trunk, the mighty boughs, portrayed in deep soft masses of grey ash, which after rain becomes an emerald fur of softest velvet moss. The tree by a natural miracle, again after long death supports a verdure deeper than in its leafy prime. Nor does even then change cease. The skeleton of the prone tree can only for a few days perhaps be visible in ash, for a few weeks in moss. It remains more durably marked in scrub. Better fed on the potash, this scrub manuka (*Leptospermum scoparium*) soon out-tops the surrounding growth and stands forth in strange arbitrary lines, a record of the past, undecipherable except to those who have watched each stage.

It is throughout the trough of the run that timber is most evenly distributed as well as most plentiful. On the marls of the east, land-slips perpetually occurring have contributed in no small degree to its disappearance. There it has been swept away, and lies buried deep below the surface. The slower-growing more durable species of tree, moreover, have not had time to establish themselves firmly; as seedlings and saplings they have been uprooted and rushed downhill in avalanches of earth.

Besides evidence afforded by timber preserved in water, buried in mud and marsh, and strewn irregularly over the surface of the run, there are other convincing proofs of an ancient forest. Many portions of the station are so honeycombed with holes and hollows, the result of rotted or burnt-out roots, that they are unsafe to ride over except at a walking pace.

The trough of the run is marked too by innumerable hummocks, their longitudinal edges running at right angles to the quarter from which blow the most violent gales. They are so numerous, and the hummock form so invariable, that it is certain these boles have been levelled by storms from the west and nor'-west. The hummocks scattered over the whole of central Tutira denote, too, a forest overblown when dead, not green, in the first place destroyed by fire, then uprooted by the prevailing winds. In green New Zealand woods great trees do not readily fall; not infrequently they are supported by neighbouring trunks, or at any rate their natural angle of fall is deflected by masses of lianes, creepers, and vines. Often they rot away standing, torn to pieces by the kaka parrot (*Nestor meridionalis*) in search of grubs.

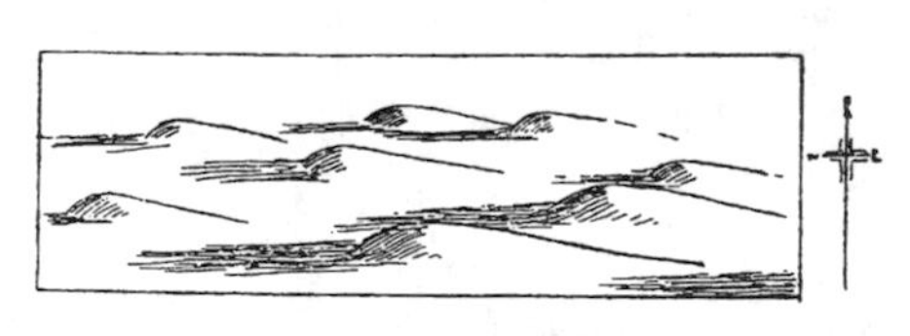

Hummocks—Central Tutira.

Lastly, there is a very pretty little bit of evidence afforded by two ferns which could never have established themselves under present conditions. Each of them is a forest species—the one, the umbrella fern (*Gleichenia Cunninghamii*), being usually found on wooded spurs; the other, a maidenhair (*Adiantum diaphanum*), on the forest floor itself.

FUCHSIA.

These survivors of an altogether different vegetative covering still manage to exist on modern Tutira though stunted and depauperated.

Although, however, there can be no question as to the existence of this former forest, its duration in time, date of disappearance, and cause of decay are problems not so easy to solve. Even taking into account the fact that subterranean soakage has stolen away the manurial values of leaves, branches, and bark—even, I say, taking that into account, it seems extraordinary that soils for any time under forest should have become so barren in so brief a time. Nine-tenths of Tutira have been unable to support ryegrass for a single season, yet it is certain that forest had not long disappeared off the face of the ground when settlement took place.

That soil and subsoil do not seem anywhere to have been thoroughly intermixed, throws but little light on the question of the duration in time of this primeval forest. As has been explained, more often than not trees do not fall when dead,—they decay upright, the great boughs snapping indeed with age and weight of epiphytic and parasitic growths, the stem as often as not mouldering away, devoured by insect life and torn to bits by birds. Admixture of humus, pumice grit, and red sand has taken place no doubt to a certain extent, yet the yellow hummock material exposed by the overthrow of a fire-swept forest shows distinctly different from the top twelve inches.

In regard to date of disappearance, the oldest natives I have questioned—men of eighty or ninety—have no recollection themselves of great forest fires, nor have the memories of such events been handed down in tribal history. It is probable that no huge conflagration has occurred, but that the disappearance of the old-time forest has been piecemeal. This negative evidence of a tardy retrogression is borne out by the amount and by the condition of timber in various parts of the run. The differences can best be illustrated by portioning the station into imaginary belts of equal width. Thus, throughout the most coastward belt, little surface timber will be found to remain even on sites favourable to its preservation. Another belt, more inland, will furnish surface timber in small quantities, bog timber and a profusion of hummocks with roots completely rotted. A third belt, still farther away from the coast, will provide a greater amount of both surface and bog timber; hummock markings are rather less worn with wind, frost, and rain, roots and stumps are not altogether decayed. A fourth belt will give us tall blackened boles, still here and there erect, also immense numbers of fallen trunks but partially decayed. This belt must have flourished as green forest within ten or fifteen years of my arrival at Tutira. Nearly one thousand acres of trees must have perished by fire about '65 or '70, for in '82—the date of my arrival—a third of the timber was still erect; thousands of boles, blackened and charred, but still branched, stood perpendicular, eighty or ninety feet high. A fifth belt would include the ranges and give us the growing bush of the present day—the last remnant of the primeval forest that once shaded the whole run.

This slow retreat towards the mountains is not likely to have been caused by change of climate. It is of too recent date to be thus accounted for; we must seek another reason for the triumph of bracken over woodland. Sometimes I incline to a solution, only the barest outline of which can be given. The latest considerable influx of islanders from outside took place, it is believed, about five hundred years ago. These immigrants from wheresoever they came probably dispossessed tribes neither so virile nor so numerous. There was no bar, therefore, to the rapid increase and multiplication of the dominant race. The ancient Maori was an excellent cultivator, keeping his crop grounds in a high state of tillage, carefully weeded, dug, and hoed.

Their earliest settlements as an island race were planted on coastal lands. Now wherever man works, one of his most helpful agencies is fire. Maybe the fires of these immigrants five centuries ago began that destruction of the forest, not yet quite complete when Europeans arrived in Hawke's Bay. In dry seasons these fires doubtless ran far beyond the limited Maori clearings; we can be equally certain that fern took possession of the rich loose mould thus opened to the sun. Furthermore, a fern crop, once established, would, every fourth or fifth season, be sufficiently thick to burn; the flames would on each occasion destroy a new breadth of timber. Even the small number of seedling trees able to compete with the bracken would never attain to more than four or five years' growth. Thus bracken would take possession of the coastal regions first, then gradnally work inland to damper, colder areas. Fire would also be largely used as a means of easy access to inland hunting-grounds. There can be no doubt that the aboriginal forest was destroyed by fire.

There is equally little room for doubt that if fires, mankind, and

Yellow Kowhai.

stock were banished, the woodlands would re-establish themselves. Twenty-five years would see the surface, so painfully grassed, once again in fern; one hundred would see forest reclothe the countryside. Within my own period, an example of this general tendency has presented itself. In '83 a part of the run known as the "Sandhills" had been fire-swept. It lay black and bare except for one patch of five or six acres of fern. This oasis of ravine or dene, evidently particularly damp even in summer heat, lay on steep slopes facing south-east, but with no other apparent defence from the fire which had desolated the surrounding lands. Unscorched in '83, it has since then again and again escaped periodical fires purposely lighted. In forty seasons it has been transformed from fern to scrub and from scrub to light bush. It contained in the 'eighties a great deal of tall fern together with a proportion of small tutu (*Coriaria ruscifolia*), koromiko (*Veronica salicifolia*), and manuka (*Leptospermum scoparium*). Later, appeared slender matapo (*Pittosporum tenuifolium*) and makomako (*Aristotelia racemosa*), fuchsia (*Fuchsia excortica*), hinahina (*Melicytus ramiflorus*), kowhai (*Sophora tetraptera*), and rangiora (*Brachyglottis Rangiora*); the original shrubby tutu and koromiko grew almost into trees; the manuka stiffened into poles; tree-ferns, lawyers (*Rubus australis*), and supplejacks (*Rhipogonum scandens*) appeared as under scrub, the fronds of the stifled bracken grew further apart. Seedlings and saplings of the larger forest species, white pine (*Podocarpus dacrydioides*), rimu (*Dacrydium cupressum*), and totara (*Podocarpus Totara*), established themselves. With the lapse of another twenty-five years light bush, the precursor of forest, would have possessed the little dene.

Tutira, then, has been at one period entirely covered with forest, bush of a lesser size and more ephemeral nature possessing the eastern coastal belt, timber of great girth and of a more durable character flourishing throughout the trough and western portion of the station. Consequent on forest fires a gradual general retreat inland of these woodlands has been traced. For two or three centuries maybe eastern Tutira has been bare of trees; on the other hand, in the central run a patch of one thousand acres has been destroyed only as recently as the 'seventies. On the far west, relics of the ancient primeval forest still grow green.

CHAPTER VIII.

TWO PERIODS OF MAORI LIFE.

THERE will be found in the following chapters some account of the bygone inhabitants of Tutira — their fortunes, their folk-lore, and their feuds. These relics of the past have been gathered from the mouths of three friends well stricken in years, Anaru Kune, now "gone west," Aparahama—for short 'Para or 'Pera,—and last but not least, Te Hata-Kani, that wonderful old man—*tauwhena*—a word meaning sometimes dwarfish, of small stature, but also used to denote a person who never grows old, but retains his youthful vigour to the end. To these three men and to the indefatigable Rev. P. A. Bennet I owe the history of the Ngai-Tatara. To be in sympathy with this *hapu* or sub-tribe and its old-world ways, readers would be well advised to shed the Decalogue, to accept for the nonce the ethics of the Stone Age, to imagine themselves bare-limbed, bare-headed, brown, the *pake* of everyday wear thrown over their shoulders, on high days and holidays clad in soft mats of woven flax, plumes in their hair and *taiahas* in their hands.[1]

[1] The generally accepted theory as to the colonisation of New Zealand by the Maoris, too definite to contain the whole truth, is that some five centuries ago a great migration from Hawaiki reached the Dominion. Mr Sidney H. Ray, to whom I applied for information on this matter, and who has kindly allowed me to use his reply, writes thus :—

"I think that whenever the introduction of an element from the west into Polynesia took place, it must have been a great deal earlier than the fourteenth century. There is internal evidence that the known Maori language is formed by the imposition of an Indonesian linguistic element[1] upon the speech of an earlier population. The words were adopted in their Indonesian derived form with no apprehension of their exact meaning. Thus a word might be *pahiwi* or *hiwi*,[2] *pahore* or *hore*,[3] *karipi* or *ripi*,[4] *karakape* or *kape*,[5] with no sense as to differences of meaning. That the prefixes are *not* meaningless is certain, but the Maori borrowers were just as ignorant of their meanings as they may be to-day of the suffix *ana* in *tariana*,[6] quoted by you, or of the *meta* in Hakarameta,[7] the *hana* in Kamupeneheihana,[8] or the *mana* in *pirihimana*.[9] Another fact is indicated by this borrowing of ready-made words. Other Polynesians do not know them in these forms, hence they did not give them to, nor did they

[1] By Indonesian is meant the original speech of the islanders of the Indian Archipelago.
[2] To jerk. [3] Peel. [4] Cut, gash. [5] To move with stick.
[6] English *-ion* in *stallion*. [7] English *-ment* in *sacrament*. [8] English *-tion* in *compensation*.
[9] English *-man* in *policeman*.

Te Hata-Kani

A preliminary word may perhaps now also be said as to the seeming redundancy in coming chapters of seemingly irrelevant names. Time to the Maori was of no account: every incident in a story was to be fully given, no detail was to be omitted. Never, therefore, must the reader be tempted to exclaim—it would not be *tika*, it would not be correct—What do the names of Te Amohia's two cronies, Mohu and Whangawehi—I give both on principle—in her escape after the captivity of Tauranga Koau, matter? What does it avail to know that Tataramoa was the father and Porangi the mother of the damsel Tukanoi—all of them, by the way, descendants of Kohipipi—in her love affair with the gallant, the gay, the red-headed Te-Whatu-i-Apiti? Why, it just matters everything; for after that fashion for ages have these stories been transmitted. It is proper, therefore, that in that exact shape they shall be crystalised in print.

It may be well now also to emphasise the anglification of place- and personal names during the brief space when heathendom and Christianity still divided the allegiance of the tribes. During that twilight interval it was that Te wai o-hinganga, for example, was changed into Bethany or—since there is neither B nor Y nor TH in the Maori alphabet—into Petane. Under the same scheme of things Te-Iwi-Whati, the grandfather of a friend who has done yeoman service in these chapters, became Abraham—Aperahama. Correlative to this change of place- and personal names was another in regard to weapons of offence—the musket was supplanting the spear. This same Te-Iwi-Whati, for instance, was desperately hurt by eight heathen spear-thrusts fighting the Urewera at Ngarua-titi. At a later period, missionaryonised into Aperahama—Abraham—he was no less badly wounded by Christian bullets at Tiekenui, again battling against the Urewera.

Of several of the lamentations, songs, and lullabies of the gallant *hapu* whose story I am about to relate, only general renderings into English are given; the older poems are not properly translatable into another tongue. I have not attempted it. There occur words so

receive them from, the Maoris. The words must have come directly from Indonesia to New Zealand in a migration which was not that of other Polynesians—*e.g.*, Samoans or Tongans. These received their share of Indonesian speech from other places and at other times. Hawaiki and Pulotu may stand for different origins, both possibly within the Eastern Ocean."

ancient that their meanings have become lost, and occult allusions almost or quite impossible of elucidation. In the folk-lore tales and tribal legends the exact Maori phrase descriptive of any striking custom or statement has been preserved. Alas! from what the writer has been able to gather from the annals of the Ngai-tatara alone, he is cognisant of the wealth of material that must have elsewhere perished.

The lands described under the designation Tutira were included in the immense territory of old, claimed or occupied by the Ngati-kahungunu—a countryside stretching from Gisborne to Woodville—from Turanga to Tamaki. Descent is claimed by the Ngati-kahungunu from Rongo-kako, whose son Tamatea arrived in the fast-sailing Takitimu, one of the most famous canoes of the great *heke* or migration from the mythical Hawaiki. In this great tribe were included the *hapu* living on or possessing interests in Tutira. Formerly it had been known as Ngai-Tatara, but later, for reasons yet to be told, it was styled Ngati-kuru mokihi: it was made up of two minor septs—the Ngati-moe and the Ngati-Hinerakai—each of which, moreover, possessed its own especial cultivation plots. The two were, however, indissolubly allied "*hoa matenga*"—friends together to the death. There were also intimate ties of blood and friendship connecting them with the neighbouring *hapus*. In the accompanying map are marked the boundaries of the lands of the Ngai-Tatara, and the names of the sub-tribes by whom they were surrounded.

Although there were *pas*—stockades—built on Tutira, yet within its boundaries the Ngai-Tatara were in great degree wanderers. At any rate they did not chiefly put their trust in stationary fastnesses; rather they relied on stout hearts and active limbs; "*Ko to ratou pa ko nga rekereke*"—"their *pas* were in their heels": that was the tribal motto. Like the Douglas of old, they preferred to hear the lark sing rather than the mouse squeak. Their temporary camping-grounds were chosen, doubtless, according to the seasons and the conditions of food supply. As another local proverb has it: "*Ka pa a Tangitu, ka huaki a Maungaharuru, Ka pa a Maũngaharuru ka huaki a Tangitu.*" "When Tangitu"—the deep-sea fishing-ground off Tangoio—"is closed, Maungaharuru"—a mountain range prolific in bird life—"opens; when Maungaharuru closes, Tangitu opens."

Man, like other animals, is dependent for his maintenance and increase on the nature of the soil in his possession. The Maori is a

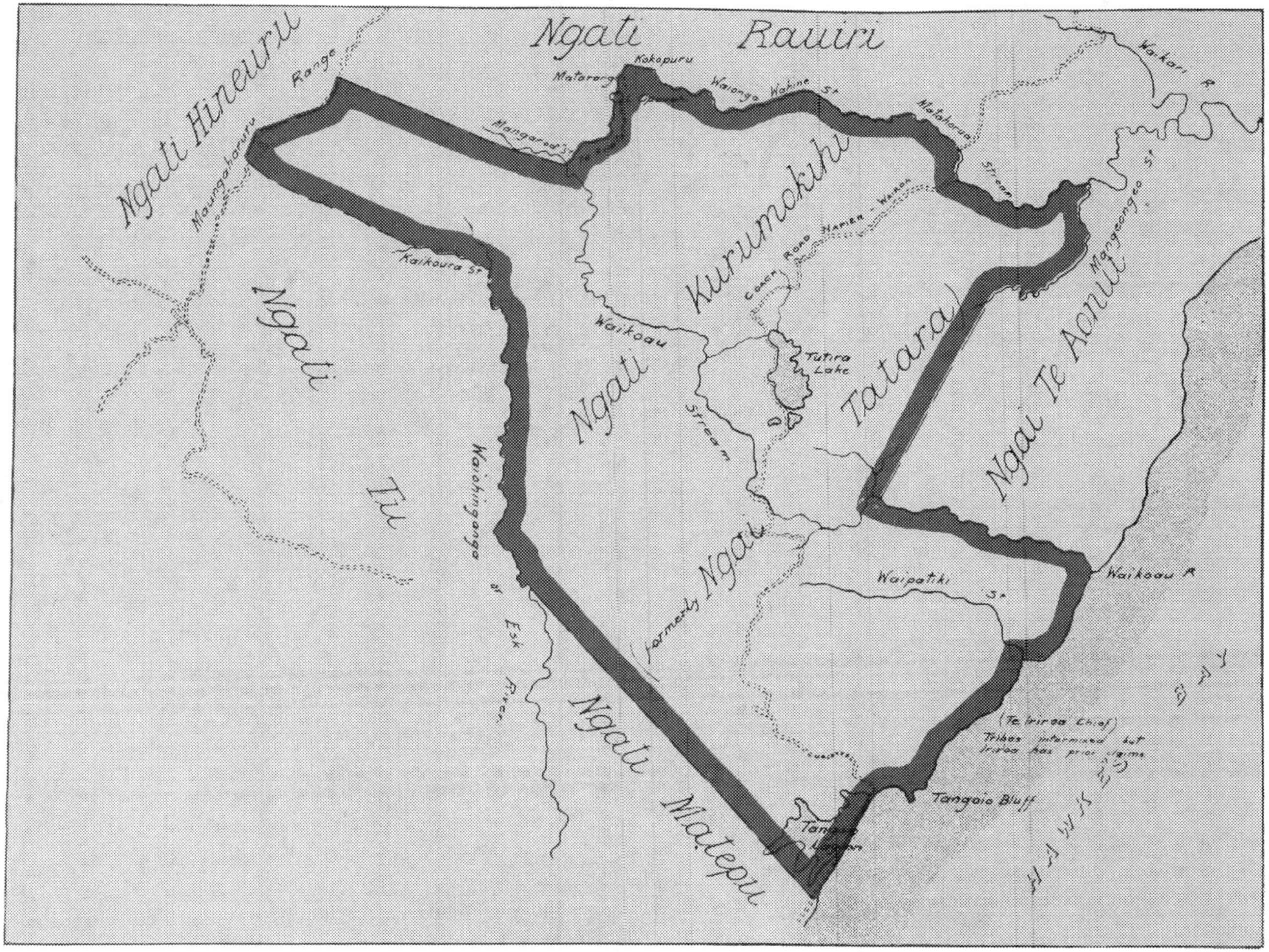

Ngati Rauiri
Kokopuru
Matarang
Waiongo Wahine St
Matahorua
Stream
Waikari R
Ngati Hineuru
Range
Maungaharuru
Mangaraa St
Kurumokihi
Coach Road
Hamien - Waikoau
Maungaongao St
Kaikoura St
Waikoau
Tutira Lake
Ngati
Tatara
Ngai Te Aonui
Stream
Ngati
Tu
Waiohinganga
or Esk River
formerly Ngai
Waipatiki St
Waikoau R
Ngati Matepu
(Te Iriroa Chief)
Tribes intermixed but Iriroa has prior claims
Tangoio Bluff
Hawke's Bay

descendant of ancestors who have travelled from warmer climes; in New Zealand he has clung to the coasts, to the thermal regions and to the northern portions of the North Island. The Ngai-Tatara during winter, and whilst planting of crops was in progress, dwelt chiefly about the estuaries of the local rivers. The climate of Tutira was rather too cold and wet, the land usually too poor for the cultivation on a great scale of such exotics as the taro (*Colocasia antiquorum*), the hue (*Lagenatia vulgaris*), and the kumara (*Ipomœa batatas*). On the other hand, the flax (*Phormium tenax*) [1] growing about its swamps was celebrated for strength, the shallows of the lake were paved with mussel-beds—kakahi (*Diplodon lutulentus*), the flavour of its eels was unsurpassed. They were speared in the lakes, they were caught in enormous numbers in eel-weirs—*patunas*—or in *whare tunas* built along the edges of streams. In the forests of the interior, pigeon (*Carpophaga Novæ Zealandiæ*), tui (*Prosthemadera Novæ Zealandiæ*), and kaka (*Nestor meridionalis*) abounded; they were captured by means of decoy birds, or snared by natives ambushed beneath selected trees. Often a superabundance of birds preserved in their own fat was bartered for the delicacies of other *hapus*. Tools of wood and weapons of stone were manufactured. These relics of bygone days—pounders for the softening of flax-fibre, adzes, eel-spears, and bundles of bird-snares hidden in rocks—are still from time to time discovered. The womenfolk by many processes worked the tall flax-blades into soft beautiful mats, or nursing their babies, sung them to sleep with such lullabies as the following:—

E hine e tangi nei ki te makariri i a ia,
Kaore nei e hine te rau o te ngahere i a taua.
Pinea rawatia ki Tutira ra;
Ki te ue pata, ki te kai rakau.
A ehara e hine i te roto hou;
He roto tawhito tonu na matou ko o nui.
Ina tonu te raro i potau atu ai e hine:
Ko Hine-rau-wharariki te hahanu noa nei:
Ko tini o hunga ki roto kakati ai e hine.

[1] Through this plant an acquaintance with the Latin tongue is the heritage of every man, woman, and child in New Zealand. All know two words of it—*Phormium*, flax; *tenax*, tough. No writer on country matters can forgo the magic words; even flax-millers attain scholarship. What *pax vobiscum* was to Wamba, the son of Witless, *Phormium tenax* is to the New Zealand settler. Wool may be down, stock may be down; he braces himself in the knowledge that *phormium* means flax, and *tenax* tough.

O maiden, who art weeping because of the cold,
We own no garments of forest-leaves, O child.
Let us gather together to Tutira
Where are eel-weirs and fruit-laden trees.
The lake, my little girl, is not a new lake,
But an ancient lake possessed by thy ancestral great ones.
It is only just now that the food has gone:
Hine-rau-wharariki is preparing the fibre:
Suppressing the hunger-pangs gnawing within.

Tutira and the adjoining lands were a sort of connecting link between the seaside villages and the ranges of the interior. The Ngai-Tatara during peace dwelt about the coastal estuaries and the lake. During war they sheltered in the forests and fastnesses of the hinterland. The glory of the *hapu* was in their continued occupation of so famous a lake, in their possession of so unfailing a food supply of the most highly-prized kind. Their warriors were active, bold, and resolute; nor, as we shall see, did the womenfolk of the sept fall short of their husbands and sons in the accomplishment of deeds of derring-do.

The annals of the tribe may be divided into two distinct periods. One is of a time when the Ngai-Tatara—when the Maori people everywhere—had attained its maximum numbers; when, on Tutira as elsewhere, every height and fastness was utilised for defence, when every fertile locality was devoted to cultivation. The other period, brief in its duration, is marked by the presence of *kaingas* or open villages with considerable areas of crop-land adjacent, by *whare* sites immediately extraneous to the fortified *pas*, such sites corresponding to the overflow in old-world cities of houses beyond the ancient walls of defence, beyond the city gates; lastly, by the appearance in the gardens and cultivation-plots of alien plants and of alien fruit-trees.

The first period represented heathendom, the second Christianity. Evidence of the former is plentiful in folk-lore and tradition. There are records of forays from the direction of Mohaka and from the regions of Waikaremoana and Heretaunga. Doubtless, according to modern reckoning, no action that could be dignified by the name of battle has taken place on Tutira soil; perhaps indeed the killing of Ti-Waewae and the vengeance of his tribe is the deed that has circulated furthest beyond the marches of the run. Nevertheless although skirmishes on Tutira have been but skirmishes, they illustrate the former way of life of its inhabitants; as part of the history of the station they must be recorded.

In '82 sites which still showed distinct traces of fortification were Kokopuru far to the west, the peninsulas Oporae and Te Rewa, and the island Tauranga-koau. There were other spots also where evidences of former habitation were discernible; one sure and infallible sign indeed of ancient Maori settlement was in the 'eighties the appearance of certain native grasses. *Danthonia semiannularis* and *Microlæna stipoides,* elsewhere smothered by fern and scrub, survived about the erstwhile *whare* sites and along the edges of the hard-trodden paths.

Kokopuru was a cone-shaped hill connected by a narrow ridge with the Otukehu range—the "Nobbies." The main defensive work of the *pa* built on its top was in '82 almost intact. Immense upright totara boles and boughs, placed circlewise about the waist of the solitary hill, then stood black and erect. Undisturbed, this heavy palisade work

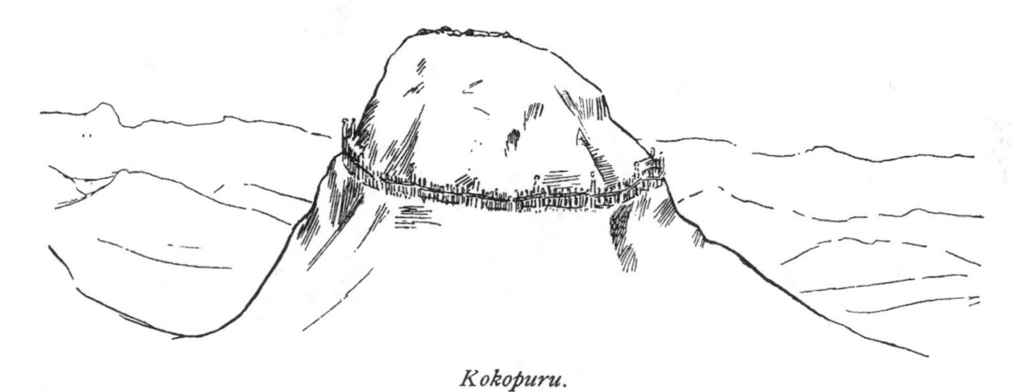

Kokopuru.

should have lasted for centuries; it was pulled down and converted, not by me, into fencing posts. This really fine example of a fortified *pa* now resembles any other peak of the neighbourhood. Signs of former use are almost gone—only ash and splintered stone tell of the ancient kitchen midden. In 1919 my daughter discovered what will prove probably the last vestige of native occupation—a fragment of totara with tool-marks still visible on its grain.

Oporae, a minute peninsula on the eastern edge of Tutira lake, also shows signs of fortification. On three sides water was its natural defence, on the fourth a bank and fosse—*maioro*—had been cut, which, though partially filled in, is still many feet in depth. On the edges of the level summit cavities remain, out of which have been burned or pulled up, or from which have decayed, the huge posts of the main defence.

Entrance across the moat was by bridge; no sign of that remains, but the narrow gap in the embankment where stood the ancient gateway is still distinct. The natural declivities also of the little peninsula have been straightened into perpendiculars. Within these defences stood on levelled ground, in close proximity to one another, the reed-thatched huts. There are faint indications still of canoe traffic on the adjacent shore.

Te Rewa, the terminal point of the spur which nearly divides

Oporae pa.

Tutira from Waikopiro lake, was another and larger fortified peninsula. Its natural defences on one side were impenetrable marsh, on two sides water, northwards Tutira, southwards Waikopiro; its fourth approach was guarded by a bank and fosse similar in principle to that of Oporae, but of greater width. Moat and embankment are now alike obliterated; they have been trodden flat by the hundreds of thousands of sheep that pass yearly to and from the wool-shed.

The pits of the ancient stockade posts are likewise worn away; only the earthen floors of former *whares* remain preserved by the matted growth of an alien grass—*Poa pratensis.*

Tauranga-koau, the island off the east shore of Tutira lake, was in the beginning a mere bare reef,—as its name signifies, "a perching place for cormorants." This natural point of vantage was built up and consolidated by soil shipped from the mainland. As late as '82, though hardly an upright remained in position, quantities of timber not yet utterly rotten lay in shallow water or on the island itself. Many of the prone posts or *take* of the palisading were still ornamented with the curious top or head supposed to be commemorative of ancestors, and dear to Maori fort-builders. Beneath the water there

Te Rewa.

were visible not only the lines of holes sunk for the main defence, but, preserved by water, even remains of the smaller innermost stakes of the breast-work—*kiritangata.* Water was, of course, the principal natural defence of this *pa*, which could only be reached by canoes, by rafts, and by swimming.

Other peninsulas have also been occupied, but of their defences little now remains saving natural declivities made more precipitous, beds of broken kakahi shell, collections of splintered stone used in the ovens, and as elsewhere levelled earthen floors. About every one of them also grew in the 'eighties the native grasses already named. On one of these juts of land, Pari-karangaranga, there remained until ten years ago a section of about twenty yards of native footpath, a trail trodden out by naked feet long prior to the advent of the booted settler.

This old-world track, slightly dished and about eighteen inches in width, used to be one of the most interesting relics of Maori life on the run. It had remained untouched on a soil of grit, dust, and powdered kakahi shell. There had been no inducement for cattle, sheep, or pig to visit this desolate little bluff with its unpalatable stunted bracken and starved danthonia. Alas! it exists no longer; like other sentimental interests dear to the writer, it has been sacrificed to exigencies of station management. Its contour has been defaced, obliterated indeed by cattle.

Such were the fighting forts and strongholds of the virile *hapu* who owned Tutira and the adjacent lands. Their way of life was similar to that of every tribe of New Zealand. Their motto, "*Ko to ratou pa ko nga rekereke*"—"their *pas* were in their heels"—was, however, only relatively correct, for until about the 'fifties, as Manning[1] says, no man slept safe who did not sleep armed and within walls. Out of their strongholds every morning marched the men, prepared for all contingencies, their womenfolk and children in the rear; into them every evening retired their owners, the women and children in front, bearing wood, water, and food for the evening meal.

About the middle of the century a change took place; an Indian summer of peace prevailed, a brief space between the cessation of tribal warfare and the struggle which from the beginning had been inevitable between the brown race and the white. Missionary influence had quenched the fires of internecine hatred, the war and bloodshed which had seemed until then the normal condition of the land. The tenets of Christianity had widely spread amongst the tribes. Instead of as formerly sleeping within the precincts of the stockaded *pas*, the natives of Tutira, like their fellows elsewhere, dwelt now "after the manner of the Sidonians, careless" in open villages. The *pa* had given place to the *kainga*; cultivation grounds lay undefended, unfenced, unhidden; there was no longer need for the concealment of crops nor for their hasty furtive gathering and storage. Heathen names of villages gave place to Christian names; Johns, Peters, Abrahams, and Isaiahs swarmed in every tribe. During this golden interval between war and war the

[1] Author of 'Old New Zealand,' a volume remarkable alike for its sympathetic appreciation of the Maori character and for its abounding wit.

principal open villages on Tutira extended secure and peaceful on the "grubbed grounds" of the Mangahinahina and on the fertile slopes now called the Racecourse Flat. There were smaller settlements also at Kahikanui and at Te Rewa; there were isolated *whares* besides, scattered here and there along the margin of the lake, the homes of outliers, each with its patch of tillage and grove of peach-trees.

A farther and a final change occurred immediately prior to the taking up of Tutira as a sheep-station. As not long before the fighting forts and heights had been vacated, now the open villages were deserted. A general shrinkage in the native population of New Zealand had drained off the inland tribes and sub-tribes towards the coast, towards warmth, richer lands, food supplies more easily won from sea, lagoon, and river-mouth. Tutira was deserted save as a temporary residence of hunting-parties.

CHAPTER IX.

TRAILS FROM THE COAST TO TUTIRA.

MAORI footpaths in olden times followed the lines of scantiest vegetation such as open river reaches, unfertile hill-tops, ridges bare of cover, lines of ingress and egress, in fact, least liable to ambuscade. There were two main trails connecting Tutira with the coast, the one from Arapawanui on the east, the other from Tangoio on the south. These, as also the tracks round the lake and the outward track to the ranges, I shall use as threads on which to string our narrative; from them I shall invite the reader to listen to the legends, folk-lore, and history of the localities traversed.

Starting from Arapawanui on the coast, the track inland followed the general line of the river Waikoau as far as the eastern corner of Tutira. The going was fairly open and level; the river, flowing only a few score feet above sea-level, had deposited along its banks sand, grit, and limestone rubble washed from its upper reaches. At its great bend, near to the several boundaries of Arapawanui, Tangoio, and Tutira, precipitous marl cliffs compelled a deviation. Almost exactly opposite the spot where the Mangahinahina stream joins the main river, our trail crossed on to Tutira soil. Immediately after passage of the Umungoiro ford a faintly defined subsidiary track followed for a quarter of a mile the general direction of the river-bed to a little clearing in a patch of bush. Doubtless it had been the home of some outlier, a residence only habitable under the conditions of the second phase of native life on the run; like every settlement of that later period, it was marked by the presence of peach-trees. Reverting to the main track from the Waikoau, it followed the line of the Mangahinahina brook until that streamlet, as streamlets do on Tutira, narrowed into a gorge. It continued along a narrow ridge, first in a northerly line, then along the ridge of another

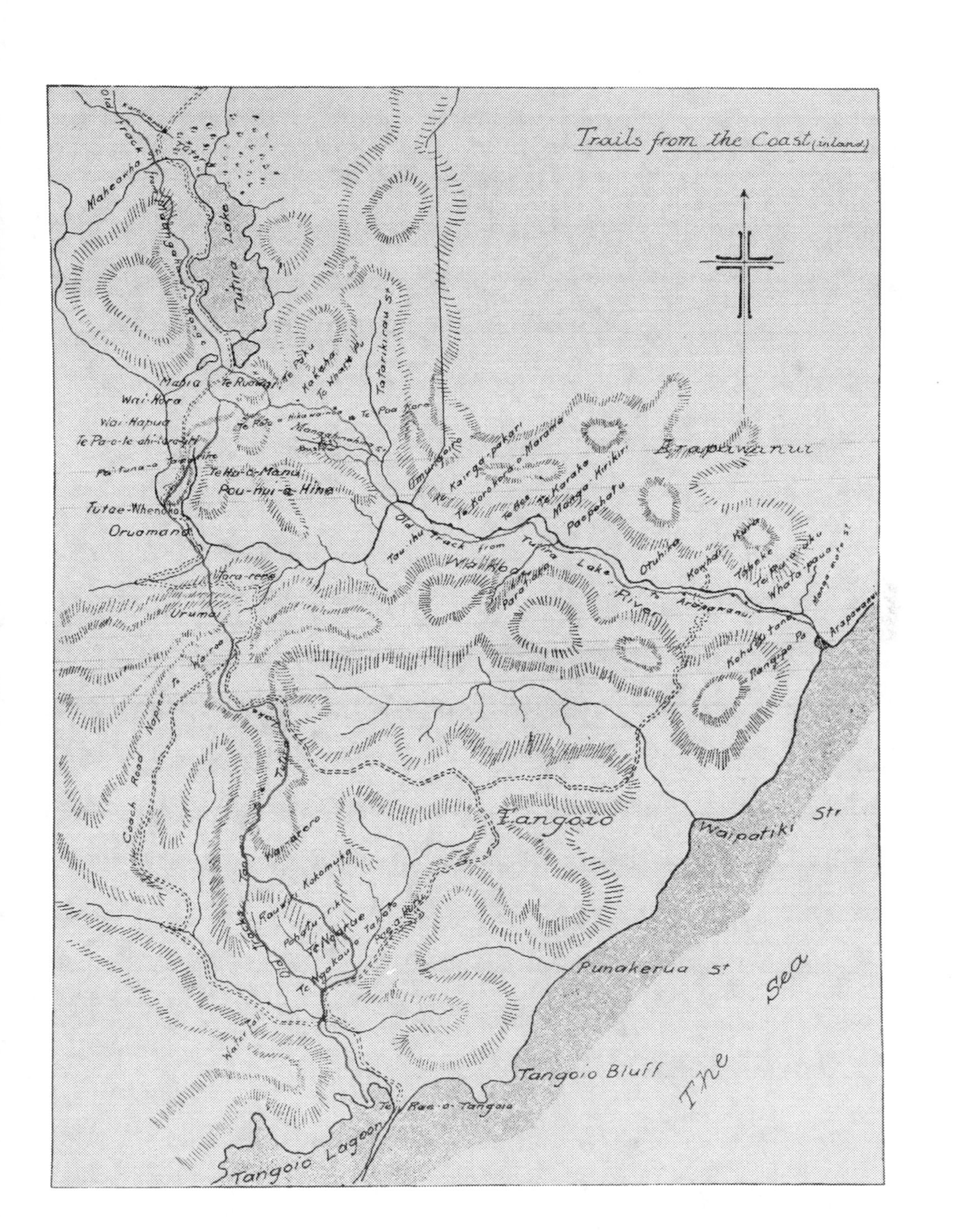
Trails from the Coast (inland)
Tutira Lake
Maheawha
Mahia
Te Ruawai
Kakaha
Tatarikirau St
Wai-Kora
Wai-Hapua
Te Pa-o-te ahi-turoa
Te Ho-o-Manu
Pou-nui-a-Hine
Tutae-Whenua
Oruamana
Omunu
Kairaa-pakari
Te Karaka
Manga-Kirikiri
Paopohatu
Arapawanui
Old Tauihu Track from Tutira Lake
Waikoau
Parataua
River
Arapawanui
Otuhira
Tabake
Whata-paua
Kohukutanoa
Rangipo Pa
Uruma
Coach Road Napier to Wairoa
Tangoio
Waipatiki Str
Kokomuka
Punakerua St
The Sea
Tangoio Bluff
Tangoio Lagoon

spur in a westerly direction. At the base of the long ascent, on which are situated the group of rock fragments called Te-Poa-Kore, it bifurcated, the less trodden path turning south towards the *kainga* of Mangahinahina. This *kainga* was perched on a rise near to woodlands of the same name. Here in ancient times grew the largest trees to be found on eastern Tutira. One of them, a magnificent totara named Te Awhiawhi, lay in the 'eighties fallen, topped, and rudely hollowed into the shape of a canoe. About the *kainga* itself were visible no signs of defensive works; in spite of this total lack of fortification the village belongs, nevertheless, to the old order of things, and is illustrative of what has been already told of the Ngai-Tatara—that their *pas* were in their heels. The kumara or sweet potato plantations here were the largest on the run, the rich ground and excellent exposure well suiting the requirements of this tropical tuber. About the sites of the old *whares* grew also in the 'eighties the usual signs of the later era—peach-groves. Surviving from the garden plots of this derelict village I have found clumps also of another alien—a species of mint (*Nepeta cataria*).

Long prior to the 'eighties the "grubbed grounds," as these cultivation lands used to be called, had reverted to a wild state. Only the name remained to show that they had been stumped by native labour. Thickly covering them, groves of ngaio, wine-berry, and manuka had sprung up, none of these natural plantations showing normal forest growth. The trees in each patch were of similar age; there was no admixture of species. They had evidently taken immediate possession of tilled ground abandoned and disused. The original vegetation of the "grubbed grounds" had probably been light bush, with just sufficient intermixture of bracken to carry a fire. The natives had burnt this growth in a dry summer and afterwards taken advantage of the favourable conditions to clear the land thoroughly. The *kainga* itself was built on just such a site as the old-time natives cared for: its clustered *whares* stood on the gentle slope of a spur studded with huge limestone crags deeply sunk into the ground. One of the most lovely sites on Tutira, it it was raised well above the damp of the wooded ravines on either side; it caught the earliest rays of the morning sun up the long rift of the valley of the Waikoau. If its inhabitants did not live happy in content and country freedom they must indeed have been hard to please. We know at any rate that at least one other person desired to be on that pleasant spot. She was a girl called Hariata, in love with Te-Iwi-

Whati. Looking downwards from Te-Karaka, a high point between Waipatiki and Arapawanui, she could see, or nearly see, the dwelling of her lover. The following is the *waiata* composed for her singing by a friendly poetess, Kowhio :—

Akuanei au ka piki ki te Karaka ra ia
A marama au te titiro ki Manga-hinahina ra.
Kei raro iho na ko taku atua e aroha nei au.
Taku hinganga iho ki raro ra ko turi te tokorua ;
Te roa noa hoki o te po tuarua e Iwi.
Oho rawa ake nei ki te ao, hopukau kahore, ei.

I will climb with the dawn to the top of Te Karaka
So that I may get a clear view of Manga-hinahina.
Just below lies my beloved one.
Whilst I slept alone, my tucked-up knees only were my bedfellow.
During the long night, twice, Iwi, I have dreamed of thee.
I awoke, I felt for thee ; thou wast gone !

Returning again to the main route, it followed in a westerly direction the ridge of a very steep leading spur passing the group of limestone rocks, Te-Poa-Kore, already named, and later the minute tarn, Te-Roto-a-Hikawainoa. Still following the hill-tops it reached the elevation Te-Whare-Pu, and lastly the high ground called in most ancient times Kakeha, but more recently, in commemoration of a gross episode of the nursery, Tutae-o-whare-Pakiaka. Here the track again branched, the less trodden portion dropping in a steep descent on to terrace levels, known in modern times firstly as the "Reserve" and later as the "Racecourse" Flat. The other branch also dropping over the brow Te Puku, and passing the group of limestone rocks also so named, followed the unbroken line of a narrow ridge downwards towards Waikopiro—this jut or headland Te Puku being known as the "head," the lakelets Waikopiro and Orakai as the "eyes" of Tutira.

For the present we can leave this path and describe the other line inland — the trail from Tangoio. From the important coastal *pa* Ngamoerangi, long since swept away by the sea, and in later days from the Rae-o-Tangoio *pa*, it followed for a considerable distance cultivated lands along the bed of the small stream, Te Ngarue, that debouched on to the flat lands from the north. At the junction of this stream with the Pae-a-Huru the trail forked, one branch ascending a steep northerly spur, the other proceeding along the Pae-a-Huru for half a mile, when it also turned north; on the first-mentioned path there are no signs of use, but forty years ago scattered peach-trees and grape-vines survived along the second trail. In early days these, and more rarely other foreign fruits, were planted by travellers as acts of good citizenship. The seeds thus dibbled in flourished extraordinarily; blights were unknown, there were no sheep to nibble, no cattle to break down and destroy.

Leaving the stream-bed when it became a gorge, the last-mentioned track rose by steep gradients up the Te Ngakau-o-Takoto spur, and followed several leading ranges of the interior of the Tangoio run in a north-westerly direction to "Dolbel's boundary gate," Kai-arero, where the two branches conjoined. Later the track descended from the range Urumai by precipitous ridges into the valley of the Waikoau. Near that river flourished in the 'eighties a couple of small peach-groves, marking as elsewhere during the second period the unfenced cultivations of outliers, sometimes aged couples whose children had grown up, sometimes solitary individuals. This locality was called Tara-rere. A few yards down-stream from the site of the present bridge our track crossed from the Kaiwaka run to Tutira. It climbed the steep spur Tutae-o-Whenako, and continued along the western side of the limestone streamlet Te Hu-o-Manu. Where this rill joins the main river is situated the cave Oruamano.

On the right below the high top called Pou-nui-a-Hine is another small cave beneath a limestone projection, in ancient times the home of a *kumi*. The story is still related by the Ngai-Tatara of a visit by a Waikato chief to Tutira. He had heard of the *kumi* at Pou-nui-a-Hine, but derided the tales that were told concerning its powers. Maybe, however, he was less of a disbeliever than he posed to be. At any rate, he was persuaded by one of the *tohungas*—wizards or priests—who had power over the *kumi*, to visit the spot. They climbed the heights, and eventually reached the projecting ledge beneath which the creature lived, in the likeness, I am given to understand, of a *tio* —a bivalve of some sort. The *tohunga* then recited the necessary incantations, with the result that the shell gradually opened, revealing a small lizard-like reptile, *moko-parae*. The Waikato man was interested but still unconvinced. The *tohunga* recited further incantations, which had the effect of making the *kumi* visibly grow. The attitude of the

Waikato man began to change. He saw with his own eyes the reptile increasing into a formidable monster. He dared not watch longer, but becoming panic-stricken, took his departure as fast as his legs could carry him. His flight was the signal for the *kumi* to give chase. Down the cliffs they hurried as fast as they could go. When they reached the "Racecourse" Flat they were seen by Hine-kino, a wise woman or priestess or female *tohunga*, who also had considerable power over the *kumi*. She saw the predicament into which, by pride and presumption, the Waikato man had put himself. Straddling out her legs, she called to him to run between them. The Waikato man—his choice the devil or the deep sea—did so, with the result that the *kumi* stayed its chase and returned to its home below Pou-nui-a-Hine. Now, in olden times, except in the case of a wife, it was not proper that a woman should pass over any part of a man; sitting at night with legs outstretched around the whare fires, a woman about to move across the circle will always for that reason give notice of her intention, the menfolk tucking up their legs to avoid contact. When, therefore, the Waikato man rushed between the legs of the priestess Hine-kino, he lost *mana*—authority, prestige, reputation,—the word is hard to translate; he had sued for protection; he had forfeited his highly-prized attributes of rank and chieftainship; no longer would he be recognised as a leader of men in the lands of the Ngai-Tatara. His travelling *mana* had undergone what the Maoris termed *tararo*—a casting down.[1]

Our track still rising, now passed on to the "Racecourse Flat." Much of these rich washings from the hills above has been worked, the Maoris having taken advantage, as in the case of the burnt bush of the Mangahinahina, of favourable natural conditions. Through its cultivation-patches the track proceeded towards Tutira lake, passing a large square rock upon which has been growing, during my owner-

[1] A well-known instance of this custom occurs in Percy Smith's 'Maori Wars of the Nineteenth Century': "Te Ao-kapu-rangi, a woman of rank of the Ngati-Rangi-wewehi tribe, being anxious to save her own people when Mokoia was attacked, insisted on going with the *taua* or war-party. She importuned her husband, and through him Hongi Hika, to save her friends. To this Hongi at last unwillingly consented, making it a condition that all who passed between her thighs should be saved. She was in Hongi's canoe when Te-Awaawa—owner of the only musket in the island—crept behind a flax bush just where the canoe landed, and fired, knocking Hongi over. Hongi's fall, though protected from a wound by his steel helmet, created a sort of panic, during which Te Ao-kapu-rangi sprang ashore and, quickly making her way to a large house belonging to her tribe, stood with her legs straddled above the doorway, at the same time imploring her people to enter the house, which they did until it could contain no more, and all these were saved; hence the saying, '*Ano ko te whare whawhao a Te Ao-kapu-rangi*'—'like the crowded house of Te Ao-kapu-rangi.'"

ship of the station, a handsome kowhai tree. This great quadrilateral, Te Pa-o-te-ahi-tara-iti, was in bygone days a favourite haunt of the village children, who played on it "King of the Castle" and other games common to children the world over.

Proceeding, the track passed on the left the locality Wai-hapua, on the right the locality Wai-hara, then on the left Mahia. Here exist several deep pits, near which used to stand a couple of boundary-stones—*pou-rohe;* these pits—*ruakumara*—which are too minute for the storage of any potato crop worth garnering, were probably, as their name denotes, used for kumara. Far to the left, distant perhaps half a mile in the river-bed of the Waikoau, lay the locality Patuna-o-Tamarehe. The low rounded spur or hillock, Te Rua Awai, the ancient burying-ground of the tribe, was next passed on the right. Near-by grew the great *ti*—cabbage-tree (*Cordyline australis*)—on whose branches the bones of the dead were exposed previous to final sepulture. The burial-grounds, the tree, and the pit Piraunui, were alike deeply *tapu*—sacred—in ancient times; nor even now is the recollection of the *tapu* entirely gone; old Te Hata-Kani, whose recollections go back some eighty years, and to whom I am indebted for many of these old-world legends, was most circumspect in his perambulations, and though he said nothing, scrupulously forbore to tread on consecrated grounds.

Here for the present, conjoined on the southernmost shore of Waikopiro, we can leave the trails connecting Tutira with the ocean and the outside world.

CHAPTER X.

TRAILS ROUND TUTIRA LAKE.

IN Maori occupation the water area of Tutira was more productive of food than its solid surface. "*Te wai-u o koutou tipuna*"—"the milk of your ancestors"—runs the local proverb, signifying the constant supply of food ready to hand from lakes and rivers. It is natural, therefore, that a larger number of place-names, legends, and traditions should have been remembered about the vicinity of the lake, about its shores, small fertile marshes, and promontories, than about the remainder of the run. Most of the traffic was by water; even in the 'eighties there were several old canoes afloat; others still intact rest to this day submerged and safe in Waikopiro. There were narrow trails of a more or less temporary character connecting *pa* with *pa*, *kainga* with *kainga*, cultivation-ground with cultivation-ground, but probably in many places no permanent route existed. The line of sparsest vegetation would be the only general description of the eastern lake path, a line that must have altered in some degree with every fire run through the flax and fern, with every flood and consequent crop of landslips.

Starting from Piraunui and following the eastern margin of the lake, the trail, such as it was, passed on the right the celebrated spring of water Te Korokoro-a-Hine-rakai, on the left the log Te Waka-o-whakairo, ere reaching the small marsh known in modern days as "Pera's Swamp." Here on a dry patch of good land stood, in the 'eighties, the remains of an old hut, its little garden plot marked with a patch or two of thyme (*Thymus vulgaris*). About the dry warm apex of the same valley flourished a considerable peach-grove. On another dry rise, rich in leaf-mould and travertine oozings, grew a single peach-tree. On the farther side of Waikopiro swamp a sharp spur runs down from the main range terminating in the peninsula Te Rewa-a-Hinetu. Upon the

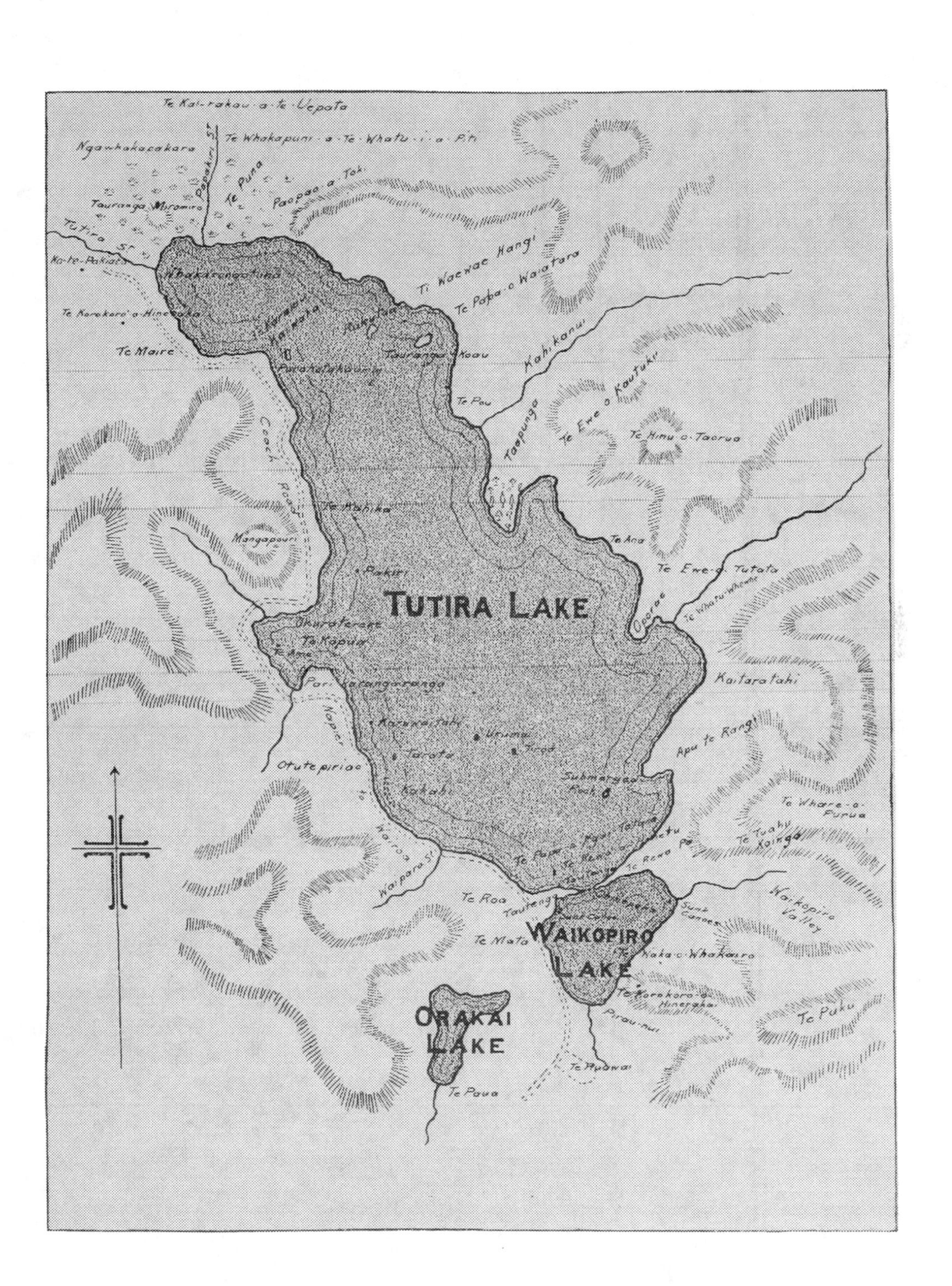

Te Kai-rakau-a-te-Uepata
Ngawhakaeakara
Te Whakapuni-a-Te-Whatu-i-a-Piti
Tauranga Miromiro
Te Puna
Paepae-a-Tok
Tutira St
Ko-te-Pakiaka
Whakarengatiha
Te Korekore-a-Hineraka
Te Maire
Ti Waewae Hangi
Te Papa-o-Waiatara
Tauranga Koau
Te Pau
Kahikanui
Taapunga
Te Ewe o Kautuki
Te Hinu-o-Taorua
Coach Road
Te Mahika
Mangapouri
Pakiri
TUTIRA LAKE
Te Ana
Te Ewe-o-Tutata
Okurateroke
Te Kapua
Kaitaratahi
Karewaitahi
Urima
Tarata
Apu te Rangi
Otutepiriao
Kakahi
Submerged Rock
Te Whare-o-Purua
Te Tuahu Koinga
Waipara St
Te Roa
Te Mata
WAIKOPIRO LAKE
Sunk Canoes
Waikopiro Valley
Te Kaka-o-Whakairo
Te Korokoro-o-Hineraka
Pirau-nui
Te Puku
ORAKAI LAKE
Te Ruawai
Te Paua

base of this tongue of land was situated the *kainga* Te Tuahu, upon its extremity was built the *pa* Te Rewa already described.

About 100 yards farther up the peninsula are to be seen the remains of a storehouse or *whata*, the land about it called Te Whare-o-Porua. Proceeding along the margin of a deep bay the promontory Te Apu-te-rangi is reached. Off the front of this peninsula exists, far beneath the water, a cavern or deep chasm, into the proximity of which no canoe would willingly venture. As my informant only knew of it to avoid, I could but learn that even to pass over it inadvertently was in the highest degree unlucky. On Te Apu-te-rangi are to be found the usual indications of occupation, naturally steep banks artificially straightened, level sites of former *whare* floors, and beds of kakahi shell, intermixed with splintered cooking stones.

To this day there flourishes on Te Apu-te-rangi a remarkably fine cabbage-tree, nourished probably on the remains of the old kitchen midden. The shore-line of the peninsula was particularly holy or *tapu*, for there in bygone days was the sacred spot—the *tuaahu*—where the *tohungas* practised their religious rites. The track then passed over the little flat Kaitaratahi, and 50 yards farther on over the larger marsh Te Whatu-whewhe, where tradition avers that in ancient days a large and valuable slab of greenstone was lost.

Farther along the lake we reach the minute jut of land on which the pa Oporae stood. It was sacked some five generations back by the Mohaka chief, Popoia, one of whose wives had misconducted herself with a stranger from Heretaunga. To rehabilitate the *mana* of Popoia two *tauas*[1] or war-parties were sent forth from Mohaka. Arriving on the

[1] Manning, in his inimitable 'Old New Zealand,' thus describes the *taua*: "Now something moves in the border of the forest—it is a mass of black heads. Now the men are plainly visible. The whole *taua* has emerged upon the plain. . . . They are formed in a solid oblong mass. The chief at the left of the column leads them on. The men are all equipped for immediate action; that is to say, quite naked except their arms and cartridge-boxes, which are a warrior's clothes. . . . As I have said, the men are all stripped for action, but I also notice that the appearance of nakedness is completely taken away by the tattooing, the colour of the skin, and the arms and equipments. The men, in fact, look much better than when dressed in their Maori clothing. Every man, almost without exception, is covered with tattooing from the knees to the waist; the face is also covered with dark spiral lines. Each man has round his middle a belt, to which are fastened two cartridge-boxes, one behind and one before. Another belt goes over the right shoulder and under the left arm, and from it hangs on the left side and rather behind another cartridge-box, and under the waist-belt is thrust behind, at the small of the back, the short-handled tomahawk for close fight and to finish the wounded. . . . On they come, a set of tall, athletic, heavy-made men. . . . They are now half-way across the plain; they keep their formation, a solid oblong, admirably as they advance, but they do not keep step: this causes a very singular appearance when distant. . . . This mass seems to progress towards you with the creeping motion of some great reptile, and when coming down a sloping ground this effect is quite remarkable."

same day respectively at Oporae and Te Rua-o-tunuku, a village near the site of the Tangoio wash-out, the inhabitants of both were slain, one man only escaping from Oporae. Considerations of why the people of Oporae should have been slain because a stranger from a district thirty miles south had insulted a chief of a sept twenty miles north, would lead us deep into the intricacies of Maori tribal custom; suffice it to say that every insult had to be expiated, if not on the person of the offender or his relatives, then on some other man or tribe, or failing that, even on inanimate nature.

Our track proceeding along the shore-line Te Ewe-o-Tutata, now passed the conglomorate cave Te Ana. "It was in the deep bay opposite that a chief named Tamairuna had cast his net for the purpose of catching eels. Tamairuna was holding one end of the net and his men the other. Presently they felt the net being dragged away from them by the *taniwha* known to haunt the bay. Their strength was powerless against the monster. Tamairuna had a wife called Te Amohia whom he had deserted some time previously, and who was noted for her prowess as a diver, and who possessed some kind of affinity or occult sympathy—it is difficult to give the meaning exactly—with the *taniwha*. So much at any rate was this the case that she was known as Te Uri-taniwha, the descendant of the *taniwha*. Tamairuna placed great value on his net. Having now lost it, his thoughts reverted to Te Amohia. He paid her a visit, and eventually succeeded in persuading her to consent to dive for his net. Preparing herself for the task by the recitation of proper *karakias*—incantations—Te Amohia dived into the subaqueous cavern and found the net rolled together and placed in front of the *taniwha*. Forbidding the monster to molest her, she pulled the net away and rising above water carried it back."

The track next passed the locality Te Ewe-o-Kautuku, situated between the edge of the lake and the great solitary hill Te Hinu-o-Taorua—the fat of Taorua. "It was so named because, when the days came for digging one of its ridges for fern-root, this man's body brought from Tangoio was eaten as a relish—*kinaki*—with the fern-root."

Reverting once more to the shore-line we reach the headland Taupunga. This headland of several acres has at one time been connected with the aforementioned hill only by the narrowest of ridges. It must then have been admirably adapted for defence. Though it is difficult to fix the date of occupation with any degree of accuracy, that there has been prolonged settlement is proved by the huge deposits of kakahi shell and splintered cooking stone, which are in places feet deep intermixed with soil. Taupunga may have been a *pa* when the Maori race was at its zenith in numbers. Except during that period it is unlikely that any population resident on Tutira could have manned so large a space. There are, at any rate, no signs of its use except as a *kainga* between the cessation of intertribal fighting and the beginning of war with the white settler.

A hundred yards inland, on the margin of the Kahikanui Swamp, and immediately beneath the western spurs of the hill Te Hinu-o-Taorua, flourished in the 'eighties large peach- and cherry-groves. The former fruit had been planted by the Maori in his last decade of occupation, the latter by the white man immediately after arrival on the run. Close to this orchard grew, in '82, three tall white pines, survivors of the kahika grove, from which the flat had probably taken its name. At this date, too, the remains of a reed-thatched whare still stood by the pine-trees. It had been for a considerable time station headquarters, one of the halting-places of the ark ere it finally rested on Otutepiriao, the site of the present homestead. Amid the then densely growing flax there existed also a clearing of several acres, the chance result of fire probably, in the first instance, but later taken advantage of and utilised for cropping, as in the case of the grubbed grounds of the Mangahina-hina, and the fertile slips and washings of the "Racecourse Flat."

Proceeding, our track passed over the point of the steep spur Te Pou. A little further along the lake lies the island Tauranga-koau, well known in east coast history on account of the death of Ti Waewae and the vengeance of the Ngai-Tatara, or, as they were later known, the Ngati-kuru-mokihi. Ti Waewae had married Hitau, a sister of Te Whata-nui, a chief of the Ngati-raukawa, a war-party of whose tribe was defeated near Puketapu. The survivors fled for protection to Ti Waewae, who was then living with the Ngati-paru at Te Putere. He entertained, then slew and ate his guests, a procedure by the way which must not shock my readers, which may indeed have been perfectly correct—*tika*,—for we cannot apply to tribal custom the standard of Christian ethics. He may have, like Fhairshon[1] in Bon Gaultier, but avenged an ancestral wrong committed generations back. Be that as it may, awaiting events Ti Waewae established himself on Tauranga-koau, and there prepared

[1] "It is now six hundred coot long years and more since my glen was plundered."

himself for the return match in true Maori fashion. "During the siege of the island *mokihis* or rush-rafts were used, and all sides of the *pa* attacked. It could not be taken, so at length a truce was called. Now Hitau, the sister of Te Whatanui, had taken part with her brother against her husband Ti Waewae. From the shore she called to him. She induced him to leave the island in a canoe laden with eels, the which eels were *ngakau*." I gather that in some way their acceptance entitled the giver to fair-play, to consideration, at the very least that

Tauranga-kaou.

he should have been done to death correctly. Not even that last melancholy consolation was accorded Ti Waewae—he was just killed, knocked on the head in the common or garden way, and with him another man Paia, who, "feeling love for Ti Waewae," was resolute to share the fate of his chief.[1]

[1] My admiration for poor, loyal, simple-hearted Paia, who chivalrously chose to share the fate of his chief and friend, met with but scant sympathy; my interpreter, the Rev. P. A. Bennett, related to both races, who had hitherto thought well of me, looked very grave. Pera and Te Hata made no bones about the matter, but burst forth with deep-chested emphatic

Notwithstanding the fact of Ti Waewae's death at the hands of Te Kahu-o-te-rangi, the defenders of the *pa* continued to present a bold front. The siege, however, endured until the Ngai-Tatara, hard pressed, decided to consult the sacred oracle—*te tuaahu*—to discover what lay in the future. Te Whitiki and Tunui-o-te-ika were the tribal deities of the Ngai-Tatara. It was the latter who was now, through the medium of the *tohungas*, consulted. He was the god of revenge, of evil passions. If any man had given offence to the tribe, if it was desired that punishment should be meted out to any individual, the assistance of Tunui-o-te-ika was invoked. It was, however, necessary before response—or, to use a modern phrase which perfectly expresses the meaning, before contact could be obtained—to lay before the god something that had belonged to the offending party—a personal ornament, lock of hair, fragment of clothing, the imprint of a footmark, spittle collected from the ground. The abode of Tunui-o-te-ika was a miniature *waka* or canoe, which was moved, as occasion called, from place to place. The avatar of the god was shown as a trail of fire, visible not only to the priests but to all members of the tribe; Pera was emphatic in the use of the words "All, the whole world."

In this dilemma of the tribe, the proper rites and incantations having been performed, Tunui-o-te-ika, taking the direction of the rocks, Te Puku, manifested himself in a trail of fire "like a comet," and here sped to earth. The interpretation of the fiery flight was plain—towards that spot the Ngai-Tatara were bidden to withdraw. Their canoes, which had been hidden in the *pa*, were accordingly prepared, though it was realised by the elders of the tribe that there was not room for all. The difficulty was surmounted by the decision that only the male members of the tribe should make their escape, and that the womenfolk should be left to the mercy of the enemy. Even the infant males were taken. *Ma ratau e ngaki te mate*—"give us all the boys, because they will be needed to seek revenge for this disaster."

During the darkness of the night, therefore, the Ngai-Tatara dragged their canoes noiselessly and stealthily into the lake, the

scorn—*porangi! porangi!*—mad! mad!—and perhaps from the business point of view it might have been wiser, as they explained, to live and slay rather than be slain. From the tribal point of view Paia had just wasted himself.

tohungas reciting ceaseless incantations so that the enemy might not be disturbed and wakeful. The manning of the canoes and the retirement were successful: no single male of the Ngai-Tatara remained on Tauranga-koau. In the darkness they escaped, passing through the narrows of Ohinepaka, landing on the east edge of Waikopiro, and there sinking their canoes in deep water. At last, safely on the heights of Te Puku, facing about and looking towards their island, they exclaimed, "*Hei konei ra e kui ma e hine ma*"—"Farewell to our women, our daughters farewell."[1]

After the departure of the warriors of the Ngai-Tatara, the attacking party seized the island and made prisoners of the womenfolk, old and young, who were taken ashore at a spot known as Te Papa-o-Waiatara. As most of the attacking party were from Te Urewera, the women were carried captive in a northerly direction. During the retreat, according to ancient custom, large fires were lighted at nightfall, for illumination and warmth. Te Amohia, a woman of high rank amongst the prisoners, visited nightly each of these fires, her purpose being to discover how her people fared, to study the situation, and to disarm any suspicion that escape might be attempted. For three successive nights this was done. On the fourth evening, when the party was not far distant from Mohaka, Te Amohia whispered her plans to her particular cronies—her "aunties" as Te Hata-Kani delighted to call them,—Whangawehi and Mohu. Towards midnight the three made their escape.

The leader of the Urewera people, whose name has been forgotten—one of the very few lapses of memory on the part of Te Hata-Kani—noticed after a time that Te Amohia and her companions had not returned to their accustomed place. He thereupon called out, "*Te*

[1] In explanation of this act of desertion I cannot but quote from a book of mine, 'Mutton Birds and other Birds,' published years before I had heard of the retreat from Tauranga-koau: "Some readers will have noted with surprise and some with pain that the conduct of the male tit during the cuckoo episode stands forth in no very noble light. Those who have done so are thinking in terms of man and not of bird. His concealment of himself in the thicket we should designate by such foolish words as 'cowardly,' 'unmanly,' and 'unchivalrous'; but the verdict of male tits would consider that his proceedings were wise, eminently proper, and that he could not have acted otherwise and yet done his duty. What man calls chivalry, which ordains that the male shall perish under all circumstances to save the female, has no place in the working of the minds of male animals. If we can imagine in a community of tits some disaster analogous to that of insufficient boat accommodation in a sinking liner, the male birds would firstly save themselves, not for themselves but for the race, for their future broods." The males of the Ngai-Tatara *hapu* were no doubt subconsciously actuated by a similar instinct.

Amohia, kei hea koe?"—"Te Amohia, where are you?" There was no response. He then shouted to the guardians of the other fires, "Do you see Te Amohia?" The reply was, that she had been seen last returning from the direction from which he called. Te Amohia and her two companions had disappeared. At break of day chase was given, and the enemy leader, by far outstripping his fellows, got on to the tracks of the fugitives. As Te Amohia and her companions were nearing a certain patch of bush, looking back they saw their pursuer not far behind. Te Amohia was equal to the occasion; bidding her friends "*kia whakanga*"—"rest and get their breath"—she prepared herself for the fray. She had previously, when crossing a stream, picked up a long-shaped stone, partly for the preparation of fern-root for food and partly in anticipation of the possibility of such a crisis as had now occurred.

The women after resting for a few moments no longer troubled themselves about further concealment, but took up positions of defence behind their leader. Te Amohia, knee on ground and body resting on her heel, crouched in front: in this posture their pursuer discovered the three women. He was armed with a long-handled battle-axe, the blade of which was steel, for by this date the change from the old régime to the new had extended to weapons of war. On approaching the women he shouted and went through the usual gestures of a warrior about to strike. Perhaps in order the more to intimidate his victims, he slashed at the boughs on his right and left, leaving no doubt in their minds about his strength and skill in management of the weapon. With boughs falling at every blow, nearer and nearer he drew to the three women. He had not taken into account that Te Amohia belonged to a warrior tribe—that the blood of the Ngai-Tatara flowed in her veins. She never lifted her eyes from the ground; she sat stolid, missing no movement, her eyes fixed upon her foeman's toes. She knew that before he struck he must first thrust them deep into the earth to obtain a firmer grip. At last he gathered himself for the blow. Lifting his battle-axe to the height he brought it down with tremendous force, intending to cleave Te Amohia's head. Te Amohia, however, leapt aside, and not only parried it with her right hand and arm, but ere the striker had regained his balance, darted up, caught him by the hair and dragged him over, calling to her companions—her "aunties," dear old venerable things!—to come to her assistance. Then ensued a fierce struggle for the axe, Te Amohia in the end obtaining possession of it at the cost of a badly-cut hand. The three ladies then pounded their enemy's head till he was senseless, when Te Amohia placed her foot on his neck, and with all the strength she could command, "once, twice, three times did she strike, and every time the axe was buried in his brain." The three women then cut him open, and tearing out his heart, still warm and pulsating, Te Amohia placed it in the palm of her right hand, and raising it above her head according to the ancient rite of "*whangai hau*," offered it as an oblation to her *mana* or *atua*.

It is interesting to note, as another example of the change from heathen to Christian nomenclature and Christian custom, that in later life Te Amohia became Elizabeth—or rather its equivalent in Maori, Riripeti; under that name, dressed in European style, and doubtless a professing member of the Church of England, she was well known to Pera and Te Hata-Kani when they were boys as a quiet, pious, elderly lady.

Well, years passed away, but the desire to seek *utu*—payment, revenge—for the Tauranga-koau mishap, and especially vengeance on Te Mautaranui, the Urewera chief, still burnt fierce in the hearts of our brave little *hapu*, which now, instead of Ngai-Tatara, was more commonly known as Ngati-kuru-mokihi, those who had been attacked by means of mokihi or rush-rafts.

Its warriors met in conclave and decided that one of their number, Hunuhunu, should be despatched as embassy to the various tribes along the route to Te Wairoa, beyond which lay the vast tract of the Urewera country, the country of Te Mautaranui. Hunuhunu accordingly set forth, carrying on his back a *taha huahua* or calabash of preserved tui. At Wai-hi-rere, in the neighbourhood of Te Wairoa, he met Te Apatu, the leading chief of that locality. To him he explained his mission and asked for assistance in seeking *utu* from the Urewera tribe; his speech completed, he presented the calabash of preserved birds to Te Apatu. That chieftain, however, did not commit himself by acceptance, but accompanying Hunuhunu, bade him proceed to Tiakiwai, the chief of the Awatere. Hunuhunu repeated to Tiakiwai the proposal already made to Te Apatu, proferring to him also the calabash. Tiakiwai, following the example of Te Apatu, also declined the dangerous gift, but accompanying Hunuhunu through his tribal lands, passed on his guest to Ngarangimataeo at Te Ruataniwha. Ngarangimataeo in his turn put aside the calabash, but forwarded Hunuhunu to Puhirua, the chief of Pakowhai, who in his turn again sent him on to Tuakiaki at Te Reinga. In presence of Tuakiaki and his people, once again Hunuhunu presented the fateful calabash with all its conditional implications. It was accepted, Tuakiaki distributing its contents to each of the other chiefs to whom Hunuhunu had previously addressed himself.

The Tutira emissary was bidden, moreover, return to his home with the message that Tuakiaki would obtain satisfaction for the attack upon Tauranga-koau, that vengeance would be taken on Te Mautaranui. Tuakiaki's method was simplicity itself: he gathered together huge supplies of pig, potatoes, and other delicacies, depositing the food at a place called Te Papuni. Te Mautaranui was invited. He came. There was a great feast, at the conclusion of which Tuakiaki pulled out a *patu* concealed beneath his mat, and with it there and then slew Te Mautaranui; again to quote the ballad of "Fhairshon," "drew his skian-du and stuck it in his powels."

The chiefs visited by Hunuhunu had in fact agreed that it would be wise policy for them to remove Te Mautaranui and so get rid of the cause of offence,—as Te Hata put it, in the language of the New Testament, it was expedient that one man should die for many. Had the Ngai-Tatara been permitted to send their raiding party through the district, one or other of the tribes through whose territory the *taua* would pass was certain to have suffered.

After Te Mautaranui had been killed, his body was cooked by a method of grilling, the dripping being caught in a miniature vessel shaped in the form of a canoe. Nothing was wasted. The more savoury parts with the tongue on top were placed in the self-same calabash that Hunuhunu had carried from Tutira, and over them the fat was poured. Finally, the mouth of the calabash was covered with skin saved for that purpose from the elegantly tattooed buttocks of the slain chieftain. The calabash was then carried to Tutira by Tuakiaki's people, together with a bundle of Te Mautaranui's bones to be used as fish-hooks: this was a very terrible indignity—the bones, it was emphasised to me, from time to time crinkling and creaking in their rage and remonstrance, "for it is in that manner that the spirits of the departed speak."

Thus was *utu* obtained for the mishap at Tauranga-koau, for Te

Mautaranui's tribe never attacked again. They contented themselves with composing a lament for their chief—a lament which, in later times, became a taunt in the mouths of the Ngati-kuru-mokihi against Te Mautaranui's people: "*Ko te papa i a matou ko te waiata i a ratou*"—"We got the victory, they got the song." The *tangis* printed are—the first, a lament for Te Mautaranui; the second, the lament of Koa for Ti Waewae.

(LAMENT FOR TE MAUTARANUI.)

"Te rongo o te tuna e hau mai ra
Kei Te Papuni kei a Wharawhara.
Nau te whakatau-a-ki nei
Te uri o Mahanga whakarere kai, whakarere waka:
A te uri o Tuhoe, moumou kai moumou taonga
Momou tangata ki te po.
Hinga nui atu ra ki te aroaro o Hineireireia:
To kiri wai kauri na Wero i patupatu.
Tarahau nga hinu, e tarahau ki runga o Mohaka:
Tarahau nga wheua e, tarahau ki runga o Tangitu,
Kia kai mai e, te ika i Rangiriri,
Tutara kauika te wehenga ka uki,
E tika ana koe mo Te Ro mo Te Apa-rakau.
Na Tikitu na te uri o Whiro-ki-te-po
Taiwhakaea-ki-te ao.
Haere ki roto o Tutira mo Ti Waewae.
Na tatou koe i tango kino.
Koa tu mai ra e Tohe i te hauauru
Ka ea ko te mate
Tenei e tai ma o tatou kape
Koi hianga i a Te Tamaki ma
I riro mai ai a Te Heketua, i mate ai Nuhaka.
Tona whakautu pahi ko Te Rama-apakura.
Haere ki roto o te Mahia, mo Kahawai mo Kauae-hurihia,
Inumanga wai te rito o Te Rangi
Te pa taea i Pu-te-karoro,
I tangi ai te umere, pae noa ki te one
I Taiwananga e I!"

(LAMENT OF KOA FOR TI WAEWAE.)

"Tenei taku toto te whakahekea nei
Rauiri rawa koe i taku rau huruhuru,
He tianga raukura no Te Mau-tara-nui,
Nau te hotu e, i riro ai ko te hoa.
E koro tu kino, te whai-kohatia

Ikapohia pea i te mata o te tao
I te aro-a-kapa, te tohu a te tane,
Nau i moumou, nau i tapae,
Ka mahora kai waho.
Ma Te Ahi-kai-ata,
Ka whakatarea koe ki 'te ika a ngahue'
Tiro hia ra te manu nui a Tiki
Ko te riu tena i whakahekea iho
Ki te wai-o-Taue, no runga nga puke
No Maunga-haruharu, no Tatara-kina
No roto i nga whanga.
Ma o teina koe e utu ki te hue
Mau e moumou te 'Ahu-a-Kuranui'
E rere kau atu sa.
Nau i whakakore te 'Whatu-o-Poutini,'
Te kahu o te tipua, te 'kiri o Irawaru,'
Te rau o te ngahere
Puai ki te whare *i*."

Although Te Mautaranui had been killed and *utu* had been obtained as far as the Urewera tribes were concerned, the Ngati-kuru-mokihi leaders felt that Te Whatanui's people must be made to suffer also, for he it was who had instigated the attack on Tauranga-koau. The five chiefs of the Wairoa district, combined with the Ngati-kuru-mokihi, made, therefore, a united raid upon Te Roto-a-tara, where some of Te Whatanui's people of the Ngati-raukawa tribe were living. During the fight Te Momo, the leader of the Ngati-raukawa, and most of his tribe were killed. The raiders then proceeded to Te Whiti-o-tu, vanquishing there another sept of the Ngati-raukawa tribe. Proceeding then to the Taupo district, they again attacked relatives of Te Whatanui living at Omakukara on the western shores of lake Taupo. There they killed Te Whaunui and Matetahora, the leading chiefs, and a large number of lesser name and fame. Thus was *utu* fully obtained for Tauranga-koau.

By the time the triumphant *taua* had regained Tutira, the tide of Christianity was spreading like a flood, tribal warfare was coming to an end. Hence arose the saying of the Ngati-kuru-mokihi: "*Ko Te Roto-a-Tara, ko Te Whiti-o-tu, ko Omakukara, ka iri te ake i te whare, e iri nei, tae ana mai tenei ra*"—"After the battles of Roto-a-tara, Whiti-o-tu, and Omakukara, we hung up our weapons in our houses, and there they have hung unto this day."

After this very long digression, once again returning to our trail and passing Te-Papa-o-Waiatara, Ti Waewae Hangi, and the shoal

Rukutoa, we reach the point of the long ridge Paopao-a-Toki, the northernmost ridge on the east shore of Tutira. About it were the usual signs of ancient settlement, levelled sites of huts, scattered tufts and patches of the native grasses already named. None of my informants know anything of the spot beyond the fact that tradition avers that men had dwelt there in very ancient times. It may have been off the shoal Rukutoa—history does not specify the exact spot—that on one occasion a man named Te Uaha set his *hinakis.* After a proper time had elapsed he returned to take away his catch. Pulling up the first

Oporae and Taupunga.

hinaki, there was no eel in it; the second wicker pot yielded no better result. When he came to the third also empty—failure in the capture of food was always a bad sign, an omen of impending danger—he muttered to himself, "*he kopunipuni pea i kore ai*"—"the presence of a raiding party must account for the absence of eels." Now Te Uaha suffered from a growth on his neck which affected his voice, giving it a peculiar guttural sound, which, by the way, my informant Te Hata-Kani imitated in a highly diverting manner. Te Uaha accordingly paddled home, and relating his ill-luck with the *hinakis,* the usual defensive preparations were made by the tribe. Well, sure enough there did happen to have been a *taua* lying concealed amongst reeds and flax on the shore of the lake. Imagining themselves detected and foreseeing the raid would fail, they took their departure. The *mahia Tutira*—the sound-carrying property of the lake surface, or, as Pera rendered it into English, the "Tutira telephone"—conveying Te Uaha's hoarse whisper had balked the foray.[1]

The lands immediately north of Paopao-a-toki close to the lake were called Te Puna. Behind this locality, also on flat alluvial ground, where the Papakiri flows into the swamp and loses or rather used to lose itself in morass and peat-bog, are the lands Te Whakapuni a Te Whatu-i-Apiti. There, ere a cut made in modern times had connected the stream with the lake, the bed of the Papakiri terminated in a string of deep blind holes, the surplus water percolating through the swamp in drought as through a sponge or evenly overflowing it in flood. It had been farther blocked by the malice of Te Whatu-i-Apiti, a leading chief of the southern part of Heretaunga, whose principal pa was at Te Roto-a-tara. Besides high birth, Te Whatu-i-Apiti had another claim to fame; his hair—a rare although not a unique occurrence amongst Maoris—was red, or as my friend Te Hata-Kani called it, "ginger." He had eaten the eels of Tutira at the large *huis*—gatherings—of the Heretaunga people, and like all men who had tasted these delicacies, cast covetous eyes upon the lake producing them. He set out for Tutira during the summer time with a large fighting force. Arriving at the northern end of the lake, and evidently fearing the strength of the Ngati-kuru-mokihi, he did not dare to attack, but decided to divert the stream Papakiri, which flowed into the great marsh, and so cause the lake to decompose—*pirau*—and as a consequence kill the eels. This he did, causing some little time afterwards a frightful stench to arise from the lake.[2]

In the meantime the local people, not much perturbed, watched his doings from a distance. At last, when Te Whatu-i-Apiti saw that the Ngati-kuru-mokihi would neither attack him nor leave their lake, he vacated the district. His embankments were destroyed, and once more

[1] My own experience of the *mahia Tutira* fully substantiates this story. In '82, lying awake at Kahikanui awaiting dawn one still morning, I heard our station cook awaking my partner in the hut on Piraunui, distant fully a mile across the lake. The carriage of his voice, every syllable distinct and clear, was the more remarkable as the reveille was uttered into the *whare* in an opposite direction to that in which I was lying.

[2] I pass the story on as it was told, but would point out for the fair fame of Tutira that its lake is fed from innumerable springs and brooks besides the Papakiri.

the Papakiri returned to its old course; its fresh healing waters stayed the process of decomposition.

Whatever may have been his methods and reputation on Tutira, Te Whatu-i-Apiti—a kinsman by the way of the Tutira folk whom he had treated so scurvily—was received in friendly fashion at Tangoio, where then stood the strongly-fortified *pa* Te-rae-o-Tangoio—"the forehead of Tangoio." Tangoio had been a celebrated chief of the very ancient Toi people who owned these islands before the time of the Maori, and upon his deathbed had requested that his *pa* should be thus named. Here on this fine foreland or forehead, the red-haired Te Whatu-i-Apiti was entertained by Tataramoa, whose wife Porangi was a descendant of Kohipipi. There he formed an attachment to Tukanoi, his host's daughter, and there he stayed a considerable time. Parting with Tukanoi—he was a man of no particular refinement of feeling—these were his good-bye words: "*Ki te whanau to tamaiti he urukehu me tapa tona ingoa ko Whakatau, ke te whanau he mangu, he tane ke nana*" —"If your boy is born with red hair, call him Whakatau; if he is born with black, I shall know you have been with other males."

As a matter of fact, Te Hata-Kani here made a slip, using the anglicised word "tariana"—stallion—instead of the true Maori word "tane"—male,—his sentence running: "If your boy is born with red hair, call him Whakatau; if he is born with black, I shall know you have been with other stallions." After all, however, as the old man insisted, the sense was the same.

Well, in due course the anticipated boy was born, and let us hope and trust, to the gratification and not to the surprise of the damsel Tukanoi, his hair *was* red; he had come true to type and was duly called Whakatau. At a later period the event proved a fortunate incident for the people of Tangoio. It happened this wise: Otua of Tangoio married a sister of Te Hiku-o-Tera of Heretaunga, a man of immense stature. One day whilst the giant lay asleep, Te Otua, his brother-in-law, particularly struck with his length from hip to knee, stooped down and began to take exact measurements, not as white men do by "hands" or "feet," but by the Maori method of clenched fists.

It was an enormous limb, a titanic limb, a limb that Porthos might have envied. In his excitement Te Otua forgot his manners and the decencies of reticence; neglecting caution in an ecstasy of delight and enthusiasm, he exclaimed to himself as he proceeded with his calculations: "*Katahi, ka rua, ka toru,*"—a free translation of which might run: "One, awaia! Two, a very tree!! Three, a sapling totara!!!" and so on. Now human leg-bones in those days were useful to others than their proper owners. Te Hiku-o-Tera perhaps may have been aware that his were dangerously valuable, he may have been unduly sensitive. At any rate, as ill-luck would have it, he woke during the operation, and, furious at the insult as he considered it, accused Te Otua of measuring his understandings with a view to converting them into bird-spears, for the longer the bone the more highly was it prized for this purpose. In high dudgeon he left the *pa*, and returning, reported the incident to his chief, Te Whatu-i-Apiti. In those times an insult to an individual was an insult to his tribe. A war party accordingly was collected—its leader, however, being warned by Te Whatu-i-Apiti that his red-haired son Whakatau, whom he had never seen, was on no account to be hurt.

The *taua* made its approach by way of the beach, between which and the *pa* lay a broad lagoon, at that season covered with multitudes of duck. Less wary and wakeful, however, than the geese of the Capitoline, they were circumvented by the following stratagem: Each warrior provided himself with plumes—*pua kakaho*—of the tall graceful toe-toe grass (*Arundo conspicua*), and thus camouflaged crept after midnight quietly round the lagoon, crossed the stretch of water—sometimes, it is said, actually touching the unsuspicious duck—and established himself beneath the outworks of the *pa*. There the reassembled warriors awaited the earliest dawn—"*Kia kitea nga turi,*"—"until it was light enough to see a man's knees."

Just before daybreak a woman from the *pa*, happening to go out, saw the *taua* just below. She gave the warning by exclaiming: "*Ko te whakaariki!*"—"hostile raiders!" Te Otua was the first man up after the warning. Snatching his bundle of pointed manuka spears, he rushed along, biting the material with which they were bound. Running thus he stepped on the spot where the refuse flax of the village was deposited. It was about a couple of feet thick with the butts of the great blades, and as Te Otua rushed forward his feet slid on the slippery surface and he landed fairly in the middle of the enemy. The gigantic Te Hiku-o-Tera, whose hip-bones had been so rudely measured, was foremost in the attacking party. Recognising Te Otua in the scuffle, he exclaimed: "*Koia tenei!*"—"This is he!" At once they pounced upon and killed him. With *utu* thus procured, Te Hiku-o-Tera called to Whakatau to reveal himself; the *taua* departed, the red-headed son of Te Whatu-i-Apiti returning with his new-found friends to Heretaunga.

It will be now convenient to return to the southern extremity of the lake, and from there follow up the track on the western side. From the flat Piraunui it passed over the ridge of land situated between the lakes Orakai and Waikopiro. Continuing northwards along the margin of the lake, it reached the peninsula Tautenga upon which the wool-shed stands. Here the lakes Tutira and Waikopiro used to be separated by what was an impenetrable morass, but is now, owing to stock traffic, a sandy bar. The peninsula, now much eroded by traffic of sheep, must have at one time been utilised as a burying-ground, for numbers of skulls and human bones have been exposed as the light top-soils have become worn away. Below its broken northern edge rests the rock also named Tautenga; and not far distant, in deep water, lies, or used to lie, the log Te Rewa-a-Hinetu. It is fifteen feet in length, a foot and a half in girth, and bears a general resemblance to a fish's head. As its name Rewa—the floater—implies, it is endowed with the magic power of moving from spot to spot, the trail of its progress being then distinct on the sandy bottom. Its approach to Tautenga was particularly ill-omened, and used to presage death in the *hapu*. Te Rewa-a-Hinetu is a branch of a tree named Mukakai, which has travelled from the South Island up the coast to Otaki; another branch rests in the Wairarapa lake, another at Tikokino, another at Te Putere. The presence of any portion of this famous tree is said to be indicative of abundance. With its disappearance the food supply of the tribe is said to dwindle and diminish.

Debouching on to the hill at Tautenga are two spurs — the one known in modern times as the wool shed ridge, Te Mata, and the other Te-roa. The latter was a guide to the shoal called Urumai; when from the surface of the lake the range Urumai on Kaiwaka station could be detected over the dip in the Te-roa saddle, the shoal Urumai could also be located exactly. The correct method of obtaining eels from this spot was to strike the paddles noisily, causing the eels below to dive into the mud, where they could be speared. Travelling northwards along the lake edge we

THE ROCK TAUTENGA

In ancient times the approach of the log Te Rewa-a-Hinetu towards the rock Tautenga presaged death in the *hapu*.

cross the brook Waipara, which used to filter through a small raupo and flax marsh. Some hundred yards farther on we reach the flat Otutepiriao, whereon is built the present homestead. On the north of this flat is a low bluff covered with deposits of kakahi shell; east of it, in thirty feet of water, projects the snag Karuwaitahi.

Still following our trail, we reach the deepest indentation on the west — the bay Te Kopua or Ngaha. On the southern edge of this bay is another bluff, lower in height, called Pari-karangaranga. Te Kopua was in very ancient days the name of this bay, but later it was renamed after the woman Ngaha. Upon her death she was buried in a cavity high above the lake. From this height the *taniwha*, whose dwelling was in deep water, carried her in her *amo* or bier. "It is true; the cave from which the body of Ngaha was torn is still on the hill-top; one of the poles of the *amo* protrudes to this day from the centre of the bay. Her little dog Pakiri, changed into a great stone, lies even now submerged in shallow water."

As amongst other primitive peoples, strange natural phenomena tend to suggest fabulous tales. In two cases cited, caves have been responsible for legends of magicians and monsters. We have now a chasm on the hill originating the story just given, a snag in the bay and a curious rock fragment substantiating the details of the legend.

Crossing a small flax swamp our trail bifurcated — one path running over low barren hill-tops until, on the far side of the hill Ko-te-pakiata, Maheawha, the ancient ford of the stream draining Tutira lake, was reached; the other track, closely following the lake edge, passed successively spots or localities of land called Okuraterere, Te Kahika, the peninsula Kaiwaka, Te Karamu, Te Maire, and the water-hole Te Korokoro-o-Hineraki. Finally, the two tracks circling the east and west shores of Tutira reunited at the outflowing stream on the lands named Whakarongo-tuna. From this last-named place — the north-westerly extremity of the lake —a deep slow-flowing creek, Tutira, runs its lazy course, meandering towards the ancient ford, Maheawha. Betwixt this crossing and the mouth of the lake it is probable that in olden times more food was obtained than from the whole of the rest of the station: sixteen *patunas*, or eel-weirs, were known and named in one short half mile of water.

At the crossing itself stood also a *whare-tuna*—an eel-house or eel-lodge.

It was not a tribal possession, but belonged to the individual upon whose land it was built—to him and to his relatives.

The size of a *whare-tuna* varied according to locality and depth of stream, but was about 15 feet long, 1½ feet high, and 4 feet wide; the sides, roof, and ends were made of manuka lashed with flax, in the same manner as raupo is bound together on the sides of a *whare-puni* or sleeping-house; there were three or four observation holes on top, sufficiently big to admit a man's hand. At the outer wall, next to the stream and away from the bank, stones were placed to withstand the force of the current. The down-stream end was also blocked and weighted down with stones. The upper end, into which the stream or part of the stream flowed, remained open. Lastly, the interior of the *whare-tuna* was made snug and comfortable by loosely filling it with water-weed—*rimurimu.* It was a permanent trap that required no watching, no baiting, and no lifting, and must have proved particularly serviceable to such wanderers as the Ngai Tatara. There the eels congregated, sometimes so thickly as perceptibly to raise the temperature of the water; to obtain them the only precaution necessary was a soft-footed approach.[1]

On occasions when eels were wanted a pliable bough or hoop—*tutu*—was attached or rather jammed against the open orifice of the *whare-tuna;* to it was fastened the *purangi* by which a secure way was made towards the huge *hinaki* or wicker-work pot, where eels required for immediate consumption were placed. When all was ready one man stood with his foot by the small end of the *purangi*, whilst his companion, inserting his hand into one of the loopholes of the *whare-tuna*, would feel for an eel and gently turn its head towards the *hinaki;* he would then give its tail a pinch or squeeze, causing the creature to rapidly shoot forward, the man at the *purangi* simultaneously lifting his foot to allow passage and immediately replacing it to prevent the escape of other eels already taken. After a heavy haul from the many *patunas* along the creek Tutira, the surplus fish were often placed in a large reserve eel-pot—*hinaki-ruru.*

[1] When asked what had suggested the idea of the whare-tuna, which seems to have been peculiarly a Ngati-kuru-mokihi institution, Te Hata-Kani replied that when groping beneath the banks of creeks and rivers eels were very commonly found in hollow logs, more particularly in the hollow stems of certain tree-ferns, *mamaku* and *ponga.*

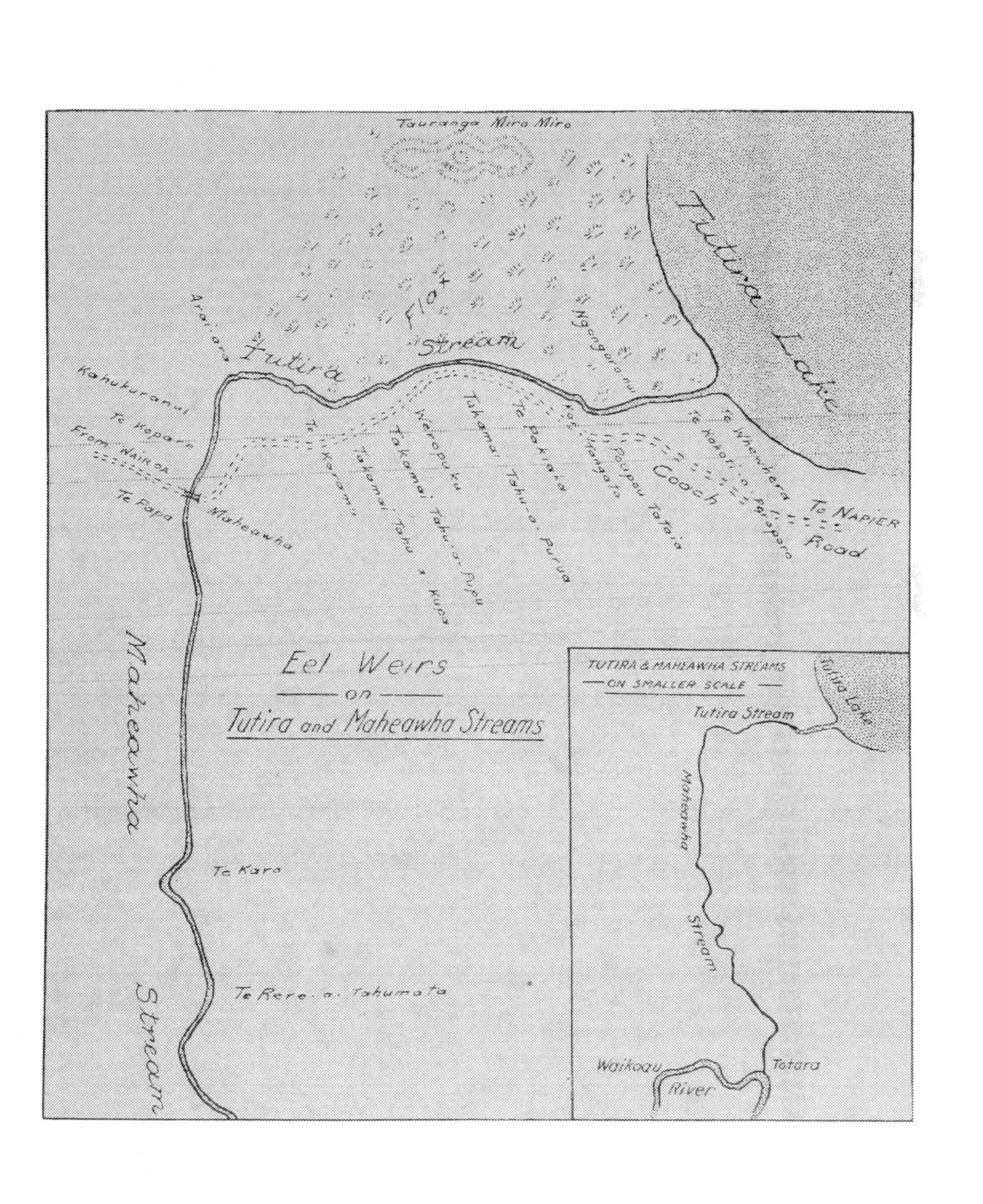

Eel Weirs
on
Tutira and Maheawha Streams
Tauranga Miro Miro
Flo
Tutira Stream
Tutira Lake
Maheawha Stream
Araiara
Kahukuranui
Te Kopare
From WAIROA
Te Papa
Maheawha
Te Karamu
Takamai Tahu a Kuao
Weropuku
Takamai Tahu-a-Pupu
Takamai
Takamai Tahu-a-Purua
Te Pakiaka
Ngangaro-nui
Tangato
Poupou Tataia
Te Kokoti-a-Paroparo
Te Whewhera
Coach Road
To NAPIER
Te Karo
Te Rere-a-Tahumata
TUTIRA & MAHEAWHA STREAMS
ON SMALLER SCALE
Tutira Stream
Tutira Lake
Maheawha Stream
Waikoau River
Totara

Conditions of eel-fishing on Tutira were remarkable, perhaps unique. As has been explained, in ancient times the waters of the considerable Papakiri stream never directly reached the lake; they soaked through a morass of several hundred acres, finally dripping into the creek Tutira, the creek that carries off the surplus water of the lake. The Maoris believe that in this great sponge of peat and root-fibre lived immense numbers of eels which never visited the lake, and which communicated with the creek by means of holes in the banks. They state, in confirmation, that although eel-weirs built on the bank require the whole width of the stream Tutira, catches as heavy are obtained in the lowermost as in the uppermost *patuna*. There were, at any rate, three sorts of eels distinguished: the common lake kind—*tatarakau;* another, also from the lake, rarely caught, much larger, and bronze in colour—*riko;* and thirdly, the eel of the creek Tutira—*pakarara*. The bellies of the two kinds of lake eels were, when taken, full of food, chiefly, I gather, a small water-snail; those of the creek eels were invariably empty. The *pakarara*, when opened up and sun-dried, would keep for four or five days, the *tatarakau* and the *riko* for as many weeks.

In view of the fact that the pursuit and capture of the *tuna* was a most important part of the life of old New Zealand, it is further worth mentioning that in one *patuna*—Maheawha,—where the waters of the creek Tutira once again begin to run violently,—its owner had to watch all night, taking each eel as it arrived, out of the *hinaki*. In all others an eel once ensnared was secure; at Maheawha only did eels seem able to find the exit as readily as the entrance.

Rights to these eel-weirs descended from father to son, but this natural transmission of property could be disturbed by force, as in the case of Tutata, or donated from the common property for deeds of arms, as in the case of Pohaki. These stories, dictated to me by Anaru Kune, are as follows:—

"Two brothers, Rere and Hongi, went down to set their eel-pots. Now in this *patuna* the waters could spread abroad in flood-time. These brothers selected the best opening and set their pots, spread like a man's fingers facing the stream. They were at work lashing on the *purangi* or guiding net to the breastwork when a man called Tutata claimed that particular spot for himself. Getting bad words, Tutata leaped into the stream, and seizing Hongi by the neck, held his head down. When Rere came to help his brother his head was also put under the water till both their bellies were well filled. Tutata then allowed them to crawl ashore. They lay for some time with their mouths open, the water flowing from their nose and throat. Tutata took the contents of their *hinakis*, and some say that Rere and Hongi never came back to Tutira. Enough! That place where the eel-pots were set was called Maheawha. It belonged to Hongi and Rere, but was taken by Tutata and remains his property to this day."

Another story, also dictated by Anaru Kune, the father of 'Pera, shows how property could be presented out of the tribal possessions to an individual, probably for his lifetime only, as the reward for assistance rendered in war.

"The Ngati-manawa were a sub-tribe of the Ngati-apa. On one occasion a party of their warriors coming by way of Maungaharuru raided Tutira. This war-party was led by Kaiawha. The only Tutira people at that time of the year living about the lake were Whai, his wife Te Rangiataahua, and their child Kupa. After the slaughter of Whai, the raiding party, carrying off the woman and child and also a quantity of *hinaki*, returned to Te Wai-whero in the Maungaharuru. The following day Kaiawha went out to hunt kiwi. His dog was restless and uneasy, and the take of birds poor; this, like the failure of Uaha to obtain eels, as already stated, was construed as an evil omen. Kaiawha returned to his *pa* and prepared to fight. That night in the dark he was attacked by Pohaki, a prominent Tutira chief. During the fighting Kaiawha shouted to his people, 'Light up the fires. *Tahuna te ahi kia marama ai a Ngati-apa te riri.*' This was wrong. The light from the blazing *hinakis* showed that Kaiawha's party numbered only eight, and the Tutira men were encouraged. The Ngati-apa were beaten and Kaiawha himself wounded. Some say he hid in a great log and escaped; some say he was never seen again. Pohaki was given two *patunas* for reward, one at the junction of the Maheawha and the Waikoau, and the other along the stream Tutira. The names of these patunas are—the first Totara and the last Te Kopare. Te Kopare has above it, up-stream, Kahukuranui, and below it, down stream, Maheawha." Later, a saying became rife on the countryside, "*upoko-pipi*"—"soft heads." It was used to denote the fate of raiding parties who visited Tutira. The exact words run, and the reader can believe they were fully emphasised when told to me, "*Tutira upoko-pipi.*" Many raids were made upon Tutira, but with the exception of the death of Ti Waewae, no other rangatiras were taken; every raiding party was beaten, hence the byword, "*Tutira upoko-pipi*"—"Tutira, the place where heads become soft."

Kupa, the child thus carried off and rescued, became a man and begat Te Umu-kapiti, who begat Parakau, who begat Aperahama, who begat Anaru, to whom, and to whose son 'Pera, I owe much of the information contained in this chapter.

CHAPTER XI.

THE TRAIL TO THE RANGES.

WHATEVER may be the value of the central portion of Tutira in the future, and personally I believe it will be very great, it was to the natives immediately prior to European civilisation almost worthless.

There existed upon its surface neither forests for birds nor suitable streams for eels. Place-names are in consequence fewer in number and records of the past scantier.

From the ford Maheawha the trail proceeded in a northerly direction. On the left of the track lay several hundred acres of flattish lands and low rolling downs by the name of Parae-ia-kai-ora. About the centre of this region rose the little hill Tamaiahua, opposite which a fairly well-defined subsidiary track branched off in the direction of Otupare, "Conical Hill." Proceeding on its way the main track rose gradually until it reached Orawaki. This height, better known as the "Image Hill," got its name from the image or *tekoteko* which at one time stood on its summit. I understand that the original, a fine piece of carving, adorned with greenstone earrings and clad in finely woven mats, was highly thought of. It was burned in one of the numerous fern fires which used to sweep the countryside; only a rude replica of the original remained in the 'eighties. The rough block of totara from which it was carved had split, but showed, nevertheless, the *moko* or tatoo pattern on the face, and the conventional three fingers crossed over the belly.[1] It had been erected by an elder brother of my old friend Werahiko in pious memory of a grand-uncle named Kupa. Here, according to tradition, a *whare-puni* of considerable size once existed, built for the convenience of people

[1] In early days the missionary bell topper was in demand as adornment for the *tekoteko*. The Rev. Mr Spencer, working during the 'forties at Tarawera, was made by the Maoris to promise—a promise he was never allowed to forget—that his discarded headgear should be reserved for this special purpose. Our "image" on Orawaki hill, naked and alarmingly masculine, clad in the ecclesiastical bravery of a top-hat, could only then have been further christianised by a bishop's apron.

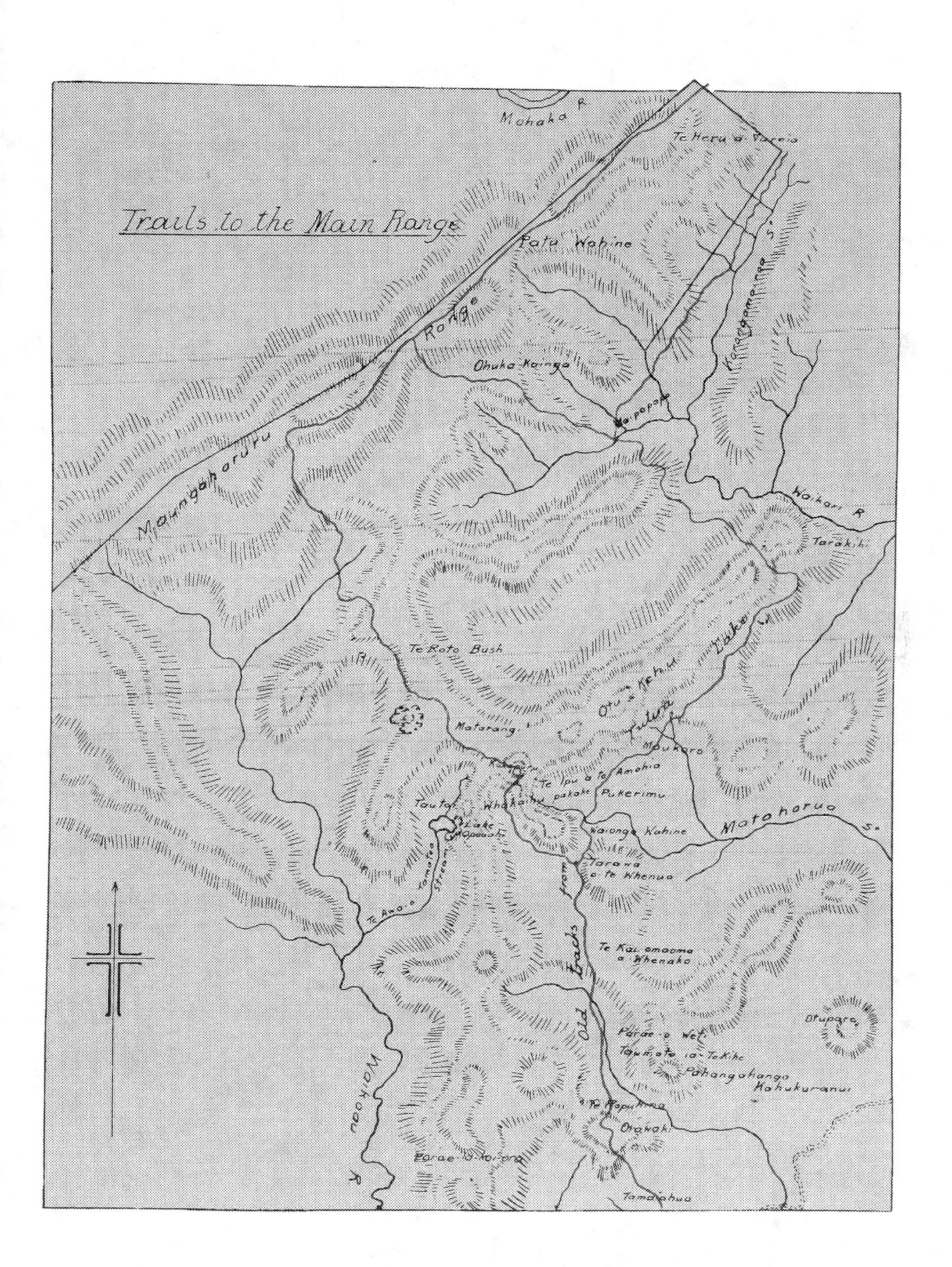
Trails to the Main Range
Mohaka R.
Patu Wahine
Range
Maungaharuru
Ohuka Kainga
Waikari R.
Tarakihi
Te Roto Bush
Matarang
Otu a Kehu
Lake Tutira
Maukoro
Te Ipu a te Amohia
Pukerimu
Waiongo Kahine
Mataharua S.
Tarawa a te Whenua
Lake Opouahi
Stream
Old Tracks from
Te Kai omaoma a Whenako
Otupara
Parae a Weti
Taumata ia Te Kiha
Pahangahanga
Kahukuranui
Orawaki
Tamatahua
Waikoau R.

working their plantations; if, indeed, this was the case, the workings spoken of must have been about the fertile edges of the lake. No man, far less a soil-wise native, would have attempted to grow crops in the vicinity of Orawaki.

Passing over this hill the trail proceeded nearly due west along the top of the narrow razor ridge Te Ropuhina. At the western termination of that ridge it descended in a northerly direction towards the barren flats and low lands of Parae-o-weti, lands which lie between the western heights of the northern portion of the "Sand-hills" and the southern slopes of the isolated hill Pahangahanga, the "Dome." Later, crossing a branch of the Papakiri, the track ran in a fairly direct line from the foot of Pahangahanga, ascended the rising ground Taumata-ia-te-hihe, and eventually reached the second crossing of the Papakiri. This crossing has always been known in my time as the "Taipo"—goblin—crossing,[1] a name probably given because of a totara block which used to lie there hewn roughly to the similitude of a man's head.

Proceeding, the track crossed the Tarawa-o-te-whenua slopes and flats situate at the foot of the western termination of the "Burnt-Blanket" range. Here the trail split, the western track rising gradually until it reached the top of the hill Whakaihu-pakake. Descending precipitously from this height it dropped into the narrow basin Te-ipu-a-Te-Amohia, at whose northern extremity lay the "Pa Hill," Kokopuru. It was on a neighbouring height, Matarangi, that a *taua* of the Tuhoe was destroyed.

Near the far-seen headland Puraho-tangihia, "Shepherds' View," the Tuhoe or Urewera people had been met and defeated by the Ngati-kahungunu, the tribe of which the Tutira people formed a sept. In this battle the Tuhoe lost their chiefs Te Mokohaerewa and Te Kapua-whakarito, whose bodies were carried off to Tangoio and there cooked and eaten. "In order to avenge the insult the Tuhoe people despatched a second war-party. It was their intention to destroy Tohutohu and Meke, the Ngati-kahungunu leaders who had been present at the skirmish of Puraho-tangihia." I have been fortunate enough to obtain from Te Hata-Kani a pictorial representation of the affair.[2]

[1] A word, according to William's 'Maori Dictionary,' used by Maoris believing it English, by Europeans believing it Maori, it being apparently neither.

[2] The old gentleman had amused himself one evening sketching on a torn bit of foolscap the meeting of his people with the Tuhoe tribesmen; afterwards on clean drawing-paper he repeated the performance, which is here exactly reproduced.

1. Kokopuru pa, showing characteristic carvings on "*take*" or main posts of palisades. Note *tewhatewha* with feather or dog-skin *puhi*, and other figure with *were*.
2. Nga-ipu-a-Te-Amohia, two little lakelets in the vicinity of the pa.
3. Opouahi lakelet, also in the vicinity, famous for the abundance of eels within it. Note the typical eel.
4. Representative warriors of the Ngai-Tatara. This sept and the Ngati-moe, it will be recollected, were *hoa matenga*, friends together to death.
5. The setting forth from Tiekenui of the Urewera foemen, evidently, to judge from their stature, inferior to the men of the Ngai-Tatara.
6. The fighting over, the enemy are invited to the great meeting-house on the Matarangi hill-top. This meeting-house was remarkable in its door at either end; there, revolving mischief, the foe can be seen cloaked in their *korowai* mats.
7. Food placed before the visitors consisting of preserved birds in calabash. Note carved wooden mouth of calabash, and woven basket around gourds and on tripods, also the kits of potatoes beneath. The guests, however, decline to partake of this food, a disinclination which, according to Te Hata-Kani, proved that they meditated treachery, and which absolved any action the Ngai-Tatara might think fit to take. The uprights of the meeting-house had, "just in case," been already prepared for these dishonourable Urewera,—almost completely cut through.
8. Talking it over, an arrangement reached by which four parties of the Tutira men show four parties of the Urewera the *Waerenga* or crop lands where the latter could gather their own food. As, however, the Urewera could not be trusted, in each of the four bands thirteen of the Ngai-Tatara, armed with spears, accompanied twelve Urewera carrying potato kits—in Te Hata's sketch the three figures on the one side and the two on the other represent for lack of space the parties respectively of thirteen and of twelve.
9. The four *Waerenga* or cultivation-grounds of differing shapes, each also showing its rubbish pit; there as a necessary precaution, to forestall the treachery of the Urewera, the four parties of thirteen spearmen slew the four parties of twelve potato-gatherers.
10. Whakahoehoe, the Ngai-Tatara leader, approaching *pa*. Note his *taiaha*, Huia feather, *were*, and mat, also his attendant on the hillside, a page or squire, possibly a kinsman of good birth.
11. Tamati Tararua thrusting *patu* into Urewera scout's temple. This was also correct —the Urewera man had failed, I understand, to appreciate properly the greatness and dignity and nobility of the Ngai-Tatara chief. In the use of the *patu* a violent thrust and slight twist were sufficient to detach the upper part of the cranium.
12. After these repeated instances of bad faith on the part of the Urewera the meeting-house is let down on to those remaining within. They are speared as they strive to emerge. Te Rangi Pumamao alone escapes. He falls in his flight and breaks the stock of his gun. He is caught up by Whakapipi. A duel with *taiaha* and gun—note broken stock—ensues, during which another Ngai-Tatara man, Whao-whaotaha, comes up behind and spears Te Rangi Pumamao through the back.
13. *Te Umu tao tangata*, the oven for cooking human flesh, showing heated stones. On this spot was the body of Te Rangi Pumamao cooked.

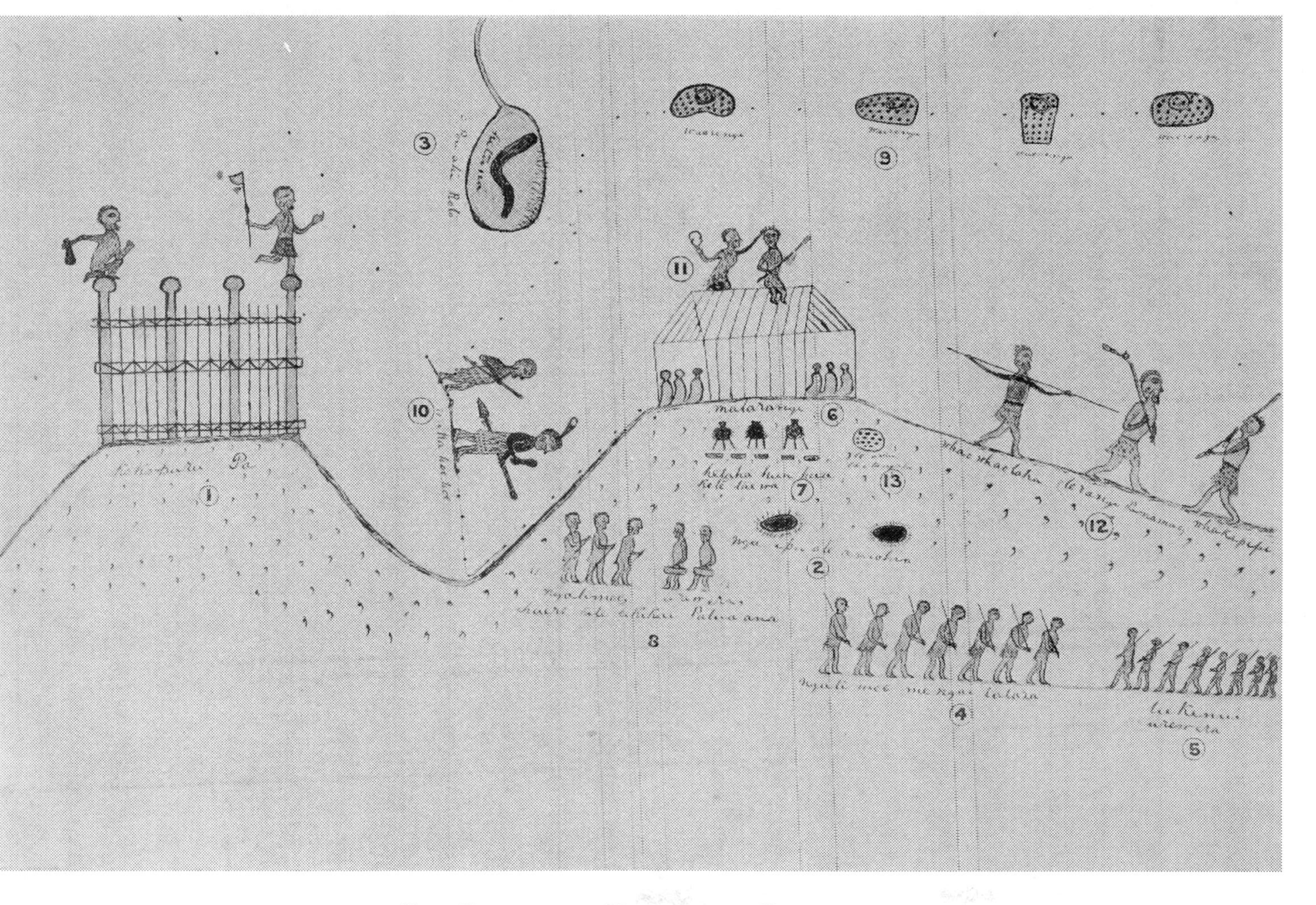

NGAI-TATARA AND UREWERA AT KOKOPURU.

Another misunderstanding on the same spot and its consequences —the visitation of the sins of the fathers upon the children by Land Courts of modern days—is to be found in the story of Waiatara and Takirau.

Waiatara was the name of a chief of the Ngati-moe, who lived at Kokopuru. His great friend, Takirau, was a chief of the tribe called Ngati-pahau-wera, whose headquarters were at Mohaka.

The district in which Waiatara lived was noted for its fat pigeons and tui. Takirau's district, on the other hand, was famous for its supply of kahawai, mango—shark—and other fish.

In token of friendship and goodwill between the two chiefs, it was their custom to make, from time to time, an exchange of food —Waiatara sending preserved birds, and Takirau returning the compliment with dried shark and kahawai.

Now it happened on one occasion that Takirau's followers made a visit to the Heretaunga district. On their return they stopped at Tutira, Takirau himself not being with the party. His followers, men of Belial, remembered the delicious preserved birds that Waiatara used to send to Mohaka. They visited Waiatara's *kainga* at Kokopuru, telling him that Takirau had sent them. Believing their tale, Waiatara readily handed over to them *taha*—calabashes—filled with birds preserved in their own fat. They carried these off to their camping-place at Tutira, but it was with covetous eyes that they gazed upon them. The temptation was too strong. They opened the *taha* and devoured the whole of their contents.

Arriving at Mohaka, and there meeting their chief Takirau, the various incidents of their journey were related, with the addition that while at Tutira they had approached Waiatara to see whether he could spare any preserved birds; not only, however, had he refused to supply any birds, but had uttered many rude curses upon Takirau and his people.

Takirau's anger was kindled at this uncalled-for insult, and he decided to form a raiding-party to seek *utu* or revenge.

It arrived at Tutira, and next day made an assault on Waiatara and his followers at Kokopuru *pa*. Waiatara was bewildered; he could not understand why his great friend Takirau should attack him in this way; finally, at the instance of onlookers, a truce was called, explanations demanded, and Takirau was convinced that he had been a victim to the covetousness and deceit of his people.

Waiatara's turn had now arrived for showing something of that *rangatira* dignity which is the peculiar property of the old-time leading chiefs. "Takirau," he exclaimed, "I have been your greatest friend for a very long time, assisting you in your troubles, and providing you with *huahua*—preserved birds—at every season. Now that you have made this treacherous attack upon me, my final word to you is '*haere*,' depart; our friendship is broken for ever."

As an evidence that the friendship was indeed not broken in vain, it may be added that when certain titles were being investigated this incident was related, and Takirau's descendants were disallowed any share in these ancestral lands.

From Kokopuru, about which so much has been said, the track proceeded nearly due west along the edge of a high ridge between the "White Pine Bush" and one of the gorges of the Waikari. This ridge was terminated by another gorge, on the far side of which lay heavy forest lands. The track then turned sharply north, and continued in a northerly direction through forest to Te-Heru-o-Tureia. Re-emerging into the open on the heights of that block, it pursued its course along the very rim of the main range above the western precipice, eventually reaching the bluff Patu-wahine, and thence proceeding out of our history to the wilds of the Urewera country.

We can now return to the lands Tarewa-o-te-whenua, where the trail had forked; the western track we have traced; the northern struck the crossing of the gorge of the Matahorua, the stream that divides Tutira from Putorino. Here at one time dwelt Titi-a-Punga. Like Rob Roy, he followed "the good old rule, the simple plan, that he shall take who has the power, and he shall keep who can." Here, also, was situated his village, and—if indeed they existed, except in the pious imaginings of an informant anxious to exaggerate the glories of the past—his plantations. At the best these can have been but of trifling extent and importance.

Probably, indeed, the residence of Titi-a-Punga on Tutira was only temporary; his permanent eyrie seems to have been established on rocky juts of the Maungaharuru range. There, encamped above the pass leading from Hawke's Bay into the Taupo country, he watched for travellers. At any rate, whatever may have been his antecedents, and wherever he may have come from, whilst on Tutira he completed a *whare-puni* or meeting-house; the building had yet to be opened, the ceremony of the laying of the foundation-stone had still to be accomplished. In lieu of the coins nowadays buried on such occasions, it was the New Zealand custom to use up a slave. Titi-a-Punga either had none to spare, or had higher ideals as to what was owing to himself and his new edifice; he had, in fact, determined on his brother-in-law, Te Rangi-nukai, as the votive offering. It was his body which was to be buried beneath the *poupous* — uprights supporting the framework of the *whare*, — his death which was to celebrate the house-warming. Friendly messages accordingly were despatched to Mohaka, requesting his attendance at the dedication of the new building. The wife of Titi-a-Punga, however, knew of her husband's intention; she warned her brother, who came, but came prepared; he arrived, moreover, by an unexpected route, thereby avoiding the ambush laid for him. It thus happened that whilst Titi-a-Punga and his merry men lay in wait on one side of the gorge, Te Rangi-nukai and his people arrived from Mohaka on the *pa* side of the river ravine. Few or none of Titi-a-Punga's band were in the village. Those few fled. The women were pitched over the cliff into the stream beneath—hence its name to this day, *Te Wai-o-nga-Wahine*, "The water of the women."

Titi-a-Punga was taken alive by his brother-in-law, and foreseeing his fate thus spake: "*Taihoa ahau e patua*"—"Kill me presently." He then uttered his farewell, still famous in the land: "*Tamai pakani a Taha-rangi toroa uta ka he i toroā tāi taratarā o Maungāhāruru ka whatiwhati,*"—"Strong son of Taha-rangi, the bird of the mountain has been destroyed by the bird of the shore; the crest of Maungaharuru has bowed itself and fallen." After that, as old Anaru quaintly put it, "he was killed—quite dead."

Crossing the ford the track passed through the locality Pukerimu, and later continued in a northerly direction through the slopes and flats east of the Otukehu range—the "Nobbies." It then swung sharp to the west between the end of that chain of hills and an isolated peak, where at one time dwelt another robber chief called Tarakihi. He, like the better-known Titi-a-Punga, also levied a toll on the track, until at last, killing some person of importance, he was himself set upon and slain.

Above the sandy ford of the upper Waikari the trail forked, one of the two branches climbing until it reached Patu-wahine and disappeared into the Urewera country. The other, proceeding roughly parallel with the Korongomairoa stream, continued through the *kainga* Waipopopo, and skirting a couple of upland tarns, also passed out of our story coastward towards Mohaka.

With it, too, is completed the history of the trails of old heathen Tutira; if they have been at times wearisome to walk, they have at any rate acted as threads upon which to string the facts; they have prevented digression in too outrageous a degree. It must have been consolatory, moreover, to the reader, that, according to its annalists, Anaru, Te Hata-Kani, and Pera, the Ngai-Tatara were always victorious, so much so indeed that the station became famous in the land as *Tutira upoko-pipi*—"Tutira where heads become soft."

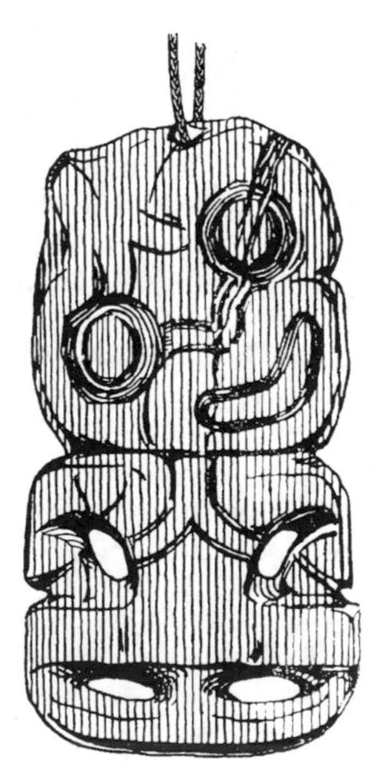

Greenstone Tiki.

(Presented by native friends to the author.)

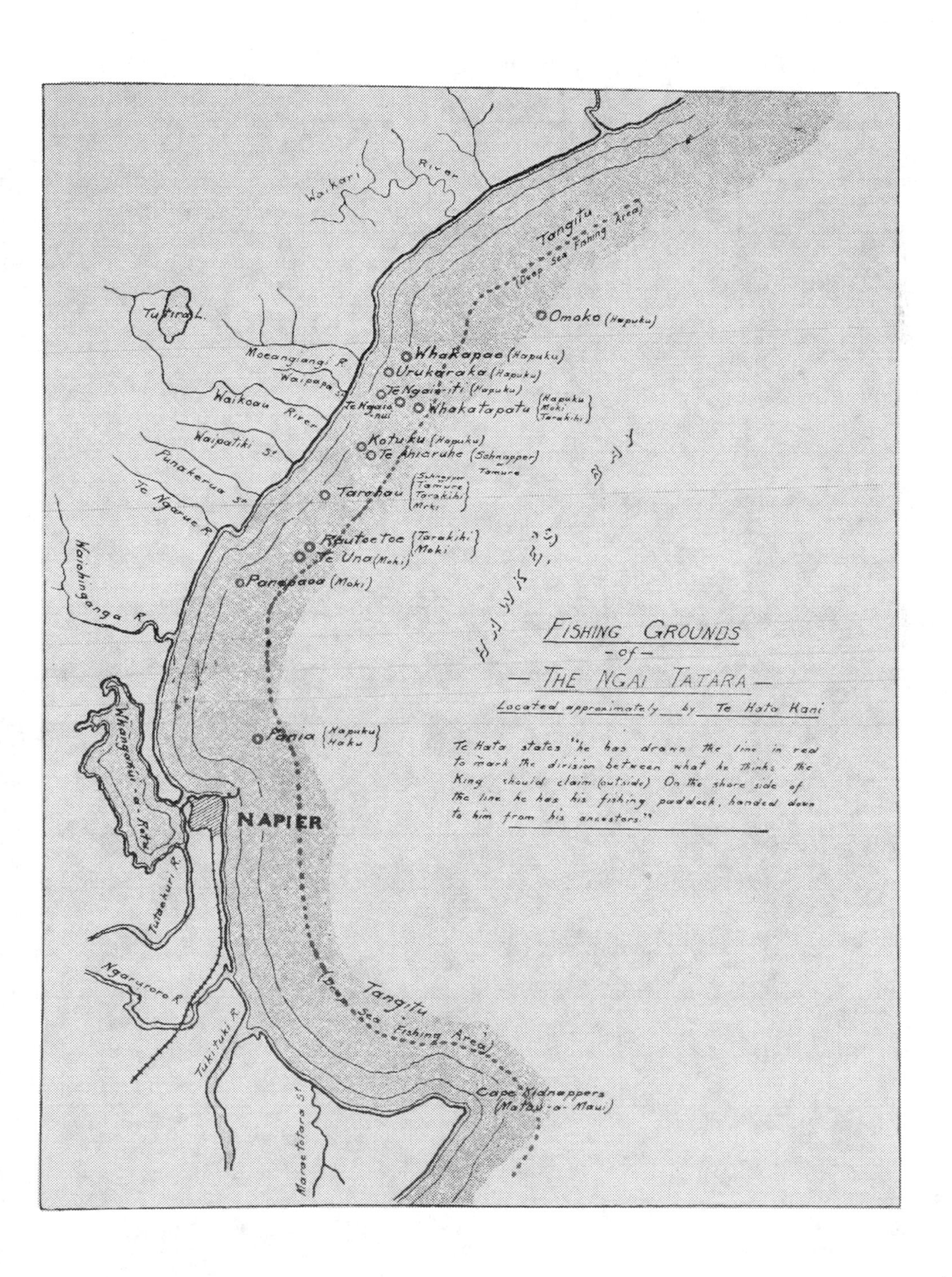
FISHING GROUNDS
-of-
— THE NGAI TATARA —
Located approximately by Te Hata Kani
Te Hata states "he has drawn the line in red to mark the division between what he thinks the King should claim (outside) On the shore side of the line he has his fishing paddock, handed down to him from his ancestors."
HAWKES BAY
Tangitu (Deep Sea Fishing Area)
Omoko (Hapuku)
Whakapae (Hapuku)
Urukaraka (Hapuku)
Te Ngaieriti (Hapuku)
Te Ngaio nui
Whakatapatu (Hapuku Moki Tarakihi)
Kotuku (Hapuku)
Te Ahiaruhe (Schnapper Tamure)
Tarahau (Schnapper Tamure Tarakihi Moki)
Rputoetoe (Tarakihi Moki)
Te Una (Moki)
Panepaoa (Moki)
Pania (Hapuku Haku)
NAPIER
Tangitu Deep Sea Fishing Area
Cape Kidnappers (Matau-a-Maui)
Waikari River
Tutira L.
Moeangiangi R.
Waipapa St.
Waikoau River
Waipatiki St.
Punakerua St.
Te Ngarue R.
Kaiahinganga R.
Tutaekuri R.
Ngaruroro R.
Tukituki R.
Maraetotara St.

CHAPTER XII.

THE VEGETATION OF THE STATION PRIOR TO SETTLEMENT.

THE two halves of New Zealand are separated by a narrow strait. At the date of their discovery, one—the South Island—was an open land fit for immediate settlement, carrying nutritious grasses; the other—the North Island—was a vast tangle of fern, of scrub, and of forest. In it there was no open country ready to the settler's hand; the pioneers of the North had to create their pasturage.

On Tutira grew a few acres of tussock-grass (*Poa cæspitosa*), a few score acres of flax (*Phormium tenax*) and of raupo (*Typha angustifolia*). A few hundred acres also of forest and woodland lay hidden in gorges and ravines. Otherwise, over the whole station stretched an illimitable sea of bracken (*Pteris aquilina*, var. *esculenta*). This plant, against which the station has been battling for more than forty years, delights in loose humus, sandy soil, and pumice grit. Into such soils—never dry, yet never water-logged—its rhizomes penetrate many feet. It is perhaps the only fern which thrives on manure. Year after year it will invade garden-plots; it will persist season after season in sheep-yards. On ploughed grounds fed with artificials its fronds spring taller, thicker in stem, and of a deeper green.

In fallen forest country, burnt and elsewhere grassed, every hollow stump eight or ten feet across, into which stock cannot reach, becomes a huge fern-vase. The fenced-in railway lines carry on either side, through cleared bushland, long ribbons of bracken. Intermingled with light open bush, I have measured fronds fourteen feet long. So situated, they develop something of the habits of a creeper—the stalks becoming finer and more pliable, the lower pinnæ aborting, the whole frond growing languorous and etiolated. In open lands on Tutira growth was most luxuriant on eastern and southern slopes.

On such aspects, in competition with tutu (*Coraria ruscifolia*) and koromiko (*Veronica salicifolia*), fern averaged five or six feet in height. On hot dry northern and western slopes it grew a foot or two less. No dry soil, however, was too bad to nourish bracken. Stunted to a few stiff inches, it covered alike the driest hill-tops and the most arid flats.

The growth of the plant is as follows: early in November myriads of minute brown-green circinate fronds begin to appear, each uplifting its own little cap of earth, as trap-door spiders raise the lids of their dry homes. Later these fronds grow into notes of interrogation, then, rising well above the old growth, each opens into the likeness of a man's hand bent back from the wrist, with fingers still curled up.

Later again the fronds develop into antlered spikes mossed with ferruginous dust. At last, fully unfolded, they assume the sombre green hue characteristic of fern country in New Zealand. On poorest soils bracken most quickly matures; on good ground weeks pass before the fronds attain completion. After its spring growth, unless scorched by fire or eaten by stock, the plant rests until the following spring. Unlike its British relative, which rots away in a single winter, six or seven different seasons' crop can be discriminated in the tangled masses of the New Zealand plant. The lowest are in various stages of fragmentary decay, others brittle and brown though sound; another is mottled with grey, but still in patches preserving its green; another bowed and weatherworn, only its tips sere; another dull green and almost perfect; the latest crop of all still erect and topping the growths of former years. Such was the appearance of Tutira in former times.

There was but little room for other plants. In fact, as mountains prove the last resort of peoples driven from their homes by conquest, so in the cliff system of Tutira plants survived which must have otherwise perished in the tyranny of fern. The reader knows the physiography of the station — an alternation of slope and cliff; a drainage system far beneath the level. Over every slope fern lay in swathes: it reached to the base of every cliff, it hung like a fringe over every precipice.

Forest and woodland covered less than two out of sixty thousand acres — forest growing in the ranges of the interior, well worthy of its name from the immense size of many of its indi-

TAWA BUSH.

vidual trees, woods flourishing on the lower-lying seaward edge of the run. Although restricted in area, this forest of the hinterland—the last shred and relic of the primeval vegetation which had at one time covered the district—was representative of both the mixed and unmixed "bush" of New Zealand. Looking downwards on to it from a higher altitude, the eye was primarily arrested by the number of very ancient grey-headed moribund totara (*Podocarpus Totara*), the very grandsires of the bush—their boles measuring 12, 14, and 16 feet in diameter. These magnificent trees live for the most part in single grandeur. They are dotted irregularly about the bush—dying, so to speak, on their feet, their short stubbed heads conspicuous in the surrounding greenery on account of the lichens glued to the dying boughs. Their great vitality has been sapped by age; their centres are hollow or choked with rotted wood, sometimes with mere dry powder. Adown their boles bark hangs loose in enormous strips and sheets. About their mighty roots lie foot-deep accumulations of mouldered wood, piles of bark already shed—for trees in the warm wet New Zealand bush thus cleanse themselves, ridding their skins of parasitic growth as birds by washing and dust-baths check lice. Considering not only the tardy growth of the totara, but its still slower senescence, I can never reckon the life of the greatest of these trees at less than one or two thousand years. Perhaps it is more—perhaps much more—for I have watched during one-third of a century certain dying branches: there has been in them no appreciable change, although that period of time is one-third of the tenth of the span suggested as the minimum duration of life. Perhaps some of these totaras on Maungaharuru were saplings when, twenty hundred years ago, Christ worked in Galilee; at any rate they must be of an enormous age. Flourishing on the spots that especially suit them are to be found also specimens of four other great New Zealand pines: white pine, kahikatea (*Podocarpus dacrydiodes*); matai (*Podocarpus spicatus*); black pine, miro (*Podocarpus ferrugineus*); and red pine, rimu (*Dacrydium cupressinum*). Other large species in the mixed bush are hinau (*Elæocarpus dentatus*), tawa (*Beilschmiedia tawa*), and maire (*Olea lanceolata*).

In the vicinity of these huge trees lie, coiled or sprawling on the ground like snakes, lianes, lawyers, vines, and clematis stems. Partly dragged up by the growth to which in youth their shoots have clung, partly drawn voluntarily towards air and light, their bare rope-like stems strike and chafe, hang and swing, against the boles like loose rigging against a mast. Seen from above, these individual trees, or little companies of trees, can easily be detected by their varying shades of green. About the middle or lower slopes stand venerable brotherhoods of tawa, grey with long pendant lichens, "old man's beard"; there are patches also of deep-green broadleaf (*Griselinia littoralis*), a species, by-the-bye, never met with on Tutira except far inland.

Another striking characteristic of this intermixed forest is the evenness, as seen from above, of the rolling contour of its ceiling of green. No tree-tops project above the general level; in this effect, however, there is nothing of blighting or blasting. The individual members of the forest community seem to have been born docile, to have acquired ante-natal knowledge of the effects of gales, never to have attempted usurpation of more than their fair share of the open commonwealth of sky. No tops are to be seen "caught and cuffed by the gale," no solitary shoots eroded and blown bare; the upper surface of the forest is as smooth in its inequalities as downlands in wheat. Conditions are somewhat dissimilar where masses of one species of tree hold undisputed sway, where narrow spurs are maned with one kind of tree as the neck of a hogged pony is stiff with hair. Such groupings of particular trees conform more or less to the shape of the locality on which they grow. They rise cone-shaped on a cone, narrow and elongated on a razor ridge. Beech of two sorts (*Fagus fusca* and *Fagus solandri*) are on Tutira the most prominent species growing thus strictly grouped; each possesses inviolate on its own territory whole spurs. Other areas are densely covered with tawhero (*Weinmannia racemosa*), others again with tall tree-manuka (*Leptospernum scoparium*). Honeysuckle (*Knightia excelsa*) is another species which, like the beech, the tawero, and the manuka, seems to revel in dry land, its long-drawn cone rising from the most arid of ridges.

So far we have viewed the forest from above; now we can take our stand beneath the trees. In forests of this sort no imprint holds its shape for long on the loose leaves; all is in process of decay, soft and yielding. The surface is cumbered with huge clumps of astelia, of species of asplenium flabellifolium, flaccidum and falcatum, fallen from above. Rotted branchlets and boughs, still encased in their husks or jackets of darker bark, lie strewn on the ground. Many of the boles rot stand-

FERN-FLOWER.

ing upright or only fall portion by portion; others prostrate are mere shells crusted with epiphytes and ferns, or clad in mosses aping in hues of softest green and yellow the forms of ferns, or stiff and erect like thickets of fairy pine. From dead trunks and boughs of harsher fibre fungus projects in ledges like lip ornaments of negro belles. Whole families of toadstools, supporting flimsy fleshy stems, their dainty parasols still rolled close, peep from beneath sheltered ledges. There can sometimes be traced in mixed forests of this sort three fairly distinct tiers of greenery: the lowest, lichen, mosses, liverwort, and ferns; the second, the massed tops of the coprosma tribe, species of which, naked below, bear their leaves on top in thin planes of foliage, thus creating a diaphanous mist, a twilight greenery, which in a shadowy way bisects the mass of trunks. Lastly, there are the tree-tops high above. In other portions of the forest there is nothing of this sort noticeable, a mere jostle of smaller and more ephemeral species competing with one another beneath the great pines, clustering about their knees and waists—fuchsia, tree-ferns, species of pittosporum, of olearia, of panax, clumps of short-lived wine-berry—makomako (*Aristotelia racemosa*)—and others.

Ferns grow everywhere, clinging like ivy to the rough stems, festooning them with elegant fronds, webbing them with veils of delicate rhizome, overrunning fallen boughs, drooping long languorous growths from matted clumps high overhead. Rooted in massy forks grow epiphytes such as Griselinia lucida, and huge rookeries of pine-apple-like astelia. Mats of sweet-scented orchids—Earina mucronata and Earina suavolens—cling with a plexus of roots to suitable sites; often a black mossy lichen exhales in sunshine a delightful violet odour. Except where massed groups of a single species prevail, and the ground beneath is bare and dark, there is a luxuriance of growth due to the great rainfall and the large number of hours of sunshine, almost unknown elsewhere. The edges of the forest exhibit a still more voluptuous profusion of tangled growth, an even thicker profusion than in its shaded heart—clematis, rubus, vine, parsonsia, and native passion-flower competing in the ampler light. Such a forest as this, typical of the North Island, is in truth tropical in all except degree, in all except latitude and longitude. The great rainfall and the full sunshine of the Dominion have created abnormal conditions. Except where massed species prevail, growing in solitary selfish gloom, an exuberance of life prevails, a luxuriance unknown elsewhere save in the true tropical zone.

The woodlands of Tutira, in contradistinction to the forest described, were confined to gorges deep and damp, gulches such as that of the Maungahinahina, where the upper soils had been washed out, where the marls had become exposed. With the exception of a valley here and there, these woodlands were bare of great trees. Their growth, compared to that of the ranges of the west—for woodland is but a preliminary step towards real forest,—was one destined on eastern Tutira never to progress beyond the initial stage. Vegetation there was dependent on two factors—rate of growth and frequency of landslips. The slower-growing pines, for example, had never time given them to find deep anchorage. Whilst still saplings they were swept to perdition by earth-avalanches following heavy floods. The surface of the ground was renewed too constantly to allow the maturing of any but fast-growing and free-seeding species. In this light bush, tawa (*Bielschmiedia tawa*), mahoe or hinahina (*Melicytus ramiflorus*), ngaio (*Myoporum lætum*)—unseen on western Tutira except after fires, rangiora (*Brachyglottis rangiora*), makomako—wineberry (*Aristotelia racemosa*), fuchsia (*Fuchsia excorticata*), and koromiko (*Veronica salicifolia*), were the most common trees and shrubs.

Nikau Palm.

Small groups of the New Zealand palm, nikau (*Rhopalostylis sapida*), and single plants of karaka (*Corynocarpus lævigatus*), grew also in the woods of the extreme eastern corner of the run. Thickets of supplejack (*Rhipogonum scandens*), entanglements of "lawyer" (*rubus* sp.), ropes of clematis and vine, were even more dense than in the forest of the west. The soils were richer, the warmth greater. Everywhere, moreover, the ground beneath these woods was ploughed and reploughed by pig in search of drupes, roots, and grubs.

A mere shred of Tutira was under marsh or swamp; such areas

were covered almost entirely with flax (*Phormium tenax*) and raupo (*Typha angustifolia*). The height of these plants varied with the drainage; on lands firm and dry each reached a noble growth; on areas of quaking bog they survived, soured and stunted with excessive wet. On dry ground grew also patches of the graceful toe-toe grass (*Arundo conspicua*). The outer edges of these marshes were rough with nigger's-head (*Carex secta*) and other coarse sedges and rushes. Sparganium antipodium also grew in certain parts, a plant remarkable in this, that it is the only native which has to my knowledge disappeared during my time on the station.

Lastly, there were on Opouahi and Heru-o-Tureia ten or twenty acres of upland meadow studded with huge, hollow, gnarled, dead, upright, broadleaf boles (*Griselinia littoralis*). On the ground lay in vast numbers totara spars and rotting trunks of other podocarps. These scraps of open upland had been under forest within sixty or eighty years, perhaps less. They were too high and cold for fern. For some reason not easy to understand, no crop of trees had sprung to possess the ground. It was grassed with yellow tussock (*Poa cæspitosa*), scented grass (*Hierochloe redolens*), one of those highly interesting Fuegian species,[1] *Poa anceps*, and other high-country grasses. Amid this rough turf many interesting species had obtained a hold and were flourishing. In their proper periods, groupings and strips of Pimelea longifolia and Helichrysum bellidioides made a brave show of blossom. On a spot most desolate and damp I have got the rare Brachycome odorata. The small terrestrial orchid, Pterostylis Banksii, was very plentiful in its season. In a sheltered nook, for the first and only time on Tutira, I have found the charming Caladenia bifolia. An interesting group of plants, including amongst its species the "vegetable sheep" of New Zealand, was represented by Raoulia australis. Other sub-Alpines of this upland meadow were Brachycome Sinclairii, Celmisia incana, Gentiana Grisebachii, Plantago Raoulii, Wahlenbergia saxicola, a delicate pale blue-bell, the barbed Acæna Novæ Zealandiæ, Spear-grass

[1] "The Fuegian element of the New Zealand flora," writes Dr L. Cockayne in the second edition of his delightful 'New Zealand Plants and their Story,' "although considerably smaller than the Australian element, has given rise to far more speculation. This arises from the fact that though biological geographers have been willing to erect a 'land bridge' between Northern Australia, Malaya, and New Zealand, many have hesitated before in imagination turning into dry land the profound depths of ocean which lie between New Zealand and Antarctica or South America. At the same time the presence of this Fuegian element so far distant from its present home has to be explained."

(*Aciphylla squarrosa*), species of Ligusticum, and species of Geranium; whilst just across my boundary flourishes safe in the rocks the lovely golden-yellow buttercup (*Ranunculus insignis*).

Other plant cities of refuge were the rock gardens of the cliffs, the sand gardens of the gritty tops, the bog gardens of the river brim and lake edge. On the dry cliffs survived two native brooms, Carmichaelia odorata and another, Vittadinia australis, Senecio lautus, Stellaria parviflora, Tillæa Sieberiana, Clianthus puniceus—brilliant in its bright scarlet racemes, and at one period, until eaten out by cattle, growing in great quantities on Heru-o-Tureia, and much more rarely on Awa-o-Totara,—Nertera depressa and Geranium sessiliflorum, both Fuegians, Pelargonium australe, Muehlenbeckia complexa, Gaultheria oppositifolia, Angelica rosæfolia, Arthropodium candidum, Daucus brachiatus, Linum monogynum, hill flax (*Phormium Cookianum*), and "blue grass" (*Agropyrum multiflorum*).[1] On the damp cliffs grew Gnaphalium Keriense, the very charming delicate Calceolaria repens, its white flowers spotted with purple, Euphrasia cuneata, Cladium Sinclairii, Lagenphora Forsteri, the native daisy—Papataniwhaniwha, Arundo fulvida, and other plants.

On aits and islands and about the river's very brim the most conspicuous small plants were Veronica catarractæ, discovered at the base of the 150-foot leap taken by the Maheawha stream, Senecio latifolius, Geum urbanum, Ourisia macrophylla, Oxalis magellanica—a fourth Fuegian,—and Viola Cunninghamii. Here and there along the lake, on the margins of springs and about damps and oozes on the limestone hills, grew a collection of miniature bog plants such as Hydrocotyle moschata, Azorella trifoliolata, Crantzia lineata, Epilobium

[1] Though now everywhere eaten out by stock, Agropyrum multiflorum was a famous grass in the early days of sheep-farming in Canterbury, its seed being considered equivalent to oats for keeping horses hard and fit. An instance of this is given by Mr George Dennistoun of Peel Forest. He writes: "On one occasion, in the middle sixties, when a neighbour, Mr Fred Kimball of 'Three Springs,' was our guest at Haldon in the Mackenzie Country, news arrived that his small son had eaten tutu berries and was dying. 'Three Springs' was thirty-eight miles distant by road, or rather by bullock-track. At once my Australian thoroughbred 'Pickwick' was run in from the block where the horses fed, country then densely covered with seeding 'blue grass.' I told Kimball, who had qualified for a doctor and was a fine rider, not to trouble himself about the horse, but to think only of his boy. I can't remember how long he took, but he said he never thought it possible to have been carried as he was. He saved his boy, and 'Pickwick,' after a bucket of gruel, later on took his oats as if he had been called on to do nothing out of the common." Readers can imagine for themselves what pace a man with medical knowledge, and a father to boot, would ride, knowing the effects of tutu poisoning; they can imagine, too, the racing-stable condition the horse must have been in to have stood without damage a forty-mile gallop over bad roads.

OXALIS MAGELLANICA.

One of several Fuegian species growing on Tutira.

nummularifolium, Montia fontana, Gunnera monoica—the red-berried form well worthy of the rock garden, — Galium tenuicaule, Mazus pumileo, Gratiola peruviana — a fifth Fuegian, — Triglochin striatum, Mentha Cunninghamii, Cotula coronopifolia, Pratia angulata, Pratia perpusilla, Lobelia anceps, Oxalis corniculata, and Spiranthes australis.

On barren crowns, arid edges, and driest of dry flats subsisted plants such as cabbage-tree (*Cordyline australis*), Gnaphalium—several species, Celmisia longifolia, Pimelea lævigata, Cyathodes acerosa, Leucopogon fasciculatus, Leucopogon Frazeri, Leptospermum scoparium, Pomaderris phylicæfolia, Echinopogon ovatus, Orthoceras strictum, and Microtis porrifolia.

At a later period, when the power of the bracken was broken, many of these plants, as will be shown, left their cliffs and deserts and rushed like eager settlers on the newly-opened land.

Cabbage Tree.

Of the sixty thousand acres of Tutira, fifty-eight, when the station was first stocked, were under bracken, less than fifteen hundred in forest and woodland, less than five hundred in marsh, less than twenty-five in upland meadow, cliff, river-bed, desert, and brims of stagnant creeks. Had, in fact, a narrow slice been shorn from the extreme west and another from the extreme east, Tutira would have been actually what it was for all practical purposes—one vast unbroken sheet of fern. Appended are the names of species noted on the station. I believe that few of the more insignificant plants have been overlooked, but since it is the nature of the writer of this volume to care for small plants rather than trees and shrubs, the list of the latter may not be quite complete.

LIST OF NATIVE PLANTS ON TUTIRA.

Ranunculaceæ.
Clematis indivisa.
" hexasepala.
" Colensoi.
" fœtida.
" parviflora.
Ranunculus hirtus.
" rivularis.
" insignis.

Magnoliaceæ.
Drimys axillaris.

Cruciferæ.
Nasturtium palustre.
Cardamine hirsuta.

Violarieæ.
Viola Cunninghamii.
Melicytus ramiflorus.

Pittosporeæ.
Pittosporum tenuifolium.
" crassifolium.
" eugenioides.
Caryophylleæ.
Stellaria parviflora.
Portulaceæ.
Montia fontana.
Hypericineæ.
Hypericum gramineum.
Malvaceæ.
Hoheria populnea.
Tiliaceæ.
Aristotelia racemosa.
Elæocarpus dentatus.
" Hookerianus.
Lineæ.
Linum monogynum.
Geraniaceæ.
Geranium dissectum.
" microphyllum.
" sessiliflorum.
" molle.
Pelargonium australe.
Oxalis corniculata.
" magellanica.
Olacineæ.
Pennantia corymbosa.
Rhamneæ.
Pomaderris phylicæfolia.
Sapindaceæ.
Alectryon excelsum.
Anacardiaceæ.
Corynocarpus lævigata.
Coriarieæ.
Coriaria ruscifolia.
" thymifolia.
Leguminosæ.
Carmichælia odorata.
Clianthus puniceus.
Sophora tetraptera.
Rosaceæ.
Rubus australis.
" cissoides.
" schmidelioides.
Geum urbanum.
Potentilla anserina.
Acæna Novæ Zealandiæ.
" sanguisorbæ.
Saxifrageæ.
Carpodetus serratus.
Weinmannia racemosa.
Crassulaceæ.
Tillæa Sieberiana.
Droseraceæ.
Drosera binata.
" auriculata.
Halorageæ.
Haloragis alata.
" depressa.
" micrantha.
Myriophyllum elatinoides.
" intermedium.
Gunnera monoica.
Myrtaceæ.
Leptospermum scoparium.
" ericoides.
Metrosideros hypericifolia.
" Colensoi.
" scandens.
Onagrarieæ.
Epilobium pallidiflorum.
" chionanthum.
" rotundifolium.
" nummularifolium.
Fuchsia excorticata.
Cornaceæ.
Griselinia lucida.
" littoralis.
Rubiaceæ.
Coprosma grandifolia.
" robusta.
" Cunninghamii.
" tenuifolia.
" parviflora.
Nertera depressa.
Galium tenuicaule.
" umbrosum.
Compositæ.
Lagenophora Forsteri.
Brachycome Sinclairii.
" odorata.
Olearia furfuracea.
" nitida.
" ilicifolia.
" Cunninghamii.
" nummularifolia.
" Solandri.

Ourisia Macrophylla—Waikoau River.

Compositæ—(contd.)
Celmisia incana.
" longifolia.
Vittadinia australis.
Gnaphalium Keriense.
" subrigidum.
" luteo-album.
" japonicum.
Raoulia australis.
Helichrysum bellidioides.
" filicaule.
" glomeratum.
Cassinia leptophylla.
Craspedia uniflora.
Bidens pilosa.
Cotula coronopifolia.
" australis.
" perpusilla.
Erechtites quadridentata.
Brachyglottis repanda.
Senecio lautus.
" latifolius.
" Banksii.
Microseris Forsteri.
Picris hieracioides.
Sonchus oleraceus.

Myrsineæ.
Myrsine salicina.
" Urvellei.

Oleaceæ.
Olea lanceolata.

Scrophularineæ.
Calceolaria repens.
Mazus pumilio.
Gratiola peruviana.
Veronica salicifolia.
" angustifolia.
" catarractæ.
Ourisia macrophylla.
Euphrasia cuneata.
Glossostigma elatinoides.

Thymelæaceæ.
Pimelea longifolia.
" virgata.
" lævigata.

Loranthaceæ.
Tupeia antartica (twice noticed on Leptospermum scoparium).

Urticaceæ.
Urtica ferox. (My rabbiter lost one dog and has had others crippled for days by this terrible nettle; a shepherd unwisely attempted to rush his well bred horse through a mass of it, the animal became unmanageable, rolled, and refused to rise: next day it was found dead.)
Urtica incisa.
Parietaria debilis.

Cupuliferæ.
Fagus fusca.
" Solandri.
" sp.

Coniferæ.
Podocarpus Totara.
" Hallii.
" ferrugineus.
" spicatus.
" dacrydioides.
Dacrydium cupressinum.

Palmæ.
Rhopalostylis sapida.

Pandaneæ.
Freycinetia Banksii.

Typhaceæ.
Typha angustifolia.
Sparganium antipodum.

Naiadaceæ.
Triglochin striatum.
Potamogeton polygonifolius.
" Cheesemanii.

Restiaceæ.
Leptocarpus simplex (edge of lake).

Cyperaceæ.
Eleocharis acuta.
Scirpus maritimus.
" prolifer.
Schœnus axillaris.
Cladium Sinclairii.
" glomeratum.
Gahnia Gaudichaudi.
Carex virgata.
" secta.
" inversa.
" Colensoi.
" echinata.
" subdola.
" ternaria.
" lucida.

Salviniaceæ.
Azolla rubra.
Lycopodiaceæ.
Lycopodium Billardieri.
" fastigiatum.
" scariosum.
" volubile.
Tmesipteris tannensis.
Orchideæ.
Dendrobium Cunninghamii.
Bulbophyllum pygmæum.
Earina mucronata.
" suaveolens.
Sarcochilus adversus.
Spiranthes australis.
Thelymitra longifolia.
" imberbis.
Orthoceras strictum.
Microtis porrifolia.
Prasophyllum rufum.
Pterostylis Banksii.
" foliata.
Caladenia bifolia.
Chiloglottis cornuta.
Corysanthes oblonga.
" rotundifolia.
" macrantha.
Gastrodia Cunninghamii.
Irideæ.
Libertia grandiflors.
" ixioides.
Liliaceæ.
Rhipogonum scandens.
Cordyline Banksii.
" australis.
" indivisa (flowers always purple).
Astelia Solandri.
" nervosa.
Phormium tenax.
" Cookianum.
Arthropodium candidum.
Dianella intermedia.
Juncaceæ.
Juncus pallidus.
" bufonius.
" Novæ Zealandiæ.
Luzula campestris.
Myoporineæ.
Myoporum lætum.
Labiatæ.
Mentha Cunninghamii.
Plantagineæ.
Plantago Raoulii.
Illecebraceæ.
Scleranthus biflorus.
Polygonaceæ.
Polygonum aviculare.
" serrulatum.
Rumex flexuosus.
Muehlenbeckia australis.
" complexa.
Piperaceæ.
Piper excelsum.
Monimiaceæ.
Hedycarya arborea.
Laurelia Novæ Zealandiæ.
Laurineæ.
Beilschmiedia Tawa.
Proteaceæ.
Knightia excelsa.
Apocynaceæ.
Parsonsia heterophylla.
" capsularis.
Loganiaceæ.
Geniostoma ligustrifolium.
Gentianeæ.
Gentiana Grisebachii.
Convolvulaceæ.
Calystegia sepium.
Convolvulus erubescens.
Solanaceæ.
Solanum nigrum.
" aviculare.
Campanulaceæ.
Pratia angulata.
" perpusilla.
Lobelia anceps.
Wahlenbergia gracilis.
" saxicola.
Ericaceæ.
Gaultheria antipoda.
" oppositifolia.
Passifloreæ.
Passiflora tetrandra.
Epacrideæ.
Cyathodes acerosa.
Leucopogon fasciculatus.
" Frazeri.
Dracophyllum (sp.).

Wild Calceolaria.

Umbelliferæ.
Hydrocotyle elongata.
" moschata.
" asiatica.
Azorella trifoliolata.
Oreomyrrhis andicola.
Crantzia lineata.
Aciphylla squarrosa.
Ligusticum (2 sp.).
Angelica rosæfolia.
Daucus brachiatus.

Araliaceæ.
Panax Edgerleyi.
" Colensoi.
" arboreum.
Schefflera digitata.

Gramineæ.
Isachne australis.
Microlæna stipoides.

Gramineæ—(contd.)
Microlæna avenacea.
Hierochloe redolens.
Echinopogon ovatus.
Deyeuxia Forsteri.
" quadriseta.
Dichelachne crinita.
Deschampsia cæspitosa.
Trisetum antarticum.
Danthonia semiannularis.
" pilosa.
Arundo conspicua.
" fulvida.
Poa anceps.
" cæspitosa.
" Colensoi.
" imbecilla.
Agropyrum multiflorum.
" scabrum.
Asperella gracilis.

CHAPTER XIII.

THE FERNS OF TUTIRA.

THE ferns of Tutira deserve special attention—a chapter, albeit a brief one, to themselves. Out of the 135 species enumerated by Cheeseman in his 'Manual of the New Zealand Flora,' or 134 if the very doubtful Davallia Forsteri be disallowed, more than one-half grow on Tutira. It is a remarkable record for one station—a record which, I am confident, can never be exceeded. Variations of altitude, large rainfall, range of climatic conditions, dissimilarity of geological formations, and careful search have each in its degree contributed to this result. The main cause, however, has been the wild and rugged nature of the country, its enormous quantity of gorges and ravines, its hundreds of miles of precipice and crag. Species ousted elsewhere maintain themselves in such spots—they afford a last foothold to fugitives; thus, clinging to the base of a low conglomerate cliff, survives a patch of Gleichenia circinata. Twice have fires almost blasted the plant to death; twice has it reappeared. A dripping precipice, otherwise usurped by Polypodium Billardieri, shelters Lindsaya viridis; crannies in a single mass of broken limestone rock high on Heru-o-Tureia afford foothold to Cystopteris fragilis; the concave base of a series of high, dry conglomerate rock-faces safeguards Doodia media; though nibbled and brushed by stock, and though endangered during every flood by landslips, it survives. The low rims of a tumbled mass of conglomerate boulders offer a last foothold to Adiantum diaphanum. It has climbed by an athletic feat from its own natural habitat—the forest floor. Asplenium Trichomanes survives on a single limestone rock broken from one of the ancient sea-floors of eastern Tutira and deeply set in the turf of the green hillside. Each of the above species has been found but on one small spot on Tutira. Another fern only to be found on rock is the slender, beautiful annual, Gymnogramme leptophylla. Thrice since 1882 it has been exceedingly plentiful on the conglomerates of central Tutira. On each occasion the plant has shown itself after periods of remarkable drought and heat, germination of its spores seeming only to happen at a temperature above normal. Except under such conditions not a single specimen has been found. It is absent or abundant, very plentiful or undiscoverable, appearing or reappearing at intervals of years. Asplenium flabbelifolium also chiefly abides on the rocks. I have got specimens of many others—burnt up, starved, depauperated—on cliffs. Their names need not, however, be given, as they grow also on sites where they thrive and which they adorn by happy growth. One other fern—Gleichenia Cunninghamii, a forest species normally—has once only been found on Tutira.

I can say of the ferns as of the grasses and orchids of the station, that they have been sought for with special care. Appended are the names of species :—

Hymenophyllum rarum.
" polyanthos, var. sanguinolentum.
" pulcherrimum.
" dilatatum.
" demissum.
" scabrum.
" flabellatum.
" Tunbridgense.
Trichomanes reniforme.
" humile.
" venosum.
Cyathea dealbata.
" medullaris.
Hemitelia Smithii.
Alsophila Colensoi.
Dicksonia squarrosa.
" fibrosa.
Cystopteris fragilis.
Lindsaya viridis.
Adiantum affine.
" diaphanum.
" Æthiopicum.
Hypolepis tenuifolia.
Cheilanthes Sieberi.
" tenuifolia.
Pellæa rotundifolia.
Pteris aquilina, var. esculenta.
" scaberula.
" tremula.
" macilenta.
" incisa.
Lomaria Patersoni, var. elongata.
" discolor.
" vulcanica.
" lanceolata.
Lomaria alpina.
" capensis.
" filiformis.
" fluviatilis.
" membranacea.
Doodia media.
Asplenium flabbelifolium.
" Trichomanes.
" falcatum.
" lucidum.
" " var. anomodum.
" Hookerianum.
" bulbiferum.
" flaccidum.
Aspidium aculeatum.
" Richardi.

Aspidium capense.
Nephrodium decompositum.
" glabellum.
" velutinum.
" hispidum.
Polypodium punctatum.
" pennigerum.
" australe.
" grammitidis.
" serpens.
Polypodium Cunninghamii.
" pustulatum.
" Billardieri.
Gymnogramme leptophylla.
Gleichenia circinata.
" Cunninghamii.
Todea hymenophylloides.
Ophioglossum lusitanicum.
Botrychium ternatum.

CHAPTER XIV.

THE AVIFAUNA OF THE STATION PRIOR TO SETTLEMENT.

In regard to the immediate past there is no reason to believe that in actual number of breeding species there has been any decrease. On a run so full of crags, impenetrable gorges, and deep river-beds, possibilities of concealment and escape are almost unlimited. The difference between now and then lies not in reduction of species but in reduction of individual birds. Undoubtedly there has been a very great diminution in the aggregate numbers. There are probably not ten birds now for every thousand there used to be immediately prior to settlement.

Male Bell-bird feeding young.

There had, however, existed—say within a century or two—other species. The older resident natives knew of them by tradition; they knew their Maori names. From hearsay they could, with a fair degree of accuracy, describe their habits. They recognised with expressions of delight their coloured representations as depicted in Buller's illustrated volumes. Thus I learnt that the Blue Wattled Crow (*Glaucopis Wilsoni*), a breed. until the forest was felled, extremely plentiful on the coastal forest between Wairoa and Gisborne, was at one time common also on Tutira. They recognised, too, the Saddle Back (*Creadion carunculatus*), a species which in my time has always been exceedingly rare on the east coast.[1] About one or two less prominently marked species, the Maoris were less confident in their identification, but I gathered that the North Island Robin (*Petrœca longipes*) had also at one time been common. These three species, it was agreed, had vanished long prior to the inroads of settlement; they had probably passed away with the passing of the ancient forest.

Young Bitterns.

We shall note in another chapter how indigenous species of birds have been affected in different ways by the development of the run—by its change from bracken and bush into grass: how some must inevitably perish, some linger in lessened numbers, and some, I am glad to say, survive and even increase.

Appended is a list of species seen on Tutira during my time:—

Falconidæ (Hawks).
- Hieracidea Novæ Zealandiæ.
- " ferox.
- Circus Gouldi.

Strigidæ (Owls).
- Athene Novæ Zealandiæ.

Alcedinidæ (Kingfishers).
- Halcyon vagans.

Meliphagidæ (Honey-eaters).
- Prosthemadera Novæ Zealandiæ.
- Anthornis Melanura.
- Zosterops lateralis.

Certhiadæ (Creepers).
- Acanthisitta chloris.

Luscinidæ (Warblers).
- Sphenœacus punctatus.
- Gerygone flaviventris.
- Petrœca toitoi.
- Anthus Novæ Zealandiæ.

Muscicapidæ (Fly-catchers).
- Rhipidura flabellifera.

Psittacidæ (Parrots).
- Platycercus Novæ Zealandiæ.
- Nestor meridionalis.

Cuculidæ (Cuckoos).
- Chrysococcyx lucidus.
- Eudynamis taitensis.

Columbidæ (Pigeons).
- Carpophaga Novæ Zealandiæ.

Apteryginæ.
- Apteryx Mantelli.

Charadriadæ (Plovers).
- Charadrius bicinctus.

Ardeidæ (Herons).
- Ardea poeciloptila.
- " alba.

Scolopacidæ.
- Himantopus leucocephalus.
- Limosa baueri.

Rallidæ (Rails).
- Ocydromus earli.
- Rallus philippensis.

---

[1] I have seen locally, indeed, but one pair in my life, immature birds in dense scrub on the slopes of the Maungahamia range in the back country of Poverty Bay

PUKEKO (*Porphyrio melanotus*) MALE BIRD ON NEST.

Rallidæ (Rails)—(contd.)
- Ortygometra affinis.
- " tabuensis.
- Porphyrio melanotus.

Anatidæ (Ducks).
- Casarca variegata.
- Anas chlorotis.
- " superciliosa.
- Rhynchaspis variegata.
- Hymenolæmus malacorhynchus.
- Fuligula Novæ Zealandiæ.
- Nyroca australis.

Colymbidæ (Divers).
- Podiceps rufipectus.

Procellaridæ (Petrels).
- Thalassidroma melanogaster.

Laridæ (Gulls).
- Larus dominicanus.
- " scopulinus.

Pelecanidæ.
- Phalacrocorax Novæ hollandiæ.
- " brevirostris (?).
- " varius (?).

CHAPTER XV.

IN THE BEGINNING.

ABOUT 1860 an immense territory was purchased by Government from the native owners of southern Hawke's Bay, the lands thus acquired being parcelled out in great runs as freehold. It was open fern country for the most part, extending from the foothills of the Ruahine and Kaweka ranges to the ocean. The homesteads of the stations thus created were connected with the port of Napier either by sea or by rough bullock-tracks following the lines of old river-beds deserted by

Packing wool pockets.

the streams which had made them, too barren to be blocked by vegetation, and offering the further inducement of sound going even in the wettest of seasons. At a later date another area of land north of Napier was confiscated from the natives who had taken part in the "rebellion" of the late 'sixties. Upon reconsideration, however, of their claims and counter-claims, it was discovered that in every tribe certain septs and families had remained "loyal." The natives, in fact, had consciously or unconsciously hit upon a device practised by many Jacobite houses in the eighteenth century,—the head of the family supporting one side in the collieshangie, a younger son as stoutly maintaining the rights of the other, the estate being thus assured whatever happened. In the case of the native lands in question, final ownership of the block seized resulted in a compromise. The Government sold outright to European settlers a small proportion of the territory taken. The residue was for all practical purposes—though I believe not absolutely—restored to its former owners.

The bush areas of Hawke's Bay were still untouched except by the hardy Scandinavians penned in their forest settlements.

The better and more accessible countryside thus taken up as freehold, later arrivals in the province had perforce to content themselves with the lands of the interior. Settlers began to push inland, and, where purchase was not permissible, to lease runs from the natives. Amongst other blocks thus taken up were Tutira, Putorino—Waikari as it was then called—and Maungaharuru.

In February of '73 Tutira was leased by forty native owners to T. K. Newton for twenty-one years at £150 per annum. The block was held in common by these natives, but it was provided that the rent—£3, 15s. per man—should be paid to each of them. Like almost every other native title on the east coast, that of Tutira was imperfect. Newton must have been anxious at a very early period in regard to one of the signatures. It is characteristic, indeed, of the tenure of the station that—the run being then in its earliest infancy, a suckling not yet three months old—there should be an entry in the Deeds Office to the effect that "William Morris, sheep farmer, husband of one of the lessors, confirms his wife's action in regard to her signature of the Tutira lease."

Newton stocked the place with 4000 sheep, and placed his brother-in-law, Craig, in charge of the new venture. Craig's headquarters during his brief residence on Tutira were near the site of the present homestead. The hummock of his clay chimney, just about the centre of my present lawn, remained for many years a monument to his memory. There is still visible the cutting whence he dug his clay. There are also mysterious excavations in the same hillock which we believe to have been his primitive dog-kennels. The 4000 sheep—merino wethers—were saved from a worse fate by the action of the notorious Te Kuiti, who at this date raided the little settlement of Mohaka, murdering impartially Europeans and "friendly" natives. His anticipated march down the coast cleared every homestead of its inhabitants. The 4000 sheep—or what remained of them—were mustered in hot haste and rushed off the place. Craig, with other outlying settlers, took refuge in Napier, and with his flight the first attempt to work Tutira as a sheep station terminated.

The Waikari run—now called Putorino—was no more fortunate in its initial stage. It also was abandoned during Te Kuiti's raid by its first owners. Maungaharuru, taken up by Philip Dolbel, was actually burnt out by Te Kuiti's band of ruffians, Dolbel and his men escaping by a fortunate delay in the delivery of certain newly-purchased stock. These sheep, which should have been ready to start from Napier on a Monday, were not forthcoming until the Tuesday; Dolbel and his drovers arrived at Maungaharuru in time to find the

Homestead of the 'seventies.

yet smouldering remains of their little homestead — the twenty-four hours' delay had saved them.

Reverting to Tutira proper, we can well believe that Craig's hasty muster was not a "clean" job. Sheep, in fact, were left on the station in considerable numbers, for until the run was again in European hands the local natives were accustomed to dog them into high fern and there shear them. Sheep, however wild, are, in six-foot bracken, helpless. They sink belly-deep into the tangled springy growth, whilst the standing fronds surround them like a wall. After shearing, the wool thus commandeered was rammed into bags and carried off by the Maoris. This first Tutira shearing must have been picturesque at any rate,—the trampled trodden wedge driven into the solid fern, the blue open sky, the wild brown Maoris, the mongrel teams of dogs, the gleaming shears, the jollity and laughter over the *pakeha's* discomfiture. Newton's

"Wild" Sheep.

experiment in sheep-farming had not been a success. He had paid two or three seasons' rent, and lost, in one way or another, probably nearly half his sheep.

In 1875 the station was sold to Edward Toogood for £5. I understand, however, from Mr J. C. Tylee, who managed the place during Toogood's tenancy, that even this sum was given, not chiefly for the goodwill of the place, but as payment for any claim Newton might have had on his abandoned sheep and their wild progeny. Again the run was stocked with 4000 sheep and 100 head of cattle. Boundaries were kept, the sheep only allowed to roam over what are now called the Natural and Reserve paddocks. The cattle lived about the swamp land round the margin of the lake. Tylee also tells me that two or three bags of grass seed were sown and that a few chains of fencing were erected.

Times were now beginning to mend a little; there were prospects of lasting peace; property was becoming more secure. The energy, moreover, of certain settlers in southern Hawke's Bay was proving that fern-runs could be made to pay, at any rate in good soils and in dry districts. Toogood, like other sheep-farmers, was beginning to "improve," and doubtless found himself fully occupied with his Tangoio property. Be that as it may, Tutira was sold by him early in '77 to G. J. Merritt for £2500—a few score pounds more than the value of the 4000 sheep delivered with the place. In March of the following year Merritt sold to C. H. Stuart one half-share in Tutira together with 3600 sheep for the sum of £2500; it had been purchased for T. J. Stuart, a younger brother not of age. As Merritt had given that sum for the full share not long before and spent nothing in "improvements," he must have cleared something by the transaction,—how much, at this distance of time, it is impossible to discover. It would depend on the age, condition, and sex of the stock delivered, and many other eventualities.

Up to this date the station had been owned by men who had not lived on it. Newton was a Napier merchant, Toogood's real interests lay in his Tangoio property, Merritt was a settler in Clive. Each of them had looked upon Tutira as a mere speculation; it had been regarded as a step-child. Its new owner, Mr T. J. Stuart, was a settler of a very different type; from the beginning he cared for the place. It was to be developed by his own labour; it was to become a home made by his own hands.

CHAPTER XVI.

THE LURE OF IMPROVEMENTS.

It has been stated that in March of '78 George Merritt sold to C. H. Stuart one half-share in Tutira. There is an underlined diary entry dated 1st April '78: "C. H. Stuart takes over the share of T. J. Stuart." There is another: "Ayson received from C. Stuart £40 to pay native track-makers." Doubtless there was some private arrangement between the brothers, for though a minor at the time, the half-share in the station actually belonged to T. J. Stuart from the beginning.

During the same year Merritt's remaining share was also taken over, no money as far as I can discover passing in the transaction. The probabilities are that there had been advances made from a bank or mortgage company. There would also be on Merritt's part responsibility for the working expenses of the place. Not long afterwards a new name appears, that of T. C. Kiernan, who entered into partnership with C. H. Stuart, the latter again, doubtless, acting for his brother.

In the hands of Messrs Stuart & Kiernan, Tutira was to undergo a vast transformation. They had bought the place to put into it their own personal work, to make a home of it for themselves. They were young, hopeful, and energetic. The earliest written record extant, that of '78, is fragmentary, but in Kiernan's diaries each day's work on Tutira is fully registered between January of '79 and July of '81. Kiernan's diaries are, in fact, items in the early history of Hawke's Bay; though written of Tutira, they illustrate incidentally the vicissitudes of every sheep-station in the province, the rise and fall of prices, the smiles and frowns of fortune.

These were the times, as in King Arthur's court, when each hour brought forth some noble deed, when each day saw some wrong to the station righted. The Stuart Brothers and Kiernan loved their run—

they were enthusiasts. They worked as young men work when hope is high. The wool-shed to be erected was a vision to dream of; it was a joy to view with the mind's eye vast stretches of green grass; their hearts leapt up when they beheld the flocks and herds of the future, the larger lambs, the fatter wethers, the heavier-fleeced ewes. In that golden age money was valued as useful only for some new improvement to the station. She was the beloved mistress for whom nothing was too good. She was to be decked with the straightest of fence lines, the woolliest of sheep, the shadiest of willow-groves; beautified with tall crops, smoothed in green grass, lawned like Arcadia. A settler gives his best love not to his parents, not to his wife, not to his little ones, but to his land.

In a former chapter I have asked the reader to shed the Decalogue and to strip himself to a Maori mat. In this I could wish that he should brace himself to the agony of perusing three months' entries from T. C. Kiernan's diary. Unlimited diary is a proverbially stodgy diet, and that of '79 is no exception to the rule, yet deliberately I intend to reproduce January, February, and March exactly as they were written, day by day, word for word, with all their repetitions, trivialities, jottings of wages, stores, mutton, tobacco, and pain-killer sold. I have thought it better to show the plain unvarnished tale as a whole than attempt to select sample days. The reader must prepare himself, therefore, for the digestion of a three months' lump of diary. Alas! that I cannot present to him the grimy original documents; then, indeed, he might forbear, or, at any rate, condone the offence. Truth to tell, they hardly bear transcription to clean paper and clear type; I feel a kind of shame in dragging to the light of day jottings pencilled in smoky huts lit by candles guttering in the draughts, the writer, with hard hands and broken nails, rising from time to time to turn the frizzling chops, to prong the simmering joint, or to pile fresh embers on the lid of the camp oven.

The play opened then in '78 with a *dramatis personæ* of owners and ex-owners: Thomas and Charles Stuart, George and Ben Merritt, a mysterious Mr Doull "from Otago,"—alone of those mentioned in the fragmentary diary of '78 honoured with the courtesy prefix,—shepherds, bushmen, contractors, and natives.

The station once acquired, improvements were not long withheld; from the beginning, indeed, they were lavished on the land with both hands: draining, track-making, and fencing were started within a few weeks of purchase. Specialisation of work had not begun—each man put his hand to the task most pressing. The kind of life led by these pioneers, the kind of work done in the early days of station life, cannot, in fact, be made more comprehensible to readers than by full citation of actual bare facts. Doubtless the work was rough and crude, but it is upon the dust and grime and sweat of these prehistoric days that the present Tutira is founded.

Well, the reader has to imagine a company of young men, living on bread, mutton, wild pork, and potatoes, in a reed-hut, garbed in little else than boots, shirt, and moleskins, the last-named garment supported by a waist-belt containing the butcher-knife, sometimes its leathern sheath, from use and wont, so warped to the wearer's shape as almost to resemble a tucked-in tail; the station itself a wilderness, unfenced and pathless, covered with bracken, bush, and flax.

The diary of '78, by whomsoever written, is broken and fragmentary, in that of '79 each day's work is entered. Making no further apologies, I shall allow this diary to speak for itself. Whether of interest or not, its pages, at any rate, portray the early days of a sheep-run. The initials C. H. S. are those of Charles H. Stuart, T. S. those of Thomas Stuart, T. C. K. those of T. C. Kiernan, otherwise the journal tells its own tale.

JANUARY—

1. *Wednesday*. C. H. Stuart and T. S. in Napier on a visit to Meanee. Kite at Petane. T. C. K. enjoying his New Year's Day by keeping the confounded cattle out of the oats. Turned them out twice, and the last time drove them over the other side of Hughie's fence. Took a look round about 6.30 P.M. and found four of them back again, so gave it up as useless. Larrikin registered from this date for a year.
2. *Thursday*. T. C. K. and Hughie fencing all day at the horse paddock. C. H. S. returned from Napier about 4 P.M., having sent Kite and Tom S. to the Kaiwaka for young bullocks.
3. *Friday*. Kite and T. S. returned from Petane and brought 2 of the young bullocks from the Kaiwaka. C. H. S., T. C. K., and Hughie fencing all day.
4. *Saturday*. Kite and C. H. S. sledged posts from Tylee's spur and cut some Kohi posts for the horse paddock. T. C. K. and Hughie fencing in ram paddock at Kaikanui.
5. *Sunday*. All hands at home.
6. *Monday*. Heavy rain all the forenoon. Cleared up about 2 P.M., and C. H. S., T. S., and T. C. K. did some fencing.

7. *Tuesday.* C. H. S. and Kite packed a load of wool to Petane. T. S. attended muster at Waikari. T. C. K. and Hughie fencing all day. I wrote to my mother.
8. *Wednesday.* Kite and C. H. S. returned from Petane with pack-horses bringing stores. Two bullocks dead at Troutbeck's. One "tuted," the other bogged. T. C. K. and Hughie finished fence at Kaikanui.
9. *Thursday.* T. C. K., C. H. S. and Hughie went to the back country to erect mustering-yards at the "Burnt Bush." T. S. still at Waikari. Kite went to McKinnon's (Arapawanui) for some of our sheep from Moeangiangi.
10. *Friday.* C. H. S., T. C. K., Hughie fencing at Burnt Bush. Kite came out about 8 A.M. and reported having got 60 sheep from Arapawanui, mostly unshorn. We are short of tucker, as Hughie neglected to fetch meat, and but very little bread. Killed young boar this evening; great rejoicing.
11. *Saturday.* C. H. S., T. C. K., and Kite and Hughie, being starved out of camp, had to return to the station. Could not tackle the wild boar. Posts and strainers for yards all split. Arrived at Tutira at 8.30 P.M.
12. *Sunday.* All hands being knocked up, took it easy. T. S. arrived from Waikari with Finlayson and brought home 40 of our sheep. Finlayson engaged to muster at 30/- per week.
13. *Monday.* Hughie and Kite started early for the Burnt Bush to finish the yards. Finlayson went to Arapawanui to try and get Atta for the muster. C. H. S., T. C. K., and T. S. put zinc on the posts of store and cleaned it out. T. C. K. and C. H. S. drove the cattle from Kakanui round to the Natural Paddock swamps.
14. *Tuesday.* C. H. S., T. C. K., and T. S. got things ready for going mustering. Left Kakanui for Tutira about 11 o'c. Martin took 4 pack-horse loads of stores to Ward's bush. Finlayson returned from Arapawanui at 12.30 P.M. Atta couldn't come. George Goodall arrived from Waikari. T. C. K. mended store door. C. H. S. killed sheep. T. S. cooked. Kite and Hughie came from yards at 7.30 P.M. C. H. S., T. S., G. G., Finlayson, Kite and Hughie went to camp at Papa Creek. Parkes arrived from Petane with six bullocks from Troutbeck at £13/10/-. T. C. K. returned to Kaikanui.
15. *Wednesday.* T. C. K. went to look for horses; couldn't find them. Went to Tutira to get the woolshed ready. Got soaked through going over. Too wet to do anything, so returned, and got wet through again. Changed things and stayed at home. Took order from Sparkes for sledge iron-work and wrote to Faulknor.
16. *Thursday.* Sparkes and Martin came to Kaikanui for sledge wheels. Sparkes took monkey-wrench to the bush, also one pack-saddle—no horse, as he packs one of the bullocks. Went to look for horses and found Charlie dead in the swamp, strangled by his tether rope. Went to Tutira at 2 P.M. and drove cattle over the Willows. Worked at woolshed. Returned and stapled fence. Very heavy earthquake at 10.45 P.M.
17. *Friday.* Found 10 head of cattle in the oats. Drove them as far as

No. 2 swamp. Looked for my horse; couldn't see him. The cattle went back to the oats, so I drove them across the lake at the Willows. Found "Tommy" in No. 2 swamp. Went round the fences at Kaikanui and stapled them where required.

18. *Saturday.* First thing I saw was a mob of cattle in the oats. Drove them across the lake at the Willows. Started for Waikari at 12.30 P.M. Couldn't find it, so returned to Moeangiangi, where I arrived at 10.30 P.M. and stayed all night. Left letters for the mail-man.

19. *Sunday.* Left Moeangiangi at 7.30 and arrived at Tutira at 10 A.M. Found Hughie in from muster for tucker. Found cattle back in the oats. Drove them to No. 2 swamp. Found Pablo fast in wire fence. Got him out after some trouble all right. Set fire to flax in small paddock.

20. *Monday.* Mustering still going on. T. C. K. at station. Found the cattle back in the oats. Drove them across at the Willows. Worked at woolshed all day. Musterers returned from back country with mob of about 2000 mixed sheep (800 woolly) about 7 P.M.

21. *Tuesday.* Drafted sheep from back country. G. Goodall got 100 of Waikari sheep; no other strangers. Docked 20 lambs, making in all 1145 lambs docked to date not including those at Waikari. Hughie cooking. Kite went for the tools at the Burnt Bush.

22. *Wednesday.* G. Goodall left with Bee's sheep at 9 o'clock and T. S. went to clear the road. Kite took 22 bales of wool to Petane, making a total up to date of 173 bales. C. H. S. and T. S. went to Ward's bush to make arrangements about the timber. Hughie cooking. T. C. K. getting woolshed ready for recommencing shearing. C. H. S. and T. S. returned from Ward's about 9 P.M.

23. *Thursday.* Kite returned from Petane with pack-horses, bringing flour, sugar, raisins, currants, and the mail bag, also parcels for T. C. K. T. C. K., C. H. S. and T. S. working at woolshed. Hughie cooking. 5 Maoris arrived this evening.

24. *Friday.* Commenced shearing with 19 rams in the morning, and afternoon, 348 of the sheep brought from the back country. Hughie started as cook for all hands and the Maoris. T. C. K., C. H. S., T. S. and Kite assisting at shed.

25. *Saturday.* T. S. went to attend Dolbel's drafting. Kite sledged firewood with "Rodney" from Reserve. T. C. K. superintending for shearing, rolling fleeces, etc. C. H. S. getting bales ready for packing to Petane. Sheared 431 sheep to-day.

26. *Sunday.* T. Stuart returned from Dolbel's about 4 P.M. with 200 sheep.

27. *Monday.* C. H. S., T. S., T. C. K., Kite and two Maori boys started mustering the reserve at 3 A.M. Got in about 1000 sheep, principally ewes and lambs, at 10 A.M. We estimate that there are 600 sheep still in the paddock. C. H. S. got wool ready for pack-horses. T. C. K. and T. S. worked the shearing. 389 sheep shorn to-day. Docked 8 lambs, making in all (with 8 at Waikari, 6 at Dolbel's) 1167.

28. *Tuesday.* C. H. S. and Kite took 22 bales of wool to Petane. On going

for the sheep for day's shearing found they had nearly all got out of paddock back to reserve during the night, 48 woolly sheep only remaining. T. C. K. went to Kaikanui to drive cattle to Natural Pad dock. T. S. attended shed. Heavy rain began at 12 o'clock. Shearers all went home. 48 sheep shorn, 3 lambs marked. Parks and wife brought the young bullocks from the bush and report them as shaping well. Took the camp oven from fencers' camp. He brought back from bush one spade, keeping one spade and pick out there.

29. *Wednesday.* T. S. and T. C. K. cleared up the shed and dried wool that got wet yesterday. Parks and wife left with 6 old bullocks for the bush, and also packed "Dan" with piece of wire and camp over, etc. He also took iron-work for sledge and gouge. C. H. S. and Kite returned with pack-horses at 7 P.M., bringing stores. Had great trouble on the road.

30. *Thursday.* Kite took grey mare to fetch stores that had to be left on road yesterday. C. H. S., T. S., and T. C. K. worked at woolshed drying wool, making up loads of wool, and cleaning shed (pressing) of wool. Kite returned and pressed. T. S. set fire to Newton. C. H. S. put stores in order, and T. C. K. fires Reserve and assisted Kite.

31. *Friday.* C. H. S. and T. S. and T. C. K. and Kite mustered Natural Paddock in the morning and finished pressing and weighing all wool in the shed. In the afternoon Hughie cook.

FEBRUARY—

1. *Saturday.* C. H. S. and Kite took pack-horses with wool to Petane. T. S., T. C. K. and Hughie drafted the sheep mustered yesterday, and earmarked all that had been missed before. Put X-breds in Newton block, and T. S. took the Merion wethers over Papa Creek. T. C. K. mended store door and secured wool in the shed.
2. *Sunday.* T. S., T. C. K. went to Tutira for the rams. C. H. S. and Kite returned from Petane bringing a few stores. T. S. and T. C. K. took 26 rams over to Kaikanui Paddock.
3. *Monday.* C. H. S. and Kite mustered cattle to take to Petane to-morrow to the sale. T. C. K. and Hughie worked at fence. T. S. went to drafting at Kaiwaka.
4. *Tuesday.* C. H. S. and Kite took 24 head cattle to Petane. Hughie came over to Kaikanui at 7.30 A.M.; sent him with pack-horse up to fencers' camp for windlass, &c., to get posts from Natural Paddock in morning. Heavy rain in afternoon. Stayed at home and wrote letters. Cleaned whare, etc. First snow this year at Cox's.
5. *Wednesday.* C. H. S. and Kite not yet returned from Petane. T. S. still at Kaiwaka. T. C. K. went to Tutira in canoe for salt, spades, etc. Hughie at the Natural Pdk. rigging windlass for getting posts out of the bush. T. C. K. returned to Kaikanui at 12 o'clock. Saw large mob of sheep on the new burn outside Hughie's fence. Went up the fence as far as the slip panel, and found the fence burnt in

several places. Counted altogether 20 posts wanting renewing. About 300 sheep outside.

6. *Thursday.* T. C. K. and Hughie putting up windlass rigging for the posts in Natural Pdk. C. H. S. and Kite returned from Petane. Cattle averaged at sale £5—2—0½ per head.
7. *Friday.* C. H. S. and Kite took wool to Petane. T. C. K. and T. S. went to Tangoio and brought the sheep (340 ewes and lambs and a few wethers) home from Dolbel's. Got the sheep in the yards at 9 o'clock P.M. Had tea at 11 P.M. Got to bed at 12 P.M.
8. *Saturday.* T. S. and T. C. K. earmarked and docked 6 lambs and all Merritt's ewes and wethers brought from Dolbel's yesterday. Hughie still at the Natural Padk. C. H. S. and Kite returned from Petane bringing stores.
9. *Sunday.* All hands at home taking a rest.
10. *Monday.* S. S., T. C. K., and C. H. S. got things ready for the muster. Hughie and new man getting firewood. Kite cooking. C. H. S. and T. C. K. put the store in order. Heavy rain all the afternoon, so couldn't start to muster as intended.
11. *Tuesday.* T. C. K., T. S., and C. H. S. went over to Tutira about 12.30 P.M.; found Neil from Dolbel's and Tangoio Joe had arrived for the muster, making in all 7 hands—C. H. S., T. S., Hughie, Kite, Joe, Neil, and new man. T. C. K. remaining at home. Musterers started this evening.
12. *Wednesday.* T. C. K. went to Tutira in canoe for tools. Spent the day at the No. 2. swamp bridge cleaning out the drains and getting material for repairing the bridge. All hands still out mustering the back country.
13. *Thursday.* T. C. K. worked at No. 2 swamp bridge. Got it finished at 5 P.M. All hands still out mustering.
14. *Friday.* T. C. K. went to Tutira; saw Hughie in from muster, they having run short of bread. Mended table, etc. Tried some burning, but the fern wouldn't burn well.
15. *Saturday.* Stayed at home and looked over ledger. In afternoon found the cattle had injured the No. 2 bridge, so had to repair it. Musterers still out.
16. *Sunday.* Musterers still away. T. C. K. at home. Cleaned saddle, stirrup irons, bit, etc. Had a visit from Tomoana this afternoon. Asked to stay at Tutira to-night. Permitted him.
17. *Monday.* Kite came over to Kaikanui at 9.30 A.M. to plough. T. C. K. went to Tutira, gave Kite stores, found several Maoris there who came to shear, Tomoana having told them the wrong day. C. H. S., T. S. and musterers arrived at 2 P.M. C. H. S. and T. S. slept at Kaikanui; the rest of musterers went up to Newton camp.
18. *Tuesday.* C. H. S. and T. S. left Kaikanui at 3 A.M. to assist mustering Newton block. T. C. K. started at 6.30 for do. Got sheep into the yards about 2.30 P.M. Good muster, about 2000. C. H. S., T. S. and rest of musterers went to camp at Maungahinahina about 5 P.M. to muster reserve to-morrow.
19. *Wednesday.* T. C. K. went to Tutira to mend gates at drafting-yard.

Musterers returned about 12 A.M. with about 2000 sheep. Hughie and T. C. K. put canvas on shed. The rest took a rest. T. C. K. returned to Kaikanui alone. Saw the bull with broken leg at whare. Also saw 12 sheep, 5 on the road over and 7 at the whare, mostly X-breds.

20. *Thursday.* All hands started drafting the sheep. Hughie cooking and rather sulky. Maoris (9) commenced shearing. 114 sheep shorn. Rained till 12 o'clock then knocked off. In afternoon all went drafting. Got through half sheep. Webb of Maungaharuru came about 5.30 P.M. to give notice of 3700 sheep coming through on Sunday.

21. *Friday.* T. C. K. and Webb left Kaikanui at 6 A.M. for Tutira. T. S., Neil and Kite out mustering Natural Paddock. Began to shear at 9. T. C. K. superintending shed. Rest of the hands finished drafting sheep. Musterers returned at 12 o'clock unsuccessful. 550 sheep shorn. Let the lambs go in Natural Padk. to-day. When I returned to Kaikanui, on going for water found the lame bull bogged in the drain where the bridge formerly was. Too dark to do anything, so left him till morning to get some help to pull him out.

22. *Saturday.* Shearing commenced at 8 o'clock, but rain put a stop to it after 25 sheep had been shorn. Only a shower. All went to drafting-yards and got through all sheep left from yesterday at 12 o'clock. Sheep dry in afternoon so recommenced shearing. 234 sheep shorn to-day. Hughie left to-day. Mistaken about bogged beast; found it was Redman the old bullock.

23. *Sunday.* C. H. S. and T. S. came over to Kaikanui in canoe. Jim and Neil shepherding woolly sheep. The three of us went over to Tutira for tea. I rode to Kaikanui alone.

24. *Monday.* C. H. S., Neil Rossell, Jim Wild and T. C. K. tried to get Redman out of creek but were unable to do so. Maungaharuru sheep passed through to-day. T. S. clearing the road. 200 sheep shorn, when rain put a stop to shearing. C. H. S. sewing bales and branding. T. C. K. went burning. Jim shepherding woolly sheep. Neil knocked off to-day.

25. *Tuesday.* Went to shoot Redman this morning, but found him dead. Shearing till 12 o'clock, when rain put a stop to it. Rained at intervals during the whole afternoon. Day very stormy. 134 sheep only shorn. C. H. S. and T. S. took 9 pack-horses to Petane this morning with wool.

26. *Wednesday.* Began to shear at 8.30 A.M. Being short of bales or bags could not press the wool, so had to pack it the best way I could. Got 482 sheep shorn, and would have done more but for the number of previously shorn sheep being mixed up with the woolly, making it necessary to fill the crush-yard more often, consequence being loss of time. C. H. S. and T. S. returned from Petane about 7 P.M.

27. *Thursday.* Drafted out all woolly sheep and took them to the shed. Shearing began at 9 A.M. Finished all sheep, 421, at 6.30 P.M. C. H. S. and T. S. drafting and putting shorn sheep out to the paddocks. Jim generally useful. Kite cooking.

28. *Friday.* Paid off all the Maoris (shearers and others) with the exception

of Spooner, who remains to press the wool. Sent him with 2 pack-horses this morning to Petane for bales. C. H. S. and T. S. mustered Natural Paddock and got about 30 woolly sheep. Cut wild rams this evening. Fine night. Paid Maoris in following cheques: Rare £3—1—0; Newton £2—8—8; Hemera £2—4—4; Winiate £2—5—0; Ne £1—11—0; Neddy £3—0—0; Tomoana £3—4—0; Honie £3—0—6; Napier £2—0—0; Jack £1—0—0; Mary £1—7—0; Mulligan 9/-, making a total of £26—10—6. This afternoon C. H. S. put store in order. T. C. K. put shed in order, sewed bales and pressed. T. S. and Kite assisting at matter. Jim taking the sheep over the Papa Creek.

MARCH—

1. *Saturday.* C. H. S., T. S., Kite and Jim went mustering the reserve for stragglers. Returned at 1 P.M. with 200 woolly sheep and long-tail lambs. T. C. K. and Spooner pressed 30 bales of wool, and cleaned up shed. Parks arrived from bush about 5 P.M. Reports getting on well.
2. *Sunday.* C. H. S. and T. S. came over to Kaikanui at 10.30 A.M. Parkes left this morning with 7 bullocks for the bush. G. C. Thompson and John McKinnon arrived at 3 P.M. Thompson started from Napier yesterday but got lost on the road and was out all night Got to Arapawanui this morning at 6 A.M. J. McKinnon left at 6 P.M.
3. *Monday.* Maoris didn't arrive as they promised to shear the stragglers. C. H. S. and G. C. Thompson went out to the bush about 9 A.M. and returned at 4 P.M. T. S. shepherded sheep in the morning and with Kite made up loads in the afternoon. T. C. K. went out burning the Newton faces and afterwards gathered up loose wool at the shed. Jim left this morning to muster at Tangoio.
4. *Tuesday.* C. H. S., Thompson and Kite left with pack-horses at 9 A.M. T. S. left for Moeangiangi at 5 P.M. T. C. K. went to Tutira and saw Spooner (presser), at 1.30 P.M., who came by himself, the Maoris not having arrived to shear. T. C. K. left Tutira at 2 P.M. and arrived at Petane at 6 P.M.
5. *Wednesday.* T. C. K. and Thompson left Petane for Napier. C. H. S. and Kite returned to Tutira with pack-horses taking stores. T. S. shepherding sheep (unshorn).
6. *Thursday.* T. C. K. in town. Warm & splendid for the young grass. All hands idle on account of weather.
7. *Friday.* T. C. K. in town to buy sheep. At the station heavy rain and thunderstorm. All hands had to keep indoors. Wet night both in town and at station.
8. *Saturday.* T. C. K. in town. Kite and C. H. S. repaired fence at Papa Creek and made a gate in the morning, and repaired fence between Natural Padk. and Reserve in afternoon. T. S. shepherding woolly sheep on Reserve.
9. *Sunday.* No work done; all hands taking a rest.
10. *Monday.* T. C. K. in town. Docked 8 lambs, making total up to date of 1244.

11. *Tuesday*. Kite and C. H. S. packed 12 horses to Petane with wool. T. S. started for Napier to meet T. C. K., in order to have a look at the "Okawa" ewes. T. C. K. went to Petane from Napier to meet T. S. Jim at Tangoio mustering.
12. *Wednesday*. T. C. K. and T. S. started for Okawa to see sheep. Kite and C. H. S. returned to station from Petane.
13. *Thursday*. T. C. K. purchased 1500 ewes from Beamish at 4s., delivery to be taken on the 27th. T. C. K. and T. S. return to Napier. Kite and C. H. S. took 12 pack-horses with wool to Petane.
14. *Friday*. T. C. K. and T. S. in town. Start for Maraekakaho and Olrig to have a look at some rams for sale. Didn't reach Olrig, so stayed at Maraekakaho Accommodation House for the night. C. H. S. and Kite returned to Petane, taking 4 cwt. wire and oats for pack-horses.
15. *Saturday*. T. C. K. and T. S. arrived at Olrig. Saw rams, but didn't like them. Most of them too old, and the young ones too coarse in the wool, so didn't purchase. Left for Napier at 11 A.M., and arrived in town at 2 P.M. C. H. S. and Kite packing wool from station to Petane.
16. *Sunday*. T. C. K. and T. S. in town. T. S. went to Meanee. C. H. S. and Kite returned to station from Petane with pack-horses and brought up 4 cwt. of wire.
17. *Monday*. T. S. left Napier for station. T. C. K. engaged two musterers, Whitehead and Rose, at 15s. per day. George and Charlie arrived from the bush to repair subdivision fence. Jim returned from Tangoio. C. H. S. and Kite took 12 horses to Petane with wool.
18. *Tuesday*. T. C. K. in town. Kite and C. H. S. returned to station, taking 5 cwt. wire, 1 cwt. staples, and stores. During absence of all hands dogs got loose, and killed nearly all the fowls, and worried five sheep.
19. *Wednesday*. Jim, Kite, and C. H. S. cleared up woolshed. Dried some damp wool, and pressed balance of wool on hand. Kite in afternoon went to Tangoio to muster. T. J. S. arrived from town; Whitehead and Rose also arrived. George & Charlie started at subdivision fence.
20. *Thursday*. C. H. S., T. S., and George and Charlie, Jim, Whitehead, and Rose, mustered Reserve and got 89 woolly sheep, 10 long-tail lambs, making total up to date 1256 lambs, and woolly sheep mustered 6378.
21. *Friday*. Mustered Reserve again, and got 20 woolly sheep and 4 lambs. All hands camped at Kaikanui for mustering Newton to-morrow.
22. *Saturday*. All hands mustered Newton. Got 170 woolly sheep and 5 lambs. Total woolly sheep mustered to date 6568, and lambs docked 1263.
23. *Sunday*. C. H. S. went to Arapawanui for shearers. Other hands drafting. Put on Newton 1500 sheep: 600 ewes, 100 wethers, and 800 cross-bred wethers. Mustered 33 rams, and put them on Reserve with 1800 ewes. Mustered Natural Paddock. T. C. K. returned from town. Did not get rams.

24. *Monday.* All hands mustered "Rocky Range." Got 62 woolly sheep. Total mustered to date 6630. Sheared 158 sheep.
25. *Tuesday.* Cut 20 wild rams. Turned out shorn wethers to the back block, and turned 52 ewes into Newton.
26. *Wednesday.* T. C. K., Whitehead, Rose and Jim started from Okawa to take delivery of ewes bought from Beamish. Charlie and George working at subdivision fence, Kaikanui. Maoris returned to shear. 142 sheep shorn. Shearing tally to date 6375. T. C. K. and men arrived at Okawa at 6 P.M.
27. *Thursday.* T. C. K. took delivery of Okawa ewes, but as it was late when drafting was finished, put off the start till to-morrow, the men in meantime shepherding sheep. Finished shearing at Tutira.
28. *Friday.* T. C. K., having seen the men start from Okawa with the sheep, went to town. T. S. burning gullies at back of Newton; fires visible to T. C. K. at Puketapu. C. H. S. inspecting George's fencing. T. C. K. purchases 60 Russell's rams, £150, and starts for Waipukurau to select them.
29. *Saturday.* T. C. K. at Waipukurau. Started at 6 A.M. to select rams, but couldn't get them in time for early train, so had to wait till afternoon. Arrived at Napier with rams at 7.15 P.M. Left them in trucks till to-morrow morning.
30. *Sunday.* Started for Railway Station at 5 A.M. expecting to meet man Miller engaged. Not finding him got G. C. Thompson to help me to drive them to the Spit. Found man there waiting. Got the rams safe across the ferry, & arrived with them at Villiers at 2.30 P.M. all right. T. and self return to Napier.
31. *Monday.* T. C. K. left town for Petane to see rams. On going to Young's stables for my horse, found Donoghue drunk in the stable. Rode over to Petane, and from there to Tangoio, and told Whitehead to keep ewes back, and come to Petane for rams. Fortunately they were all safe. Stayed at Petane all night.

Here we can conveniently close our chapter—sufficient matter has been given to show the normal daily life of a station in the making. I know it has been prosaic; I know it has been heavy. I cannot but be aware that its stolidity must have even veiled and obscured the glories, the delights, the ecstasies of improvements, for there is no fascination in life like that of the amelioration of the surface of the earth. For a young man what an ideal existence!—to make a fortune by the delightful labour of your hands—to drain your swamps, to cut tracks over your hills, to fence, to split, to build, to sow seed, to watch your flock increase—to note a countryside change under your hands from a wilderness, to read its history in your merinos' eyes. How pastoral! How Arcadian! I declare that in those times to think of an improvement to the station was to be in love. A thousand anticipations of happiness rushed upon the mind—the emerald sward that was to paint the alluvial flats, the graded tracks up which the pack team was to climb easily, the spurs over which the fencing was to run, its shining wire, its mighty strainers; the homestead of the future, the spacious wool-shed, the glory of the grass that was to be.

It was a joy to wake, to spring out of your bunk half dressed already,—there wasn't a nightshirt north of Napier then,—to glance through the *whare's* open door at the clear, innumerable hosts of stars, in the huge fireplace to open up the warm cone of soft grey ash piled carefully overnight, to push into its heart of glowing red the dry kindling, to see the brief smoke ascend, to hear the crackle of the rapid flames. Oh, those were happy days, with no cares, no fears for the future, no burden of personal possession, when every thought was for the run, when every penny that could be scraped together was to be spent on the adornment of that heavenly mistress.

"Bushrangers," white and black.

CHAPTER XVII.

HARD TIMES.

In the chapter just closed the reader has enjoyed without alloy the delights of land improvement; without thought of the morrow and without anxiety as to finance, he has contemplated the beginning of a sheep station; but, alas! pleasure and profit do not always march hand in hand. Local knowledge, experience, judgment, acquaintance with stock, each plays an important rôle, each is necessary to final success. There were other reasons, too, militating against our young men about which nothing yet has been said.[1]

Well then, harking back to the last days of March, it will be remembered that Kiernan bought sixty rams, for which he had presumably given a station cheque of £150. On 1st April I find recorded these words: "*Was surprised cheque for rams had been returned by Bank of New Zealand; had to give my private cheque for* £150 *to meet it.*" 8th April: "*C. H. Stuart returned from town with news of hard times and the likelihood of having to sell the run. Everybody down on account of bad news.*" 9th April: "*At Kihekanui, very miserable on account of bad news—no heart to do anything.*" 10th April: "*Went to town to*

[1] It is never pleasant to speak plainly on certain subjects, especially in a book of this sort, open to the general reader, yet surely, after the attainment of a certain age and of a certain amount of experience, warning becomes a plain duty. Reading, then, between the lines of certain entries in these diaries, it cannot be concealed that our pioneers had got into very dangerous company. It is not my place to preach—readers will resent anything in the way of a sermon; I suppose, too, that young men will be young men to the end of time—yet, to be frank, there were persons to be met with in the streets of Napier—and pretty openly too—whom it would have been better our pioneers should never have known. There were parlours, too easy of access altogether, where nothing but harm could happen. If the presence of a certain class, if the trade they ply, cannot be eliminated, its conditions, at any rate, should be regulated. In the 'eighties nothing of that sort had been attempted; bankers and managers of mortgage companies might charge what rates they chose. Legislation in respect to this matter came later, when the New Zealand Government began itself to borrow and relend cheap money to struggling settlers. In the 'eighties it was otherwise. In those days young fellows like the Stuart Brothers and Kiernan were, in spite of themselves, so to speak, forced into bad company.

arrange matters connected with run." 11th April: "*Returned from town owing to not being able to do business till Thursday next.*" 12th April: "*Trying to kill time.*" 17th April: "*In town waiting to see Miller.*"

Thrice fortunate those who have not passed through the dreary stages of having "*no heart to do anything,*" of "*trying to kill time,*" of "*waiting to see Miller.*"

Trusting that these entries from Kiernan's diary will prepare the reader for the sad sequel, we can go back many months—to the date, in fact, of the partnership of Stuart and Kiernan. The reader has, in fact, seen but one side of the operation of breaking in a run. If, however, he has been in any degree deceived, it has only been as Messrs Stuart and Kiernan were themselves deceived. He has intentionally been allowed to look at things as they themselves viewed their own affairs. The truth is, that from the beginning these pioneers were doomed—they were predestined—to failure. Conditions in the interior were in those days quite unknown; knowledge of local conditions—the most important knowledge of all—had to be purchased. Settlers in the fertile districts of southern Hawke's Bay may have been but little wiser or more careful, they always had this in their favour—that their soils were sufficiently rich to redeem the owner's faults, even making full allowance for the fact that in those days fern was fern, that none could tell in the 'seventies that a plant, easily destroyed in a dry climate and on warm rich soils, would prove almost ineradicable on porous land in a wet district.

The master and main difficulty was lack of sheep-feed. Eliminating the leaves of tutu (*Coriaria ruscifolia*), edible only to salted stock, and the growth of fern fronds, which ceased altogether for six months of the year, there were not 100 acres of sheep-feed on Tutira, there were not 100 acres of grass on Tutira when Newton stocked the run with 4000 sheep. In early times, not only on that station but on every property in Hawke's Bay, the sheep had to create his own pasture, himself to grow his own keep. Now, to understock is the secret of all successful sheep-farming, but action on the lines of this axiom was denied to the run. Irreconcilable contrarieties in nature had to be reconciled. The pioneers of Tutira had at one and the same time to "make" their country and to consider the welfare of their stock; it was to solder impossibilities and make them kiss.

In an earlier chapter a general description of the indigenous vegetation of the run—bracken, forest, woodland, and marsh—has been given.

With the influx of Europeans into New Zealand and the importation of stock and of alien plants suitable for stock, the natural spread of grasses had begun in a small way on Tutira. It grew on pig-rootings, on deserted native clearings and cultivation-grounds, on landslips and along the bases of the marl and limestone outcrops. These patches and spatters of grass were scattered over the 20,000 acres of the station. They were sometimes hundreds of yards, sometimes miles apart, linked with one another by narrow tracks or rather bores through high fern and tutu. In addition to these self-sown alien and native grasses, sheep-feed was obtainable, as has been already mentioned, during certain months of the year by the burning of bracken. Of this plant the circinate fronds are on good land fairly nutritious; sheep can, during summer, be maintained on them in fair store order. These scattered patches of grass, this fern growth, together with the leaves of the tutu, were the original sum-total of sheep-feed on Tutira. The Children of Israel had to make bricks without straw, the pioneers of Tutira had to produce wool without grass.

The first care of the settler was to increase his area of grass by the operation known through Hawke's Bay as "fern-crushing" or "fern-grinding,"—words ominous of the part played by the unfortunate sheep, and which will be described later. It is sufficient now to state that after fire had cleared the tangled bracken growth, the ground was surface-sown and kept clear by browsing sheep. As the greatest growth of fern took place during late spring, it was then impossible to have too many sheep. Every squatter in Hawke's Bay was in the 'eighties "fern-grinding," so that in those times sheep could not be bought at that season of the year. The result was that every sheep likely to survive the winter was kept, however old and however fleeced. It was at least a pair of jaws, a beast that could bite bracken.

Fern-grinding, however unavoidable in the progress towards creation of the large flock—that distant goal upon which the eyes of the run were fixed—was nevertheless a process utterly incompatible with the ownership of properly-fed stock. The early years of the run were, in fact, a compromise between murdering the sheep and "making" the country. The run was in the position of having to wrong its stock because no other course of action was feasible. It had to transgress the first and greatest of pastoral commandments: Thou shalt not overstock; there was no remedy for the evil.

Another difficulty, also insuperable and unavoidable, lay in the violation of the golden rule of stock purchase, the base of all sound buying. It is, never to move stock from richer on to poorer ground, never to move stock from a drier into a wetter climate. In these times, however, there was no worse or wetter country from which sheep were obtainable. Drafts purchased for Tutira had to be drawn from the drier climate and warmer soils of southern Hawke's Bay. Furthermore, these bought sheep had been done at any rate comparatively well on the runs where they had been bred.

On Tutira they were expected to act as fern-scythes and mowing-machines. Even stock removed from bad to good conditions requires time to settle down; purchased stock on Tutira changed contrariwise from good to bad, loathed their new environment, the grass contained less nutriment, there was less of it, their fleeces were oftener wet on their backs. They had to be acclimatised to wet country after dry, to bad land after good, to semi-starvation after a sufficiency of grass.

There is always a tendency for purchased stock to stray. On Tutira it was made easy by an unlucky geological condition, it was aggravated by the nature of the breed—merino—then on the run. The natural boundary to the south—to the quarter, that is, from which the purchased stock had been brought and to which they wished to return—was the only river stretch on the station not contained by cliffs. The Waikoau, though blocked and barred with vast limestone quadrilaterals, between and around which rushed and swirled the rapid stream, offered passable though highly dangerous fords. Swimming, distasteful to sheep, and especially to merino sheep, was, however, the comparative of dislike; the superlative of distaste was habitation of Tutira.

Each newly-purchased mob had therefore to be watched, until, after weeks of dogging and checking, the bulk of the newcomers accepted the inevitable and began to settle on their new abode. During the period of most marked restlessness the shepherd in charge watched his boundary-line day and night. Every dawn, as certain as clockwork, sections of the newly-bought sheep would trail in long lines down leading spurs to be as regularly checked and "barked" up again. A proportion, however, out of every mob would beat the best man. Trouble at one end of the line might give a chance to sheep at the other extreme; bright moonlight was a curse; a native pig-hunting might drive the sheep down the whole length of the line, making it impossible to check simultaneously animals but too willing to run in the wrong direction, for it is horrible how sheep resemble mankind in this, that ever such a small favourable chance will incline them to evil. Small lots, too, might be overlooked in the river-bed scrub and at their convenience cross unobserved. Through the cut manuka blocking approach to the easiest fords pig might have bored, thus opening an avenue of escape. Then, again, long after the bulk of the mob had resigned themselves to their fate and the boundary keeper had been withdrawn, leakage would still occur. Revived, I suppose, by misery and semi-starvation in winter-time, the old longing for home and comfortable quarters would again prompt the idea of escape; small lots would succeed in crossing the river, others would be drowned. In early spring, too, a considerable number of old ewes in twos and threes, anxious to lamb where they had previously lambed, would also attempt the river.

In one way or another hundreds of sheep thus straggled from the run. Some were secured again at neighbours' draftings; others died or were bogged; a small percentage probably succeeded in crossing the intervening stations, eventually to reach their original home. Thus, patrolling beats in the manner described, fetching back stragglers from neighbours' draftings, on the run itself dogging sheep from oases of grass such as the old Maori cultivation-grounds on to burnt fern lands, consumed time out of all proportion to the size of the early flock. In the diaries of this period, day after day occur such entries as "*dogging sheep from flat*," "*attending draftings*," "*bringing home stragglers.*"

All sheep suffer from nostalgia, but the merino is perhaps the most miserably home-sick beast on earth. In Kiernan's diary of 1879 I find a note to this effect: "*The newly-purchased wethers persist*"—he underlines the word persist—"*in lying against the new fence.*" Liberated in strange country, a mob of merinos will lie against the barrier—cliff, river, fence, whatever it may be—blocking their homeward route. Night after night, day after day, week after week, there they will camp resigned to starvation. They will hug the fence-line that debars them from return to their old haunts till their droppings are inches deep, until their lank frames reveal every bone. When they rise, it is to "string" up and down till the ground is worn bare, till not a bite of foodstuff remains.

From this sketch of the psychology of the merino some conception may be formed of Stuart and Kiernan's trouble with their first purchased stock. The result of this restlessness was by no means, moreover, covered by loss in stragglers and drowning. In a dozen ways besides, death gathered the wretched beasts with both hands. I find in Kiernan's diary the following item: "30 *per cent loss in stock between* 1*st April* 1877 *and* 31*st March* 1878." It was an entry that must have given pause even to our pioneers.

The reader will recollect the first draft of 4000 sheep planted on the station by the intrepid Newton, and almost at once removed in consequence of Te Kuiti's raid. It was the earliest mob delivered on the run, but my own experience of similar occurrences has been so prolonged that this first mob may be taken as a text to illustrate the melancholy processes of a "30 *per cent loss in stock between* 1*st April* 1877 *and* 31*st March* 1878." The preliminary leakage in droving would be small; a few sheep, however, would probably have been drowned at the crossing of the broad estuary of the great rivers, Tutae Kuri and Ngararoro. There was no bridge then; the sheep crossed in punts, the drovers swimming their horses behind with the offchance of an attack by sharks. I find in an early diary that on one occasion when the punt was filled with rams, its plug was kicked out and sheep and shepherds alike had to reach shore as best they could by swimming. The Petane river, too, would possibly claim a few victims; the "wash-out"—that dangerous break in the beach through which, under certain conditions, the tides passed to and fro—a few more; a handful or so might have managed to drink salt water; a few poison themselves on tutu, a shrub exceedingly dangerous to unsalted stock; a few drop out from lameness, or be lost in under-runners and pitfalls. It would not, however, be until the 4000—already depleted perhaps 2 per cent or 3 per cent—reached their destination that losses on a serious scale would begin. After that would commence the long conflict between sheep determined to return to their own pastures and owners determined to hold them on the station. The most careful collies will be rash at times; their shepherd masters had to walk by faith at least as much as by sight. The doings of their dogs were hidden by dips of the rugged land, by patches of intervening scrub, by belts of woodland offering harbourage to the leg-weary sheep, by deep bands of low charred tutu stems, by alternate tongues of dense bracken and of open ground, the whole countryside in addition pitted with under-runners and seamed with narrow gorges. The decencies of high-class shepherding were impossible in such broken lands—in such entanglements of scrub. Holding stock on to new ground in those days broke the hearts of the men and wore the frames of the merinos to greyhound lankness. With stock unharrassed and feeding leisurely numbers must have been trapped and lost, but with sheep "stringing" or hurried, companies of tens and twenties were swallowed at a gulp. The animals themselves did not know where to go or what to expect. The country was as strange to them as to their owners. What happens in every paddock "worked" by sheep for any length of time had not then occurred, the first action of a mob in a strange enclosure being to map it out, to explore it, that is, by lines radiating from established camps. In time tracks turn aside and thus cease to reach crossings discovered to be impracticable because of bogs; soft spots, localities mined with under-runners, blind oozy creeks, cliffs and so forth are avoided. Neither man nor beast had purchased experience then, however; it had yet to be bought by lives of sheep and money of pioneers—to paraphrase Kipling, by the bones of the sheep of Tutira, Tutira has been made. These early losses were inevitable; they were as unavoidable as the mistakes of travellers exploring lands of unknown races of men, of unknown diseases, of unknown climates.

Conditions were not ameliorated by the nature of the breed of sheep then run in Hawke's Bay. They were merino, and there is something maddening to the merino in the sight of its fellows escaping to fancied freedom. There were in early times—many of them since hardened by processes to be described later—numerous stretches of narrow marsh, firm enough to bear the weight of the foremost dozen or score of sheep, yet insufficiently sound to withstand the puddling and poaching of hundreds of hoofs. The leaders of the mob would safely traverse such a barrier. It would then become a quaking slough, the original narrow line of traffic marked by bogged animals.

The wallowings of the wretched sheep in their mud baths would deflect the line of travel by a few feet until another parallel track would undergo the same process and in its turn also become a bog. Where hundreds had crossed, dozens remained—their carcases sinking into the morass or remaining half submerged; if discovered at all, advertised by the presence of the harrier hawk (*Circus Gouldi*), which from the date of the stocking of Tutira began greatly to increase in numbers. Another type of trap taking from this conjectural mob its two or three or four, day after day and week after week, was the crevice typical of marl formation. Sheep stringing closely to one another, especially if alarmed, are apt to blunder in the leaping of what appears an insignificant crack. Out of Newton's 4000, death would have reaped its harvest piecemeal in other ways. Fires constantly lighted to open up the surface of the country would have destroyed a certain number, some blinded by the flames, others losing their hoofs in the scalding heat. A few would have been snared by their wool in thickets of lawyer (*Rubus australis*), a few would have been caught by the foot, or, like Absalom, by the neck, in forks of low stiff scrub. Some would have died from the effects of ergot on certain of the coarse native grasses. During spring and early summer many would have poisoned themselves on the shoots of the tutu (*Coraria ruscifolia*). Landslips would have accounted for not a few, some actually caught in the moving masses, others stuck in the glutinous streams that exuded from them. With the arrival of winter, conditions would have become increasingly adverse. By reason of change from a dry to a wet locality, from rich to poor land, and because of constant dogging and shepherding, these conjectural 4000 sheep even in autumn would have lost condition. The grass about the old Maori cultivation-grounds, the slips, the marl outcrops, would have been eaten bare by mid-winter; stock would have been forced by hunger into spots where hitherto they had not ventured, spots where there were still additional risks to be run.

To make a long story short, if Te Kuiti's raid had not caused the clearance of Newton's 4000 sheep, 1200 or so would have died in rivers, pitfalls, slips, under-runners, cliffs, deep pot-holes in the ground, marshes, boggy crossings and ravines, or would have been poisoned, trapped, or burnt; about 200 of them would have become what we used to call "bushrangers"; from 500 to 700 would have straggled off the run, most of which would never again have been seen. Out of this first draft, in fact, not much more than half would have passed through the hands of the shearers.

In early times there were similar difficulties with horses and cattle, shortage of feed in winter and absence of sufficient fencing always, but just as the run had to be forced to carry stock without grass, so pack-horses and bullock-teams had to be somehow kept alive to work the place. The former fed on the rich marsh-land that extended along the margin of the lake. Thereabouts in summer-time grass was plentiful, for at that season of the year the merino, startled by every outbreak of barking, kept to the upper slopes and hill-tops. In winter, when eaten out by sheep, body and soul could still be held together on rank sedges and giant grasses like toe-toe (*Arundo conspicua*), yet, when forced by hunger into dangerous places, horses too perished in numbers.

The station bullocks in one way were less well off than the horses: a horse can bite as close as a sheep, a beast requires a ranker growth. On the other hand, there was ample scrub for cattle. When not in work they were indeed expected to wander and fend for themselves. Risks had to be taken in any case; if kept in hand about the alluvial lands they ran the risk of bogging; if allowed to wander in the scrub a proportion poisoned themselves on the shoots and fruit of the tutu. There were, moreover, in those days herds of wild cattle, more or less "salted" to tutu, roaming everywhere on the hills, and although sober team bullocks as a rule held aloof from these unbranded beasts, yet an odd worker would occasionally join them. When that happened he was lost to Tutira. Though the jangling of the bullock-bell worn may have revealed his whereabouts, it was usually impossible to follow on horseback into scrub through which heavy cattle could scarcely burst their way.

When not in use these bullocks, and others bought to supplement the team, were for ever straying. Though from time to time rounded up and driven back to the lake as headquarters, they were perpetual wanderers. I find by Kiernan's diary that on one occasion they got away from Tutira, crossed Dolbel's Kaiwaka run of 30,000 acres, and were discovered "*on Troutbeck's near the coast.*" They had been bred on that station, and having nothing particular to do had walked home, tinkling and jangling their route through three runs, no doubt smashing down and lumbering over the single fence between them and the coast. Other diary entries prove them to have been almost as great a nuisance at home as abroad. Attempting no doubt to remedy the shortage of winter horse-feed, a single-furrow plough had been packed into the place and a patch of crop sown for oaten hay, sufficient for the one or two horses kept handy to run in the scattered team. This bit of delightful green must have been highly appreciated by the bullocks, led probably by "Dan," who would take bread from the hand and allow himself to be packed. I find many entries such as "*keeping cattle out of oats,*" "*turned them out twice,*" "*at nightfall found four back again.*" At dusk, no doubt, with every indignity they were again hurried from the premises. Their triumph came with dark. Each bullock bears a bell suspended by a leathern neck-strap, so that when feeding in high scrub or flax his whereabouts can be readily determined. In the hours of light when searching the hills for a lost animal the tinkle of a bullock-bell is a pleasurable sound; at night it is not. Just as the camp is dropping off to sleep the far-distant faint jangle of the grazing beasts is heard. With aggravating slowness the sound approaches, until at last a man leaps up in drawers and shirt, and muttering in the gloom pulls on his boots, snatches his stock-whip and lets loose the dogs, who know the game well and have been yelping and howling in anticipation of the treat. The bullocks are hounded off, their bells performing mad music, momentarily half-choked when swept round to horn a heeling dog, clanging dull as the beasts swing away in an elephantine gallop, or merrily and clear as they file out in a rolling trot. With a final hounding on of the collies and a pistol practice of stock-whip, the sweating, dew-drenched rescuer of the crop returns. In early diaries there are very many entries regarding the bullocks. They were a necessary nuisance, whether about the homestead or away from it.

These were evils, but a still greater misfortune to a growing run like Tutira was lack of sufficient credit and lack of sufficient time, either of which would have saved our pioneers. Stuart and Kiernan had by hard labour and energy managed somehow or other to make the station carry 8000 sheep, or at any rate begin the winter with 8000 sheep. They had cut tracks, they had drained swamps, they had sawn timber—but none of these improvements had yet had time to produce any beneficial result on the station bank account. The repayment to the station of a line of fencing may fairly be spread over a score of years, whereas a cheque has to be given then and there for the wire. Reimbursement to the station for grass seed sown might also be reasonably spread over decades; the merchant, however, has to be paid at once. The same may be said of track-cutting, swamp-draining, the sawing of timber for house and shed; nay, the very increase of stock, surely an improvement of the first importance, spelled at first a large overdraft; it also had to be paid for with borrowed money. The benefits were to be perennial, the payments, however, for these perennial benefits had to be paid instantly in coin of the realm. Thus it came about that an immediate debt was run up for improvements which could not at once bear full fruition.

Properties in the transition stage, their improvements paid for, but the financial results of these improvements not yet apparent, are the first to feel the pinch of bad times; for what is a line of fencing to a banker, or a drain or a bag of grass seed? Simply damnable items on the debit side of a balance-sheet. From enthusiasm then, from inexperience, from want of good advice—there was nobody to administer the last, for no man had worked the light lands of northern Hawke's Bay at that time,—the obvious dangers to themselves do not seem to have troubled the brothers Stuart and Kiernan.

There could at any rate have been but little forethought in the financing of the run; indeed it is to be feared that a diary entry of a very early date was typical of the financial methods then in vogue: "*Bought from William Villers, one team of eight bullocks, waggon and all complete, for the sum of* £135. *Terms, to pay when able.*" In the diaries, entered amongst shearing tallies, lists of washing sent to Napier, inventories of chattels, as the pots and pans of the station are rather grandiloquently termed, appear also from time to time financial calculations, figures enow in all conscience, but often lacking items to which they can be attached. These reckonings have apparently been jotted down hot from the writer's brain and then left high and dry, stranded and never retouched a second time. There is yet extant also a little note-book whose perusal will raise a sympathetic sigh in the bosom of every Hawke's Bay pioneer. Lined in columns for the months and weeks and days of the year, it is nothing less than an attempt at the daily registration of lost sheep, cattle, and horses. This melancholy volume, however, like other New Year resolutions, made only to be broken, seems to have been discontinued after the deaths of 31 sheep, a drowned horse, and a bogged bullock.

More carefully kept and deliberate calculations do nevertheless exist. For instance, though in the diary of '78 there is no mention of shearing—probably the flock, consisting wholly of dry sheep, was clipped at Tangoio on the coast,—there exists the catalogue of the earliest Tutira clip, 218 pockets, which, at the rate of 18 or 20 fleeces to the pocket, would roughly correspond with the following figures carefully written out and repeated on another page:—

Sheep received with station	3600
Put on since, before shearing	{ 1389 685
Put on since, after shearing	{ 1211 1081
	7966
Less 30 per cent, leaves	6460
Lambs	300

30 per cent from 1st April to 31st March 1878.

I give the figures as they are, though I cannot follow them; they are doubtless approximately correct. On another page are further calculations:—

On hand from shearing	4200
Received since shearing	2282
	6492
Shorn by Mackinnon	8
	6500

We may take it, therefore, that the shearing of '78 totalled 4200. At the beginning of the winter of '79 the number of the flock had been by further purchases brought up to 9999. Of these devoted beasts 7164 were shorn. At the beginning of '80 there were running on Tutira 8324, of which 6344 passed through the shearers' hands.

Altogether apart from these losses, however—losses which were perhaps inevitable—the finances of the station had never been on a sound footing.

The City of Glasgow Bank in Scotland had failed some years previous to the date we are considering. Between that bank and certain New Zealand land companies there had been close connection; its fall had already reacted disastrously on all New Zealand securities. It was likely, therefore, that a drop in the wool market would seriously affect the already weakened system of colonial credit. It was certain, moreover, that should such a condition of affairs occur, the first and foremost to feel the pinch would be owners of partially-developed sheep stations. Wool did drop, the mournful rumour circulating that so-and-so had "offered his clip at 5d. for the next six years, and couldn't get a taker, mind you"; the usual squeeze began of those least able to bear pressure. Among them were the brothers Stuart and Kiernan. There is indeed something almost pathetic in the naïve surprise evinced in Kiernan's diary entries of this fatal April of 1879. That the price of wool could possibly fall, that bankers could conceivably tighten their purse-strings, seems never to have entered the heads of our pioneers. The necessary knowledge had, I suppose, to be paid for, and although the considerable number of thousands of pounds lost in its acquisition may not—as Miss Wirt, in 'Vanity Fair,' was wont to tell of her father's financial transactions—have convulsed the exchanges of Europe, these sums were all their owners possessed. The hole in their resources, though neither as deep as a well nor as wide as a church door, was sufficient. 'Twas enough, at any rate, 'twould serve.

After this ill-starred month of April there appear few further references to finance. Immediate difficulties seem to have been tided over. Improvements, moreover, proceeded, although work done now was doubtless the completion of work already begun, which could not have been stayed even from the point of view of an uneasy banker. Nearly 200 bags of good ryegrass and cocksfoot were sown; delivery of pit-sawn timber, begun in happier times, was proceeded with, and wool-shed built. From April of '78, however, owners worked with the sword of Damocles suspended over their heads. With anxious eyes they scanned that fatal barometer of hopes and fears—the wool market. I find, for instance, this entry: "*No good news—wool market showing no signs of improvement.*" Kiernan's diary is, nevertheless, as ample and careful as ever. Details are given of the first station garden, of the planting of eucalypt, willow, and pine. I think no fact could more clearly prove how its owners must have cared for their station than its adornment under these tragic conditions—the adornment of a bride about to be ravished from their arms. These eucalypts, willows, and insignis, planted on a promontory jutting into the lake, have now been for forty years an ornament to the station. I never look at them on the fine headland Taupunga without thinking of the sad circumstances of their planting, how in the joy of labour chilling thoughts of the future must have obtruded themselves, thoughts that take half the energy out of the settler's arms. The days, weeks, and months of the year passed away. Again shearing-time arrived without an improvement in the price of wool. I suppose it was recognised that matters were desperate; at any rate, I find that in February of 1880 "*C. H. Stuart left the run.*" In October 1880 "*the first piles of the new wool-shed were in the ground,*" but, as with the planting of the trees, the impending calamity must have taken the heart out of its erection. Hope was almost gone, and without hope no man can put his best into his work, the labour of his hands can no longer be what it should be—pure delight. On 27th May 1881 the entry occurs: "*T. C. K. and T. J. S. transacting station business all day with Bank of New Zealand and Loan and Mercantile Agency Company; made arrangements to tide over everything till shearing.*" Alas! alas! "*transacting station business,*" or any other business at any time, is a loathsome task, but how much aggravated this renunciation of an incomplete labour of love to an unemotional bank or soulless mortgage company. What pangs of disappointment, what heart-searchings as to the past! What disgust of self and all concerned! What a sickening void of interest! I can picture the poor wretches overwhelmed with abominable figures, signing mechanically, their minds idly wandering to green Tutira, its ranges and lakes.

The end was rapidly approaching, for on 28th May "*T. C. K. left for Melbourne.*" In August he was again in New Zealand, for in that month, there being then a debt on the run of £8600, he sold his half-share to C. A. M'Kenzie for the sum of £160. There is then a blank in the station diary until the 1st September. Upon that date appears in a new handwriting—no doubt that of M'Kenzie—"*Left Napier for Tutira after having squared up everything; gave Matthew Miller bill at three months for amount due him by Tutira.*" The station was now worked entirely by two men, Stuart and M'Kenzie, with W. Stuart, a younger brother, acting as cook. I have never gathered that W. Stuart was a brainy man. I find in the station diary kept by M'Kenzie, 10th September: "*Willie trying to make bread.*" Three days later the diary was again an outlet to the feelings of the writer, "*Willie trying to make bread,*" and later this entry, almost with the ring of tears in it, "*Willie wasting good flour and yeast.*" When a man can confide his sorrow to a diary, he must indeed have suffered. Conditions were now desperate; in M'Kenzie's diary, from time to time, there are ghastly reminders of bills about to fall due at briefer and briefer intervals.

By a deed of July 1882, the place then owing £9000 to the Loan and Mercantile Agency Company, C. A. M'Kenzie transferred his share to T. J. Stuart for ten shillings. W. Stuart now, I believe, took over a half-share and put some money into the place. The brothers, at any rate, shared the few score pounds remaining after final sale. In September 1882 Stuart & Stuart sold the property to W. Cuningham Smith for £9750—the price then owing to the Loan and Mercantile Agency Company. It had been purchased on behalf of H. Guthrie-Smith and Arthur M'Tier Cuningham, at that time minors.

PUAWHANANGA (*Clematis indivisa*).

CHAPTER XVIII.

THE RISE AND FALL OF H. G.-S. AND A. M. C.

It was upon the 4th of September 1882 that the new owners of Tutira took delivery of their sheep-station. They were wild with anticipations of sport, of riding, of the mastery of animals, of life in the wilds. At least, by one of them, every hour of that golden day can still be vividly recalled. He remembers wakening at dawn and rushing out to forecast the day. It was a perfect Hawke's Bay spring morning, and be it said, no weather in the world can beat a fine September in Napier. The sky was cloudless, the faintest crisp suspicion of frost mingled with the salt tang of the beach. Behind the town rose the magnificent snow-clad ranges Ruahine and Kaweka, in front heaved the Pacific's vast expanse.

What magic there is in possession! What a pleasure the sight of the hacks! They were not quite like any other horses in the world; they were our own, they belonged to us, an earnest of that glorious sheep-station which was to provide after a few seasons easy enlargement of our minds and fortunes, endless rivers, moors, and forests in Scotland.

We rode through the picturesque town—our horses' hoofs sounding loud on the quiet streets—where half the inhabitants were still asleep. We passed through Port Ahuriri, crossing the newly-built wooden bridge which linked the northern and southern portions of the province. We followed the beach road along the western spit, peopled then only by a few fisher-folk, some of them living in homes built of biscuit and kerosene tins in lieu of iron sheeting. We passed the Petane Hotel, then run by the redoubtable William Villers. We forded the Esk river, and, riding through the *kainga* of Petane, scanned with deep interest the reed-thatched *whares* of the owners of Tutira.

O Ananias, Azarias, and Misael! that morning our happiness overflowed the world. We even loved our landlords—after all, they had been heathens until recently; they had never read Henry George; they knew no better. We rode along the shallow sandy turf of the Whiranaki Flats and over the Beach Hill to the County Boundary Peak. We were then in the county of Wairoa, one of whose divisions was the Mohaka riding, within which lay Tutira; it was another step towards our new possession. Farther on we reached the Tangoio Bluff, and turning inland at right angles passed the homestead and wool-shed of the Tangoio run. Tangoio and Tutira marched with one another.

We followed several miles inland, but parallel to the sea, the switch-back track, the old coastal pack-trail, so lost to common-sense as to think its deplorable grades good going. We struck the First Fence, and saw for the first time the Tutira station mail-box. To ordinary eyes it might have seemed, as indeed it was, a kerosene case nailed to the top of a strainer-post. To us it was much more; that box, simple and unpretentious to the outward eye, had been the receptacle of communications about Tutira wool, about Tutira stock, about Tutira interests of a dozen sorts. We viewed it with a kind of reverence.

We turned sharp inland and followed up and down over the hill-tops, the trail faintly marked by the station pack-team. Three miles farther on we struck Dolbel's Boundary Gate, and saw in the distance the Delectable Mountains of our pilgrimage—the ranges of Tutira. Shortly afterwards we looked down upon the Waikoau tumbling along amongst its boulders. We led our horses over the "shoot," the almost perpendicular drop, down which the pack-team used in muddy weather to slide with stiff legs and unlifted hoofs. We zigzagged down the steep trail of Dolbel's Face, disturbing mobs of wild cattle, each of them raising pleasant anticipations of future huntings. With the delight of Scott crossing his Tweed at Abbotsford, we splashed across the unsung ford of the Waikoau. We trod Tutira soil. We viewed for the first time our own sheep. They were merino ewes,—skin and bone, scrags, their

Blue Duck—Waikoau.

wool peeling off,—anxious to escape yet balked by the river, the kind of stock always in the very worst of condition. Such was our fatuous folly, that we believed against the evidence of our senses that they were not so very, very, very wretched, that not every single solitary bone in their lank frames was visible.

We climbed from the river-bed to the Reserve—long afterwards rechristened "The Racecourse Flat," and rode very quietly through the lambing ewes. We could hardly bear to tear ourselves away. If the sight of the scrags at the Waikoau ford had thrilled us with pride of possession, our hearts exulted at the sight of these lambs—lambs that apparently came from nowhere, but were even now swelling the numbers of our newly-purchased flock—as if thrown in gratis, a gift from a beneficent Heaven to H. G.-S. and A. M. C.

We rode along the shelf of the flat until suddenly, in an instant, the lake lay at our feet. The feelings of one of the new owners were those of Marmion's squire at sight of Edinburgh. Had the grass-fed pony permitted the feat, its rider, like Eustace, would have "made a *demi-volte* in air" at joy of the prospect. Before his eyes lay the whole length of the lake, picturesque in its wooded promontories and bays. Along its steeps grew breaks of native woodland brightened at this season with the deep yellow blossoms of the kowhai. The silky leaves of the weeping willows were in their tenderest green, the peach-groves sheets of pink. The south-westerly breeze that blew stirred the flax blades, making them glitter like glass; west of the lake the land was dark in shadow, the eastern hill-tops were bright in sun. I have looked at this lovely sheet of water a million times since then, but have rarely seen it more fair.

There are some spots on earth that seem to inspire in their owners a very special affection, as if perchance there might exist an occult sympathy betwixt the elementals of the soil and those who touch its surface with their feet. A race so eager in their appreciation of natural objects of beauty as the Maori could not but have felt thus towards Tutira; we know they did so,—I have heard its native owners a score of times rejoicing in their possession; the lines of the *waiata* cited on the first page express it. The diary entries of its pioneers bear witness to the pangs with which the place was relinquished; if there is anything of value in this volume, it is because of the author's affection for the spot where he has lived so long.

Dropping from the Racecourse Flat we reached the primitive homestead of the 'eighties. It was situated then on the Piraunui flat at the southern end of Waikopiro. The buildings were a weather-board hut 15 feet by 12 feet, divided by a partition reaching half-way to the roof. At one end was built the usual clay fireplace and iron chimney. Camp ovens, go-ashores, and billies stood on the floor, or were slung from bars above the empty hearth. Hung by wires from the roof, and thus immune from rats, was suspended a stage on which lay flour and sugar bags, currants, and other necessities of those Spartan days. Outside the house a small lean-to sheltered from the elements a barrel of pickled wild pork. Bottles of yeast stood on the smoke-stained mantelpiece. The architecture of our mansion was Noahian—a door that

Homestead of the 'eighties.

is, with a window on either side. The door, I remember, was open when we arrived, for inside were several foraging fowls, some of which fled into the huge unswept hearth, stirring up the ashes in clouds, whilst others attempted the window, several of the panes of which had already been broken and were mended with brown paper or stuffed with rags. The other buildings of the primitive homestead were an ark, 6 feet by 9 feet, a *whata* or store-house on piles, empty now in the station's dire extremity, and containing straps and pack-saddles. In front of the door lay the wood heap, that adjunct to all homes of early settlers; alas! that its litter of chips, fresh and white, or mouldy and grey or brown with age, its ever-blunt axe, its larger logs so comfortable for seats, are vanishing before new-fangled ideas of tidiness.

This hut, where for nearly a year we continued to live, was afterwards cut into sections and rafted down the lake to the site of the present homestead. There for many seasons it served as our kitchen. Later, when a newer and larger kitchen was made, our original domicile was again moved, and became part of the shearers' hut. Opposite the wool-shed to this day it stands, sound as a bell, and likely to last another forty years.

Meantime, whilst Stuart, who had ridden up with us, was preparing the immediate luxury of flapjacks, mixing flour and yeast for the morrow's bread, and fishing salt pork out of the barrel for the evening collation, we rushed off to inspect our wool-shed and yards, thence proceeding to Otutepiriao—the little valley where stands the present homestead.

In that September of '82 we could barely pass along the edge of the lake, so dense was the growth of flax and fern. On the flat itself grew huge scattered bushes of the former plant; groves of tall manuka marked the site of Craig's former garden; otherwise the surface of the ground was entirely rooted by pig. With instantaneous decision we settled this spot should be the homestead of the future. I recollect, too, that we agreed how disgusting, how disgraceful, how abominable must have been the mismanagement that could have wrecked such a splendid property. That day, in fact, we were very, very happy, and very, very foolish.

The second day on Tutira was still an ecstasy. We rode out to inspect the Back Country. We viewed its illimitable wastes of fern from the top of the Image hill. My present recollections are that we viewed them about as intelligently as an infant looks from its perambulator on to the world, and with about as little foreboding of the ills it might inflict.

Now, in 1920, at a distance, alas! of forty years, I am amazed at the hardihood of the pioneers of Tutira, Puterino, and Maungaharuru. What, I wonder to myself, could have been the inducement to attempt the handling of such runs. Tutira lands were as I have described them. Putorino contained no limestone or marl land whatsoever. Maungaharuru was thirty miles from the coast, its wool-clip packed out on bullocks. The tenure of these runs was leasehold, and native leasehold at that; without exception the titles were flawed; the land was devoid of grass, the climate was wet, the access bad, the soil ungrateful and poor. There was no compensation for improvements. It seems impossible now that any reasonable soul could have believed there was either money or reputation to be made out of them.

The truth is, that their owners were not reasonable, that they did not think at all. Most of them were new chums hardly out of their teens, of the sort moreover who welcomed physical toil as a delight, who preferred manual labour to any kind of thinking. To this day indeed I am not sure whether we were splendid young Britons, empire builders, and so forth, in a small way, or asses of the purest water. We bored inland for freedom, for adventure, for the chance of dealing with stock and soil, in obedience perhaps to an instinctive desire to push further back. Only for very brief intervals, and only in very careless fashion, did we think about the pound, shilling, pence aspect of our work. There were no proper books kept. Jottings in the station diary represented the Italian or double-entry system of book-keeping, as taught by Dominie Sampson to Lucy Bertram. Figures were doubly distasteful after a hard day's work—work, of course, was physical work. The idea of wasting even a wet day on accounts never seemed to have entered our heads. The sole excuse for such distasteful idleness—we called it idleness—would have been ill-health. Nobody ever was ill in those glorious days, so the accounts were left undone. The result was that the finances of the station were never properly known. It was a disability not decreased by the New Zealand habit of purchasing 30s. worth of property with 20s. worth of cash.

We enjoyed a perfectly happy open-air life in the present, convinced that everything, of course, would turn out all right. We split posts, we erected fences, we mustered sheep, we killed pig and cattle—less from any particular reason in connection with money-making or even benefit to the station than from an insatiable appetite for exercise; we lived, I may say, to gratify the calves of our legs. We enjoyed to the full a giant appetite, a slumber unperturbed, that anodyne, too, which keeps the labourer content—the delightful physical feeling of relaxation after prolonged muscular toil described by Tolstoi in certain passages of 'Anna Karenina.'

It was the delicious reward of a real good day's work—"real good" meant daylight to dark; "work" meant manual labour—riding or packing or mustering, or pig-hunting or fencing or bush-felling. We cooked for ourselves; we lived on porridge and water, bread baked in camp ovens—there is no better bread in the world,—mutton, potatoes, and duff. It was a delightful existence; there were no cares, there were to be none.

From the time they had forgathered at Rugby—H. G.-S. on some bug-hunting business, of course out of bounds, A. M. C. with dogs—he used to rat with retrievers, I remember—equally of course forbidden by the school authorities,—these two young gentlemen had determined to live the simple life. The simple life as they envisaged it was to preclude all thought on disagreeable subjects such as in the past, say, Latin and Greek, Euclid and Algebra, and in the future anything connected with figures or business. They were of the type to whom the Hypothetics taught in our Colleges of Unreason were particularly odious and distasteful. The writer's scheme of life was to work hard in the daytime, and in the evenings—must he confess it?—to write verse, a fatal habit which his relatives deplored. If he must have a vice, they argued, let him rather drink; many drunken flock-masters had prospered, but never a one who perpetrated verse. Drunkards, too, had been known to reform, but the verse habit was ineradicable; they regarded him as a lost soul, predestined to the pit.

Enthusiasm for the poets on my part, enthusiasm for football—he was a magnificent Rugby forward—on that of my partner,[1] were assets not particularly likely to assist in the development of an up-country sheep station.

From the start things went wrong. To begin with, the place had been purchased for us in late August, a time of the year when it was impossible to muster on account of the approaching lambing. Instead, therefore, of collecting the sheep from every part of the run and counting them in the yards, we took delivery "off the books"—that is, we accepted the flock on the previous shearing and lambing tally. 9000 sheep had been shorn, 1500 lambs had been docked; there should, therefore, have been 10,500 sheep on the property. From a South Island —Canterbury—point of view, a 10 per cent mortality was an almost inconceivable death-rate. It was thought perfectly safe to reckon the losses on Tutira during the previous twelve months at a tenth of the whole flock. On this basis of calculation the station was bought and

[1] My partner, like D'Artagnan, hated verse as he hated Latin. "An Address to my Banker," paraphrasing Goldsmith's lines, and beginning "Sweet source of all my joy and woe, thou found'st me poor and left me so," was, however, considered "not half bad."

paid for. As a matter of fact, we did shear something over 7000 sheep. The shortage in our flock, therefore, was nearer 3000 than 1000, nearer 25 per cent than 10 per cent.

Perhaps the reader may marvel how Stuart and Kiernan could have in so brief a period brought to book at shearing-time 9000 sheep. They were thus carried: 3000 wethers ran on the "back country," living until late autumn on the fern fronds that spring up after fires lighted purposely, and during winter feeding on tutu leaves, vast groves of which shrub, commingled with bracken, covered the whole of the central and west. In these regions there was literally no grass whatsoever, not one single acre, for the sheep camps were each season ploughed and reploughed by innumerable wild pig; a further 500 sheep would be stragglers raked in at various draftings from the neighbouring stations of Arapawanui, Tangoio, and Kaiwaka. The remaining 5500 were able to survive during the brief North Island winter, because the merino is a small sheep and can subsist on little, and because the surface of such country as was then in grass was virgin land, and grew feed with an exuberance altogether unknown after a few years, but principally because these 5500 ewes and hoggets were very badly done, because the country was very grossly overstocked. According to modern lights perhaps 3000 instead of 5500 could have been properly carried on the newly-grassed area.

The first brood of martyrs to the cause had emulated Bret Harte's hero, Briggs of Tuolumne, "who busted himself in white pine." If any particular factor in addition to loss of stock may be said to have given the mace blow to the three Stuarts, Kiernan, and M'Kenzie, it was their expenditure on heart of totara and work in connection therewith.

H. G.-S. and A. M. C. chose a new road to ruin. They knew that 9000 sheep, such as they were, had been shorn on Tutira, and reasoned, perhaps not unnaturally, that what had been might be again. They were mistaken.

For explanation of this dictum we shall have to revert to the work done by Stuart and Kiernan, as recorded in the station diaries. They had burnt out the countryside; they had scattered broadcast large quantities of grass seed until the high water-mark of grass expansion had been reached. Conditions were then at their best: the sheep, running over a great area of open ground, obtained a larger pasturage; the surface of the ground newly burnt was clean of parasites; pitfalls, holes, under-runners, bogs, lay exposed. All this had been accomplished, but, alas! much of the improvement was of a temporary character—Stuart and Kiernan had cut off a larger chunk than they could chew. A process of contraction, of ebb—to be fully explained hereafter—had set in over eastern Tutira, where all the improvements had been lavished. The flock was insufficiently large to eat off wholly the spring rush of bracken. The consequence was that along the lower slopes, about the outlying corners, over the cold damp spurs facing south and east, upon the poorer portions everywhere, bracken began to sneak away, to unfold itself, to recover its hold, once more to overrun the ground; during the later part, in fact, of Stuart & Keirnan's occupation of Tutira a process of general contraction in the feeding area had begun; towards the end even of their brief day the sum-total of winter feed had diminished.

When H. G.-S. and A. M. C. purchased the run the flock had been squeezed on to the upper slopes, tops, and sunny faces, the balance of the whilom grassed lands having reverted to bracken. It was a process which even the Stuarts and Kiernan could not have anticipated, still less was it comprehensible to the new owners of Tutira. They were aware only that a certain number of sheep had on one occasion been shorn: they decided that numbers must be kept up by purchase of fresh drafts. They bought 3000 ewes, which, added to the 1000 lambs saved from wild pig, brought the total up to 11,500—of which number 500 culls were sold to George Merritt. H. G.-S. and A. M. C. began, therefore, their first winter with 11,000 sheep. Of these they clipped 7400 at the following shearing.

The losses of the two previous seasons had not unnaturally shaken the confidence of the National Mortgage and Agency Company, with whom the station banked. They advised us to sell. It was good advice —the difficulty was to find a buyer, the number of fools in the district being limited.[1] Again large purchases of ewes were made, and a dry season helping us, we managed to shear 9200, and to reduce the death-rate to a little over 10 per cent. It was but a respite, for the following year again there was an enormous loss. The clip, too, was very light, for a starved flock grows a miserable fleece. The adage, "feeding is half breeding," was unknown to us; we chose to believe that the station had hitherto

[1] As reason for this sudden desire to sell, it was given out that the climate did not suit our constitutions—this to stock and station agents!!

been using inferior rams, had been breeding from badly-woolled sires. We imported our rams—high-priced Vermont sheep—from the South Island. They died wholesale. They were two-tooths, and could not stand the change in the quality of the grass; each year we lost about three-quarters of them. During autumn they did inferior work as sires, during winter they scoured themselves to death.[1]

Even such shreds of knowledge as we had acquired—a little knowledge is a dangerous thing—hurt us. Our three months' cadetship on Captain—afterwards Sir William—Russell's Tunanui station had taught us that indiscriminate burning of fern was unwise. This teaching, sound in itself, was applied by us on Tutira without discrimination. No fern-burning was done, therefore, because the work was to be done perfectly at a later date. From this determination not to burn fern until the countryside had become rough enough to ensure a "clean fire" arose other evils which, although unavoidable in themselves, were accentuated by mismanagement. "Lungworm," which broke out in Hawke's Bay in the early eighties, everywhere ran its course with all the virulence of a new importation, whether blight, weed, or beast.

In a wet locality like Tutira the disease could not but have affected young stock otherwise than seriously. On the good runs of Hawke's Bay the losses were considerable. Everywhere sheep-farmers were dosing their young sheep with turpentine and oil, or attempting the smoke-cure with sulphur. In a flock like that of Tutira, jammed by the process of contraction already explained on to foul camps, overstocked tops and clearings, three-quarters of our weaners, station-bred, and therefore by far the most valuable section of the flock, perished.

Another trouble of these times was footrot. With the increase of English grass it was becoming impossible to keep the merino on his feet. The breed was unsuited to the soil and climate of the province. The area of marl land in grass on Tutira was small, but upon that area footrot was rampant. Like lungworm, it too found a congenial nidus in dirty sheep camps, crowded grounds, and wet grass. For several years after our purchase of Tutira, an average of 25 per cent of our ewes were lame dur-

[1] Young stock, two-tooths, however well done, are particularly liable to suffer from change to a wetter climate. Twenty years later than these troublous times, again an experimental lot of two-tooth rams was bought for Tutira Notwithstanding that they were run on ground more fertile by far than that upon which they had previously depastured, more than half died. Doubtless other wet district sheep-farmers who have purchased stock from dry country have experienced similar results.

ing one part or another of the year. There was endless labour in paring the hoofs of the limping brutes, in running the lame drafts through the arsenic troughs.

Everywhere there was wastage and leakage; the old sheep died, the young sheep refused to live. The lambings were affected by the poor condition of the rams, by the age of the ewes, to a lesser extent by pig.

Every one of these adverse factors admitted of a cure,—a cure, however, only to be discovered by experience. Lacking that empirical knowledge, A. M. C. and myself stood amazed at the ills meted out to us. Our efforts at originality, such as purchase of young rams from the South Island, had failed. As I have pointed out, we were aware that 9000 sheep had in the shearing of 1881 passed through the shed. We did not know, we could not know, of the contraction in grazing area. We did not know of the importance of fires and clean feed. If the station had carried —thus we argued—9000 sheep, it could be made to carry them again. Of course it could; it must be made to do so. Every year, therefore, sheep were purchased to replace those fallen in the fight.

It does not require demonstration that farming on these lines cannot be continued for an indefinite period. The gross income derived from the place was a poor few hundred pounds' worth of wool. A considerable proportion of our flock appeared on the shearing-board with bellies, sometimes with sides too, bare of wool,—"Pareperries"—bare bellies—joyfully the shearers hailed them in the catching-pens. Their fleeces had been worn off by wandering through fern and scrub or peeled off through fever and poverty. The wool of the wethers especially, stock that lived almost entirely on tutu and scrub, was often not more than a couple of inches in length, and black with the sand and dust that stuck in the dense merino fleeces. It was no rare sight, during a spell of hot, wet, autumn weather, to see sheep come into the draftings distinctly green on the back with sprouting grass, their wet fleeces, plus the animals' natural warmth, forcing the seeds as children grow mustard and cress on wet flannel in a nursery. I blush when I think of our flock of the 'eighties.

The return from surplus stock was likewise pitiable. The younger generation, who nowadays grumblingly receive a pound and twenty-five and thirty shillings for sheep sold, will hardly credit the prices in the 'eighties. Sometimes sixpence and sometimes ninepence per head was the price obtained by Tutira for its first, second, third, and fourth draft of old ewes and hoggets. They were purchased by George Merritt, who fed them to his pigs at Clive. My recollections, moreover, are that after the first transaction he was not keen for our old sheep. More or less we had to work on his better nature, to demonstrate that he was morally bound to buy. He had been a former owner of Tutira and yet survived—he had escaped the wrath to come. We flung our skeletons at his head; he was a coy buyer; much correspondence at any rate would pass in regard to the annual sale, the station claiming that the draft was quite unusually prime and well worth a shilling, Merritt asseverating that his pigs could hardly digest the last lot, and that he absolutely could not go beyond sixpence. The station gave delivery of the brutes at Petane. After that they were "Merritt's sheep,"—the shame of their ownership had passed for ever from Tutira. The bargain, however, was by no means concluded then; the sheep had still to be paid for. The station would generously grant three or four months' grace, and would then write a friendly letter, as from man to man, hinting that when quite convenient it would be pleased to receive payment. Our mail-box, the open case pegged to the top of a fencing-post in the heart of the Tangoio run, was half a day's ride distant, and only visited at intervals. Opportunities of delay, therefore, were not wanting. First of all, Merritt would be obliged if we could wait till the pigs were fattened; then till they were sold; then till he himself had been paid for them. At last the station, becoming ravenous for its twelve or fourteen hundred sixpences, would have a "lawyer's letter" despatched intimating that unless cheque reached Tutira by next mail Merritt would be persecuted with the utmost severity of the law—or words to that effect. Even then, on one occasion, I remember that although the cheque duly arrived, the signature had been omitted. Merritt doubtless had also a banker jumping on him, and these delays were regarded as part and parcel of the deal, a comedy to be re-enacted the following season. Merritt, indeed, was regarded by us with very high respect,—we reverenced him as needy Hebrews reverence Rothschild; he had touched pitch and had not been defiled; he bore the unique distinction of having owned Tutira and yet escaped ruin. A man who could accomplish that could squeeze blood out of pumice. Merritt had another claim to consideration: he was the only buyer who could by prayer and supplication be induced even to look at our cull sheep. When he would not take them for his pigs, we had ourselves to kill and skin the wretched beasts. Once, I remember, they were boiled down for soap,—my recollection is that sevenpence-worth of soap was extracted from each sheep. Merritt, however, was our stand-by; he never went beyond ninepence per head, but that sum was more than the station could obtain otherwise.

With a gross income of about £1300, the partnership of H. G.-S. and A. M. C. only existed as long as it did because the owners spent nothing on themselves, because there were almost no wages to pay, because the price of wool remained high.

Certainly the shortage of sheep at shearing-time, the miserable clip, the more miserable annual draft of surplus stock, gave us momentary pause, but I do not recollect that on one single occasion we talked matters out or realised the danger towards which we were drifting. Were it not for the entries of another and previous owner's diaries set down in cold blood, and still to be seen as quoted, I should have said that the idiocy of myself and partner was of a unique brand; I should have said it was impossible that one station should have carried so many fools—in shepherd's language, a fool to every 4000 acres! We realised the condition of our affairs no whit more clearly than in the past had the Stuarts and Kiernan.

The still extant station ledger is a model of original thought. I remember its inauguration a few weeks after the purchase of the place. A. M. C., who, by the bye, always breathed deeply through his nose when excited, was the book-keeper of the firm, but I stood by ready to assist, and to see that in this important matter everything was done properly and correctly. I recollect the breeze my partner blew—it was like whistling—whilst we debated whether the price of the place, £9750, should be entered on the debit or the credit column. There seemed to be sound arguments for either course. What the devil!—if we had paid for the place, how on earth could it be chalked up against us; it would have been better never to have started sheep-farming than to have landed ourselves straight away with a £9750 debit. We might just as well have gone to Oxford after all. Yes, but damn it all, we had not paid wholely for the place—we had only paid down £6000, unless, of course, we had made a regular bargain, and gained £3750 straight off the reel. Well then, why not compromise the thing, why not put down £6000 as a credit and £3750 as a debit? That didn't seem right either, so the £9750 was accordingly written down first as a debit, then as a credit, and each time a fresh start was made, a page was taken out, for we knew it looked rotten not to have tidy well-kept books. Finally it was fixed as a debit—partly, I think, because a five-pound note lent us for current expenses by Captain Russell immediately before we rode up had also to be entered somewhere or other. As we had just borrowed that fiver, there could be no doubt that it at any rate must be a debit, and as the £9750 looked so absurd by itself facing the pitiable little insignificant £5, we jammed the two entries into the same category. We never got further. They are the only entries in that station ledger—except numbers and sex of wild pig slain by H. G.-S., lists of soiled linen sent to Napier, and the dates of letters despatched by A. M. C. to his father in India.

National Mortgage and Agency Co.	£9750	0	0
Captain Russell	5	0	0

We were great on the nothings—they were safe sort of figures, and filled up the page.

With the debit and credit question thus still unsolved, the rain, perhaps, began to clear—we never did accounts, of course, except in storms—and we rushed forth to some delightful labour which kept the brain entirely inactive, and produced the nirvana of muscles relaxed in rest, deep sleep and enormous appetites. Book-keeping had no place in our conception of the simple life, though, perhaps, to paraphrase Wolsey's lament, if we had loved our banker as we had loved our legs, he would not have thus left us to perish miserably.

Partly, then, by an unwise purchase, partly by unacquaintance and inexperience of conditions that would have puzzled wise men, and partly by ignorance of business and of business methods, the finances of the run passed from bad to worse. The end came with the same unexpectedness that has been revealed in the pages of a former diary. There occurred a crisis in the wool market—the most unstable market in the world. A sudden drop in prices precipitated our fall—a fall which could not have been in any case long postponed. A. M. C., for the sum of £600, paid into the station account, was allowed to quit Tutira. H. G.-S. took over the derelict half-share for the sum of five shillings. He survived, a melancholy illustration of the "martyrdom of man," of the theory that each individual of the human family, if he stands a little higher in the scale of civilisation than his predecessor, does so through the sacrifice of that predecessor—that our civilisation, like a coral island, has been built by generations of workers who have used themselves up in the process and passed away. At any rate, whether this theory be tenable or not, the writer of this ower-true tale stood with head barely above water on the carcasses of those who had fallen in the fray—Newton, Toogood, Charles Stuart, Thomas Stuart, William Stuart, Kiernan, M'Kenzie, and Cuningham.[1] They had spent all and gone under, each adding, however, ere financial death took him, his accretion to the coral island, his contribution to the future of the station—one timber, another ewes, another cattle, another rams, another grass-seed, another drainage of swamp land, another fencing, another—his mite to the general sum—that "*team of eight bullocks bought from William Villers, waggon and all complete, for the sum of* £135, *terms, to pay when able.*"

The derelict half-share thus forced upon him for five shillings, the writer became sole owner of Tutira—*Tutira upoko-pipi*—Tutira, the place where heads become soft.

[1] To the best of my belief, every one of these adventurers did well in later years; New Zealand of all countries in the world certainly is the land where after a stumble a man can most easily pull himself together again. In Hawke's Bay, at any rate, I can hardly think of a prominent settler of early times who has not been at one time or another on his last legs.

Pack-horses crossing stream.

CHAPTER XIX.

FERN-CRUSHING.

To attempt the portrayal of the successive processes by which, over the whole run, its lighter lands have been trodden and trampled into some sort of utility, would be to work on too large a canvas. A multiplicity of details, unavoidable even in the annals of a single paddock, related of many would confuse and mystify. One will suffice: the enclosure called the "Rocky Staircase" is in soil typical of every block on central

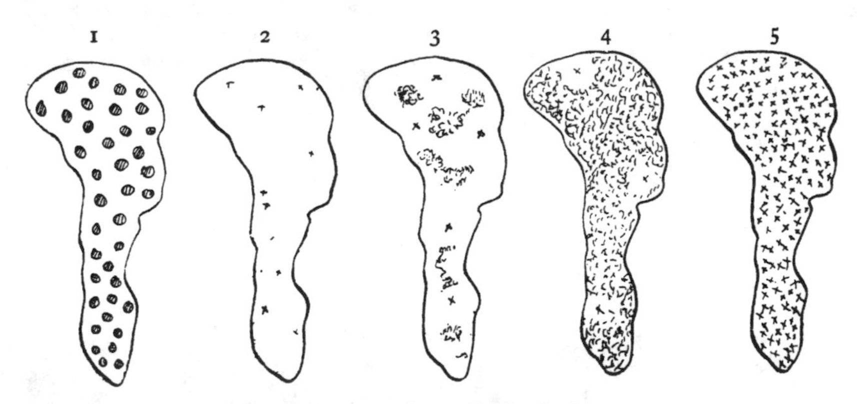

Successive growths on Rocky Staircase.

1. Fern and tutu (black). 2. Fern and native grasses (X). 3. Fern, native grasses, and manuka. 4. Fern, manuka, and native grass (X). 5. Fern and native grass.

Tutira. It will serve to show the struggle between bracken, manuka, and danthonia, and to demonstrate the discomfiture of alien fodder plants, the ultimate triumph of native species.

Perhaps the chapter may possess another interest. During its perusal no great stretch of fancy will be required to note in our little world of plants a process not altogether unlike that now taking place amongst classes and individual members of the species *homo sapiens.* Armageddon, in truth, has been raging as fiercely in the Rocky Staircase as in the Old World. The ancient régime in both has been overthrown. The good things of life have been opened to all capable of taking them; the selfish sway of capitalism and landlordism—call them brackenism and tutuism—has been broken. There has occurred a revolution which, however personally distasteful to Pteris aquilina and Coriaria ruscifolia, has proved quite delightful, I should imagine, to humbler members of the community who had hitherto been half-starved in breathless slums and barren crofts—bog brims, arid tops, and precipices—and who now for the first time could breathe fresh air and sate their appetites. The reader will see in Chapter XIX. the Rocky Staircase "made safe for democracy."

Allusions have been made from time to time to fern-crushing, to the ebb and flow of sheep-feed, to the contraction and expansion of feeding areas; they will now be fully explained. Pteris aquilina, var. esculenta, is a form or sub-species of the British bracken. Its roots were in ancient times of a certain value to the Maoris for food; at a later date its circinate fronds were moderately palatable to stock. The normal growth of the plant is that of its English relative, a single even crop of fronds in spring-time. Unlike the bracken of England, however, which rapidly withers and disappears, the fronds of the New Zealand pteris endure for years.

Briefly described, the science of fern-crushing on rich stiff land is as follows. A section rough enough to carry a fire is selected proportionate to the number of sheep forthcoming to crush it, the tangle of fern is burnt off in autumn, whilst immediately afterwards the land is surface-sown with grass and clover seed. A week or fortnight after destruction of the old stalks and stems young fronds begin to appear. Sheep are then poured into the paddock, the number required per acre varying with the fertility of the land, and, equally important, with the weather conditions. Drought means cessation of growth; sharp frost, temporary destruction; heavy warm rain, stimulation of the rhizomes. The stock used has also to be carefully shepherded. Sheep "hanging" in corners, or against fence lines barring them from the paddocks where they have been bred, have to be driven elsewhere or skimmed off. Feed, and sheep to eat that feed, should be exactly balanced. Without a big enough mob, the bracken fronds uncurl and become uneatable; on ground too heavily stocked, sheep fall away in condition. At the end of the second season, if all has gone well, if the sown grasses have sprung up dense enough to cover every bit of open ground where otherwise there might have been germination of undesirable weeds, if the land has been rich enough to support a heavy head of stock which will have trampled down unwanted plants, or devoured them unnoticed in mouthfuls of grass, then the work has been permanently done,—grass has taken the place of fern. On first-rate land these results are obtainable without injury to stock; on first-rate land the work is permanent.

The soils of Tutira, except for a few hundred acres, were, however, not first-class; they were not, except perhaps for a couple of thousand acres, even second-class. The great trough of the run—the vast bulk of the station—was third-class.

Fern-crushing on Tutira was accomplished on its few acres of first-class land as described—that is, with a minimum of trouble to man and beast; on the few hundred acres of second-class land, with rather less good results and a considerably greater output of labour to shepherds and injury to stock. On eighteen out of its twenty thousand acres, it is no exaggeration to say that the surface had to be stamped, jammed, hauled, murdered into grass. It was only the low price of sheep that made such procedure possible, for the stars in their courses fought against the station. The rainfall, double that of southern Hawke's Bay, stimulated this terrible growth of fern against which we warred. Weather conditions militated against the station in another way too—they immensely prolonged our shearings. Not infrequently a break in the weather would occur immediately after the gathering in of sheep from a block in process of crushing. These sheep, having once been mustered, could not be put back; in the first place, because the weather might have cleared at any time, and the fleeces become dry and fit for shearing; in the second place, because sheep dogged overmuch grow callous and sulky; they will not run well and give a second clean muster. I have known stock in this way kept for a fortnight or three weeks away from a paddock, where every day the fern-stems were lengthening, where every day the fronds were uncurling.

The soil of the trough of the run has been described. It was spongy, porous, and relatively unfertile, as well fitted to the requirements of bracken as unsuitable for grass. The alien fodder plants sown, nowhere amalgamated into anything that could be termed a sward. Between the isolated plants of the miserable "take" of seed there was ample space left for the germination of undesirables. We shall see, in fact, that as the station began to get the better of bracken, its place was taken by another and a worse plant. The grassing of nine-tenths of Tutira has not been—it could not be—a fine art run on scientific lines of husbandry. It has been accomplished by brute force at the expense of owners and stock.

Between '82 and 1917 the story of this paddock will divide itself naturally into seven distinct periods, each prefaced by a fire, each showing a maximum and minimum of carrying capacity, each demonstrating the ebb and flow of sheep-feed, the contraction and expansion of land over which stock was able to graze.

In the spring of '82, when I first saw the paddock, it lay a blackened waste, strewn with a tangle of tough stems—the ropy, parboiled stalks of the latest, greenest, and therefore least combustible of the many layers of fern that had covered the ground. On the damper aspects—the southern and eastern slopes—stood extensive groves of tutu, black, stiff, and stark. Mustering this block a few weeks after arrival, I remember my astonishment at these miles of desolation unrelieved by a single green blade. I had seen nothing like it before. Of manuka, excepting a compact fire-swept patch of ten yards square on the main top, there was none. The rood or so of sheep camp on the main range had been ploughed and reploughed by pig in search of grubs and roots. Immediately beneath the great conglomerate cliff from which the paddock takes its name, also beneath lesser lines of cliff, lay narrow ribbons of open land, also uprooted and grubbed and regrubbed by pig. These strips of overturned sods contained survivors of such plants, native and alien, as Microlæna stipoides, Danthonia semiannularis, ryegrass (*Lolium perenne*), white clover (*Trifolium repens*), suckling (*Trifolium dubium*), Cape-weed (*Hypochœris radicata*), Geranium sessiliflorum, Pelargonium australe, &c. On the very edge and rim of the cliffs hung scant tufts of Poa anceps, and of Deyeuxia quadriseta, a species which even to this day on Tutira seems terrified to venture into the open. Here and there, too, on the rocks grew plants of blue grass (*Agropyrum scabrum*). Except for these pig-ploughed shreds and these rock surfaces the Rocky Staircase was in October of '82 a blanket of blackness. Save in the immediate vicinity of these strips of uprooted turf there had been no germination of any seed whatsoever. The paddock had nevertheless then attained its zenith of food supply, its maximum of expansion. Certainly there was no grass in it, but the tutu groves sprouting again from their burnt stools were open. Sheep running in the paddock could at least wander over every acre of it. With the advance of the year the initial stage of contraction began; by November millions of brown circinate fronds had appeared; they grew tall and strong, their pinnæ expanding until the ground was once again overspread with fresh crops of bracken, which season after season lay thicker on the ground. On this abomination of desolation some 300 merino wethers managed to survive, their fleeces black with dust from bracken and tutu thickets, their wool stripped from belly and side through constant contact with scrub and fern.

In the autumn of '89 began the second period. By that date the bracken, which had been until then carefully conserved from fire in accordance with the general plan of fern-crushing was again fit to burn. The paddock became for the second time in our story a wilderness of charred stalks and stems. Now, however, on its surface for the first time began to appear cotyledons of certain aliens. They were sprung from plants which had by this time obtained a firm hold on the grassed lands of eastern Tutira and had begun to move inland. Some like the Prickly thistle (*Cnicus lanceolatus*) and the Cape-weed (*Hypochœris radicata*) had taken advantage of the wind to extend their range and to increase their numbers; others like Mouse-ear chickweed (*Cerastium glomeratum*), Silene (*Silene gallica*), and the native Houtawai (*Acæna australis*), had by different contrivances fastened their seeds to the legs and fleeces of sheep. Seedlings were still, however, rare compared with later germinations, the more so that winds blowing from the south and east, from the sea, over the grassed lands damped the feathery pappus of winged seeds. Their appearance was in any case of relatively trifling import, for the hand of man was about to interfere on a great scale with natural conditions. It had been, in fact, determined that the block should be crushed, sown, and stocked. Instead, therefore, of a few score of sheep running, as formerly, haphazard, many thousands were decanted into the paddock. Many hundred bags of grass seed were also sown. The first result of this stocking was the annihilation of every tutu thicket throughout the block. The bracken also was severely checked both by hoof and tooth, especially on the tops and along the upper portions of the slopes, positions dear to all sheep and especially to the timorous merino.

As might have been anticipated from what the reader has already been told,—a knowledge, however, then unattained by the writer and his partner,—this attempt to grass the Rocky Staircase was a comparative failure. On the worst portion of the paddock the seed failed to germinate; on rather better soils it held out feebly for one or two seasons, but everywhere its grip weakened with lapse of time. It was discovered, too, that on large areas of the paddock, sheep would starve rather than eat the fern fronds. On other portions, only with difficulty could they be forced to crop the shoots.

The result was that the worst parts of the paddock immediately relapsed into bracken; that another immense proportion became overgrown with hardly less rapidity; that only the steepest—that is, the best—portions of the block remained open to the sun and air for any length of time. A fresh process of fern expansion in fact recommenced in spite of us, its progress synchronising with the gradual failure of the sown grasses and clovers. At the termination of this second phase in the history of the Rocky Staircase, the paddock appeared to have reverted almost to its original vegetation, except indeed for the establishment of certain sheep camps, oases of deepest, most luxuriant green.[1] There had occurred, nevertheless, certain great, albeit hidden changes. The tops had been heavily trodden by stock; light had been allowed to reach the ground—in some parts for a season or two only, in others for three or four years. The seeds, though few and far between, of plants already enumerated as having been blown by wind or borne by stock, had before their final smothering managed to mature and sow themselves abroad. There had taken place, too, a slight permanent widening in the strips of native grasses beneath the cliffs. On hard hill-tops and narrow ridges and knobs, danthonia and microelena had grouped themselves in little companies. Where pig had tunnelled

[1] The continued enrichment of these camps month after month, year after year, decade after decade, where night by night thousands of sheep concentrate on small areas, has been for forty seasons very distressing to Harry Young and myself, constantly scheming as overseer and owner how to provide more grass for our sheep. The waste of ammonia on soil already enriched beyond all reason has been the more vexatious in view of the glaring needs of the arid areas around. Picturing the surrounding land as it might have been but for the conservative habits of stock, almost involuntarily have the words rushed to my lips, "I say, Harry, wouldn't it be grand if sheep only wouldn't always pee on the same place?" and often have I heard his responsive sympathetic sigh, "Ah, sir, Tutira would be like heaven then."

and terraced the caved hillsides, scattered representatives of these native grasses had also appeared. The paddock, moreover, was now lined from end to end with innumerable sheep-paths, on the edges of which the above-mentioned grasses had also managed, very thinly and very sparsely, to establish themselves.

The thickets of tutu had been annihilated, whilst here and there could be seen isolated manuka plants — forerunners of the coming invasion. The bracken itself had suffered in the long engagement with man and beast; its exuberance of vitality was gone. The treading of the lands for several years with thousands of sheep had given it a preliminary shock; nor, from the point of view of the station, had the crushing operation been an entire failure. Though it had not realised our hopes, yet there had taken place a substantial increase in the number of sheep carried — even though that number had been wintered but for a couple or three seasons. During the maximum period of expansion the Rocky Staircase had wintered 1900 head, a total, however, diminishing, with the increase of bracken growth, to 300.

The third period in the annals of the Rocky Staircase began in '96. When in the autumn of that year the paddock was again fired, it was found that progressive movements had taken place along each of the lines noted formerly. The surface no longer remained altogether void and black; hundreds of thousands of cotyledons opened fresh and green in the vicinity of the heavily manured, densely grassed sheep-camps, on slopes beneath the narrow strips of native turf, and along the winding stock-tracks. Especially had Thistles, Suckling clover, Cape-weed, Mouse-ear chickweed, Houtawai, and Manuka multiplied themselves. There was an increase, moreover, in species as well as in numbers of individual plants. This great multiplication of other vegetative life than bracken was owing partly to the less fierce fire consequent on the less thick growth of fern. Seeds lying on the surface had not been wholly destroyed by heat and flames. Winged seeds, moreover, had been blown in greater profusion from a larger area of handled land on eastern Tutira. Lastly, pig had been destroyed; the camps were no longer wastes of overturned sod.

In addition to increase in cotyledons, a considerable number of plants also had survived the fire. They were chiefly grasses — such

PRICKLY HEATH (*Leucopogon Fraseri*).

as Micrœlena stipoides and Danthonia semiannularis, species seemingly created for such lands as Tutira, able to survive alternate smothering by fern and blackening by fire. On points, peaks, tops, and ridge-caps, surfaces where the ground was hard naturally or had been stamped into solidity by traffic, these native species appeared to have established themselves. Where formerly two or three plants had been gathered together, now small congregations remained to pray.

During this third period there began, in fact, an insurrection of aliens and natives alike against bracken, their ancient oppressor and tyrant. Among the insurgents, manuka (*Leptospermum scoparium*) was not the least forward. Its seed-capsules mature when the plant is but three or four years old; they are produced in enormous profusion; the seeds germinate freely; the plant is able to draw nourishment from the most arid of soils. Often it is rather scorched than utterly destroyed by fires that consume the bracken. If it be not true that its capsules, like those of some of the eucalypts, open only after fire, it is at any rate noticeable that they expand then most freely. The plant's rapid growth offers this further advantage,—no small benefit either,—that seed is blown abroad or shaken out from an elevation of six or eight feet. Manuka is, in a word, a plant preeminently fitted to survive on lands such as those of the trough of the run. It now began to colonise the paddock, straying from its original sites, appearing about pig-rootings, along sheep-tracks, but especially taking possession of open ridges and peaks now clear of bracken. Other less rapacious settlers also appeared. A small densely rooting heath, patotara (*Leucopogon Frazeri*), during this third period began to colonise suitable localities. A native Carrot (*Daucus brachiatus*), a little Chickweed (*Stellaria media*), and a low-growing Michaelmas Daisy (*Vittadinia australis*), stepped down from their banishment on the cliffs. The Fern-flower or Sundew (*Drosera binata*), the little orchid (*Microtis porrifolia*), vacated the barren ridges on which perforce they had been confined. Other species like the native Thyme (*Pimelea lævigata*) and the alien Horehound (*Màrrubium vulgare*) selected small holdings about the camps. A Broom (*Carmichælia odorata*) made a brave bid too for certain special sites. Rat's-tail (*Sporobolus indicà*) proved itself able to thrive on a light diet. Hare's-foot clover (*Trifolium arvense*) appeared here and there. A species of Groundsel (*Senecio *canadensis*) also temporarily overran the Rocky Staircase in vast profusion.

It is not again necessary to describe the processes of stocking and crushing. Suffice it to say that once more the old operations were re-enacted, once more a certain number of bags of ryegrass and cock's-foot were sown, once more the worst parts of the block relapsed into bracken, until at last stocking of the paddock was altogether discontinued. The sheep were wanted elsewhere, for as one block began to fail, station policy arranged that another should come into use. After the third or fourth season it was in fact an advantage to allow a paddock to become again overrun with fern-growth, to become again fit for firing. The maximum number wintered during the maximum expansion of the third period was about eighteen hundred. Except over an insignificant area of camping ground, English grasses and white clover had disappeared. Sheep were chiefly, if not altogether, wintered on suckling clover, an invaluable plant which from this date became our mainstay on the pumiceous area of the station. The minimum number carried dropped back as usual to about a couple or three hundred.

The fourth period in the history of the Rocky Staircase was particularly marked by the failure of fern to maintain its ancient sovereignty. The plant was weakening under the long warfare waged against it; although it covered the ground still, the covering was less dense and matted. The "burn" of 1902, consequently, was not what is technically known as a "clean" fire. Unlike previous conflagrations that had swept the Staircase from stem to stern, this fire left unburnt the ridge-caps, the tops, sometimes even the upper slopes. There had been a lack of herbage to carry the flames; they had died down for want of material. On these localities manuka had already made a lodgment. On all of them it remained now in the fourth period, green, flourishing, unburnt, five or seven feet high, its infinitesimally minute seed shaken abroad in every breeze, spread by the hoofs of stock, in wet weather sticking to every dislodged pebble, washed downhill in every sheep-path runnel. Otherwise, as before, save on unburnt portions and on the bright green sheep-camps, the paddock was a blackened tangle of fern stems intermixed with scorched manuka growing singly or in twos or threes.

During this fourth period, however, it remained a blackened tangle only for a few weeks. Seedlings which had appeared, so to speak, singly after the first fire, in hundreds after the second fire, in hundreds of thousands after the third fire, now during the fourth period germinated in hundreds of millions. As before, but in far larger numbers, they had self-sown themselves or been blown from other parts of the run or carried in by stock. Thus in one way or another enormous numbers of Cape-weed, Suckling clover, and Mouse-ear chickweed seedlings appeared on the freshly-burnt surface. There was the usual though diminishing recrudescence of the Thistle (*Cnicus lanceolatus*); certain tracks were more thickly overrun by Leucopogon Frazeri, the prickly heath already mentioned; Houtawai (*Acæna australis*) obtained, too, its share of the fern-vacated ground. Fresh arrivals also, such as Pomaderris phylicæfolia and a couple more heaths—Cyathodes acerosa and Leucopogon fasciculatus—took up permanent quarters in a small way. Rat's-tail, though increasing slowly, occupied the spaces overrun with a hirsute mat, ousting all other growth. Lastly appeared Clustered clover (*Trifolium glomeratum*) and Suffocated clover (*T. suffocatum*). From this time forward, in fact, wherever conditions were favourable, aliens and natives alike struggled with the moribund bracken and with one another for possession of the soil.

As before, the Rocky Staircase was at first heavily stocked and the failing fern again heavily punished by sheep. As before, too, a certain number of bags of English grass were scattered abroad, but in this fourth period seed was scattered only on the steepest, best parts of the paddock. Even on them it made so poor a show that English grasses, such as rye-grass and cock's-foot, have never again been surface-sown on this type of land—such elements of virtue as may have been in the soil had been used up.

A conspicuous feature of the fourth period was the multiplication of manuka, its rise illustrating the law of progressive increase of new plants in units, hundreds, hundreds of thousands, and millions. The spread of this plant now began to cause serious uneasiness. On the upper portions of the hill-slopes from which fern had been worn out by the trampling and nibbling of sheep, manuka during this fourth period increased year by year. On the middle slopes where the fern-growth was becoming thin and short in stalk, single manuka plants or little groups were also to be found not far apart from one another. Even where slopes merged into flattish land, individual specimens appeared. It dispossessed danthonia and microelena from the hard bare tops where they had seemingly established themselves; practically these grasses disappeared. If not wholly destroyed, so weakened were they by shade that only sparse spindly blades showed life was not quite extinct.

The threatened seizure of the whole paddock by Leptospermum scoparium now modified our policy in regard to the bracken. For the first time we were careful in our stocking not to overweaken it. Our ancient foe, now humbled and subdued, had become an ally in the war to be waged with the rising power—manuka. Sheep, therefore, were run more lightly on the land. The Rocky Staircase was allowed almost without let or hindrance to clothe itself once more in fern; its growth hastened the date when fire could again be run over the ground, when the manuka could be destroyed once more. The small amount of sheep carried augmented another change: it allowed other invading weeds to sow themselves more freely. This was the more important, because one of them, Suckling (*Trifolium dubium*), had become a fodder-plant of prime importance. Its spread had more than compensated for the loss of ryegrass, cock's-foot, and white clover, grasses which had been sown and failed. Weeds of low growth, foreign or native, were indeed during this period rather hidden than obliterated by the bracken growth. On the ridge-tops it had almost disappeared, on the uppermost portions of steep slopes, especially on the warm west and north aspects, it had retreated far down the hill-sides. All these spots, nearly bare or sparsely covered with dwarfed, depauperated fronds, were now at the end of the fourth period of the paddock heavily sprinkled, some of them packed, with manuka bushes. Even in parts where the fern-growth still retained something of its pristine vigour, scattered plants of manuka topped the fronds. The reign of bracken—a sovereignty of centuries—was in truth passing away; the day of manuka had dawned. Alien grasses, except on the camps, had completely disappeared; native grasses, light and air denied to them, barely evaded death; on the other hand, an enormous spread of suckling clover had compensated for their loss. The maximum head of stock carried during the maximum expansion of the paddock was again about 1700 sheep; the minimum again about 200.

The fifth chapter in the history of the Rocky Staircase included the years between 1907 and 1913. As related, our paddock had been swept bare by fires of the first, second, and third periods; after the fourth fire small portions only of top remained unburnt. Now, after the fifth fire, the Rocky Staircase was parti-coloured, striped and patched like Joseph's coat. Where fern had predominated it was as of yore, black; in other areas the prevalence of scorched manuka produced from a distance a grey, sere hue. The tops, peaks, and ridge-caps, clothed in the same growth, remained green. The autumn of 1907 had been wet and cold, the admixture of growing manuka amongst the fern had furthermore acted as a damper. The accumulated growths of bracken were lesser in bulk, they were no longer capable of producing the raging, roaring conflagrations of early days.

On the blackened portions of the paddock conditions likewise had altered. Seedlings germinated in millions on the dark ground; there was the usual reappearance of Cape-weed, Mouse-ear chickweed, Houtawai, Groundsel (*Senecio canadensis*), and Pelargonium australe. There was the customary waning recrudescence of the "Scotsman" (*Cnicus lanceolatus*), a plant which, whatever its name might seem to infer, does not thrive, and eventually ceases to germinate on hungry soils. The three heaths named had extended their range, especially Leucopogon Frazeri. Pomaderris phylicæfolia had settled in small dense colonies on suitable localities. Besides this vast general increase in seedlings, there was also a vast increase in the numbers of the plants themselves that had survived the fire. In many parts the last crop of bracken-growth had not been dense enough to smother the established roots of Cape-weed and Houtawai (*Acæna australis*), Leucopogon Frazeri, and other plants. Amongst the manuka, by some miracle, danthonia and microœlena still survived, each etiolated plant still throwing forth a few meagre green blades. Though always apparently on the verge of extinction, these species just managed to exist. Their growth was sparse and meagre; to be seen they had to be searched for. Nor, as we shall see afterwards, did these invaluable species—invaluable at this period—content themselves with passivity.

Stock debarred by reason of manuka-growth from the crests and crowns of the paddock had developed on the upper slopes new series of traffic lines, parallel below parallel. Along these, native grasses now also lodged precariously, inconspicuously, breathlessly.

The increase in the number of other plants and seedlings was, however, as nothing compared to the increase of manuka. The heights everywhere were now crowned and crested with its dense thickets and winding shrubberies. Seedlings appeared in millions of millions of millions. After the heat of a fire which had rather scalded and withered than burnt the shrub, its berries opened fully and shook forth their innumerable tiny brown seeds. On the dry surface, in company with charred morsels of stick and stem, mingled with dust hardly more minute than itself, manuka seed was whirled downwards in nor'-west gales and eddying whirlwinds. In wet weather it was everywhere transported in sheeps' hoofs. In deluges and tropic showers it was poured downwards along the hard stamped tracks. On every wet pebble that rolled from the conglomerate slopes the little seeds clung fast. Plants did not appear one here and another there as in former periods; they germinated, sometimes in tens, sometimes in hundreds, sometimes in thousands, on every acre of burnt ground. Over certain portions of the paddock they sprung up like hay-seed round the edges of a stack. The bracken, crippled and weak, now endured the sufferings it had formerly inflicted on other plants; in the company of this virile newcomer it was squeezed to death, throttled, denied the right to air and light. So completely, during the last years of the fifth period in the history of the Rocky Staircase, had manuka dominated bracken that in spring-time great sections of the paddock, areas of hundreds of acres, appeared at a distance of miles as if sprinkled, appeared even as if laden with snow, the snow of manuka petals. It looked as triumphant in 1912 as tutu and bracken had looked in '82. The paddock had changed between these dates from fern to manuka—Pteris aquilina had fallen before Leptospermum scoparium. Throughout this fifth period in the history of our paddock no attempt was made to crush fern. From a foe it had, in fact, become a friend and ally. Without intermixture of its fronds further fires would have been unobtainable. Our paddock would have become a vast manuka thicket with a permanent carrying capacity of nothing at all.

There was again during this fifth period but little change in the maximum and minimum of sheep carried. On parts where native grasses had formerly thrown a certain amount of feed, green growing manuka now held sway. This loss of feed was, however, more than made up by the wonderful spread of suckling clover; stock carried during the fifth period subsisted, in fact, wholly on this invaluable annual.

During the sixth period of the paddock an extraneous factor for the first time came into operation. It was this, that after a quarter of a century the writer had been granted a sound title to his holding.

Work which could not formerly have been undertaken with any hope of return, now became at least worth the risk inseparable from any improvement. The crests and crowns of the paddock were cleared of manuka by axe-work; several hundred acres of manuka were also felled on certain slopes and valleys. Another innovation, now also for the first time determined upon, was an alteration in the date of firing the paddock. Until this sixth period fires had been lighted in autumn, weather permitting, late in February. This custom had been followed for two good reasons: to provide autumn food whilst another block elsewhere was "spelling," and to break the exuberance of frond-growth during the following spring. Now, however, that manuka had overrun the paddock to such a dangerous degree, a clean burn had become all important. Vegetation, such as fern and scrub, is never so dry as in late spring, when fresh fronds of bracken, new shoots of manuka, that damp the matted mass with sappy growth, have not appeared, when the rays of the sun have once again grown fierce. It was determined that the paddock should be burnt out in spring.

Partly owing to an extraordinary dry day in an abnormally dry spring, partly owing to the extra heat of many hundred acres of fallen scrub, the Rocky Staircase was swept as bare of green stuff as in the early 'eighties. There was this difference though, that the paddock then had been black; now it took its colour from the fire-swept manuka. In spite of the extra heat of the spring fire, wide areas of the paddock had been rather scorched and scalded than burnt. The harsh small leaves of the manuka had fallen, the bark hung in grey frayed tatters. The plant had so increased during the preceding six-year period that the general colour of the paddock was greyish, not black as in '82.

It cannot be maintained that Tutira generally has been helped by its weather; on the contrary, climatic conditions have been malignantly unkind. The summer of 1912 was an exception to the rule. Had it been wet, had even a fair proportion of rain fallen, huge areas of the block must have permanently reverted to manuka; instead, the summer proved to be a long series of terrific gales interspersed with half-inch showers. These rains, falling from time to time on the baking surface, temporarily made the ground a hotbed. Seeds germinated as if forced under glass. Renewed gales then blew from the hot nor'-west and scorched the tender cotyledons. Weed seeds, grass seeds, manuka seed, and suckling clover seed, that summer, shared a like fate. Each time Harry Young and myself rode through the paddock, we searched on hands and knees for the well-known and dreaded manuka seedlings. There were none to be seen; they were destroyed that summer by alternate warm rains and arid gales. That otherwise hundreds of millions of cotyledons must have germinated on every rood of the paddock we were assured of, for about the rims of damp spots on the hills, along the edges of the winding oozy creeks, they sprang up like grass on a wet seed-bag.

The fern, no longer a necessary ally, once more became an undesirable, and now for the first time Microelena stipoides and Danthonia semiannularis leaped on to the vacated stage. In descriptions of former periods I have been cautious to show that though these hardy grasses had been reduced to a fraction of their proper growth, and that although they were an inconspicuous factor in the herbage of the paddock, yet they had not been utterly destroyed.

Period after period in the progress of the paddock they had survived under cruel deprivations; now, stimulated by freedom to breathe, their recuperation was a marvel. On every ridge and spur cleared by the axe, appeared a broad band of native grass. In other localities where dead thickets of unfallen manuka stood stiffly impenetrable to stock, danthonia and microelena, guarded by dead lateral branches, rushed into being and seeded freely. What had appeared formerly to be moribund stools on the sides of paths and about pig-rootings, as if by magic multiplied themselves. The magic was but light and air; there had in truth been, at the termination of the fifth period, more native grass than had been reckoned; stunted, dwarfed, depauperated, throttled, only a few spindly blades showing from every crown, it had been passed over. During this first season of its triumph on the Rocky Staircase, I feel positive that no grass seedlings appeared. Conditions that had withered the manuka cotyledons had also destroyed all other germination; the sudden show of native grass was altogether from old plants rejuvenated by light and air.

After this first season of the sixth period these hardy natives continued rapidly to increase. The seed-stems of both species grow with uncommon rapidity, and attain maturity even in heavily-stocked paddocks. From the hill-tops and ridge-caps their seed was blown by the wind, poured down during summer thunder-showers in short-lived rivulets of grit and sand, or glued by wet to pebbles displaced by stock. Both species, moreover, when close cropped, possess the remarkable habit of sending forth culms perfectly flat to the earth. The caul of grass originally confined to the tops spreads each season like a mantle lower down the slopes.

The second winter after the fire the paddock was carrying 2500 sheep. It was not until the third season that manuka, reintroduced on the feet and wool of sheep, again began to show itself. By this time, however, all danger had permanently passed away. Time only now was requisite for the establishment of a turf, over which fires could be run every two or three years, fires that would scorch the low bracken fronds and short manuka. The Rocky Staircase had been grassed.

Of the several points to be noted in the annals of our paddock, one is the failure of some and the success of other aliens. Three times—other seed was in early times practically unprocurable—ryegrass, cock's-foot and white clover have been surface-sown: the first time with a certain temporary success; the second with less benefit to the paddock; the third with no satisfactory results whatsoever. In spite of three sowings, therefore, after thirty-five years' work, only the highly-manured sheep-camps grow a turf of English grass. On the other hand, chance comers such as suckling (*Trifolium dubium*), and in a less degree clustered clover (*Trifolium glomeratum*), and now, last of all, *Trifolium arvense*,—in books dealing with fodder-plants passed over or classed as worthless,—have each and all done yeoman service on the Rocky Staircase. All these plants have a certain future on this type of land. On the whole, however, the aliens so far have failed, the natives succeeded. To the latter, bar ploughing and manuring with fertilisers, a great proportion of the trough of the run will always belong. Each time a plant has overrun central Tutira it has been a native. Thus bracken has in my time reigned on the Rocky Staircase for twenty-five years, manuka twelve, danthonia two. Tutira plants have competed for Tutira soil; two species, Pteris aquilina and Coriaria ruscifolia, did hold the station; others, Danthonia semi-annularis, D. pilosa, and Microlœna stipoides, do hold it.

A second point worth noting in this progress towards pastoral utility is the spread downhill of new species. Manuka, Danthonia, Microlœna, Leucopogon Fraseri, Suckling, Clustered, Suffocated, and Harefoot clovers, have each and all first appeared on tops and ridges. There was the bracken soonest cropped and killed, there was the surface of the soil first open to sunlight.

An important factor, too, in this settlement of the run by new-comers has been consolidation of the ground. There were localities on Tutira where no plant life has appeared, ground so porous and spongy that horses used to sink dry-bogged in it to their girths. Save for a sprinkling of stunted blue grass, such spots were bare of vegetation. It was sponginess rather than poverty, nevertheless, which had caused these bare patches. Consolidated in later times by the tread of heavy horse-teams and rollers, these spots have proved not less good but better than the average land. Perhaps, therefore, of all changes, consolidation by trampling and treading of stock has been the most vital to the needs of plant life other than bracken. Perhaps there has occurred a physical alteration in the nature of the soil unrecognisable save by vegetative results, marked only by the appearance of grass; certain it is that the presence of native grass was in the 'eighties an unfailing mark of old *pa* sites and Maori cultivation-grounds; where the land had been trodden hard, Microlœna and Danthonia had been able to root themselves in firm ground. Perhaps this process of consolidation on a huge scale accounts for their triumph in later days on the Rocky Staircase, and indeed generally throughout the whole trough of the station.

To reiterate: our paddock originally was a thicket of bracken, intermingled with vast groves of tutu; consequent on the first crushing by sheep, the latter plant was utterly destroyed; at that date manuka was unknown except for a single small patch; later, when the vigour of the bracken had been quelled, native grasses appeared sparsely, then seemed to die out, whilst manuka (*Leptospermum scoparium*) mastered the bracken and overran the paddock. Native grasses had, however, been rather dominated than utterly destroyed; at the first chance they reasserted themselves, and have now taken possession of the paddock.

Notwithstanding the efforts of man, the Rocky Staircase has grassed itself in its own way, selecting and rejecting, and clothing itself finally with the fodder-plants best suited to its particular requirements.[1]

[1] The reader, however, must not cease the perusal of this chapter in the belief that the present plant-covering of the Rocky Staircase is to be its last. Danthonia semiannularis and Microlœna stipoides, natives as they are and fitted as they may be to the soils of the trough of the run, are less well adapted to them than an alien thrown by mere chance on to the shores of the Dominion. Chilian grass or Rat's-tail (*Sporobolus indicus*), according to the late Bishop Williams, "made its first appearance at the Bay of Islands in 1840, shortly after the arrival of a ship called the *Surabayo*, which, while on a voyage from Valparaiso to Sydney laden with horses and forage, put into the Bay of Islands in a disabled state, and was there condemned and her cargo sold." From the Bay of Islands the plant spread south to Auckland. There, whilst on a holiday in the 'nineties, it was noticed by Harry Young flourishing on light sandy lands. He gathered a palmful, and upon his return scattered it on the Staircase. Later, when assured of its value and suitability to the local environment, thousands of pounds weight were purchased and sown broadcast. The plant, though a wretched germinator and therefore slow in taking possession, is proving on light lands fully exposed to the sun of incalculable value. Nothing at any rate is more certain than that on northern and western hillsides Rat's-tail will completely oust other grasses, and indeed all other growths, except perhaps temporarily after fires Suckling clover and possibly Trifolium glomeratum and T. suffocatum; even these clovers, however, will be hard put to it in the company of this virile castaway. On southern and eastern slopes, however, Danthonia semiannularis, D. pilosa, and Microlœna stipoides will maintain themselves, though they too are sun-lovers. There will be found on these colder, damper aspects, besides weeds, some of no worth whatsoever and others affording an aromatic bite, several members of the clover family—grasses such as Sweet Vernal (*Anthroxanthum odoratum*), Fiorin (*Agrostis alba*), Fog (*Holcus lanatus*), Crested Dog's-tail (*Cynosurus cristatus*), Meadow Grass (*Poa pratensis*), and others of lesser value.

CHAPTER XX.

THE CHARTOGRAPHERS OF THE STATION.

If the principle of the martyrdom of man has held in regard to the pioneers of Tutira, twice over is it true of the Tutira flock, each generation of which has been decimated for the benefit of its successor.

The sheep of the station are, in sober truth, working out their own salvation. They are returfing the naked windblows, hardening the erstwhile dangerous fords, drying the bogs and marshes, building viaducts, shaping sleeping-shelves, exposing pitfalls and chasms. They are remodelling the run to suit their peculiar requirements. In this good work of reclamation other stock have participated. It is the sheep, however, that has borne the burden and heat of the day; it is owing to him that for his race the run is more easy to perambulate, more safe to traverse.

The first newcomer, however, to score a mark on the station was the pig. Swine, however, are but poor surveyors; they lack all sense of grading, climbing indifferently the steepest slopes, zigzagging in their ascent like man, in descent charging downhill like landslips. At the utmost, pig may perhaps have discovered to us some half-dozen narrow precipitous gorge crossings. Before the advent of sheep, and therefore before the establishment of sheep-camps growing grass and clover, there was nothing to tempt pig from the low grounds. There they lived and bred, trenching and terracing the hillsides in search of fern root, their staple food. Their runs, bored through the overarching scrub, were at intervals punctuated with wallowing pits—pig baths; not infrequently these runs lapsed into mere wedges in the soil, so narrow as to be scraped smooth by the sides of the animals using them. In the clearance of pig from the station I have travelled miles of these abominable tracks on all-fours. Pig tracks, in a word, have been useless in the opening up of the run.

Cattle have been equally unhelpful in the mapping of the station. They are creatures of the plain and wide river-bed, unsuited to country like Tutira, where the streams flow confined in narrow gorges. Cattle tracks, moreover, usually lead to trouble, the hoof formation of the great beasts enabling them to negotiate ground where neither sheep nor mounted shepherd dare follow.

As animals gone wild, horses have left no trace. Driven in pack-teams, they have done work that will be described later.

It is sheep that have surveyed Tutira. In the early days they worked the tops and upper slopes. Later, owing to the destruction of fern, tutu, and koromiko, it became possible for them to tread a middle course; at length they were able to circle the bases of the hills.

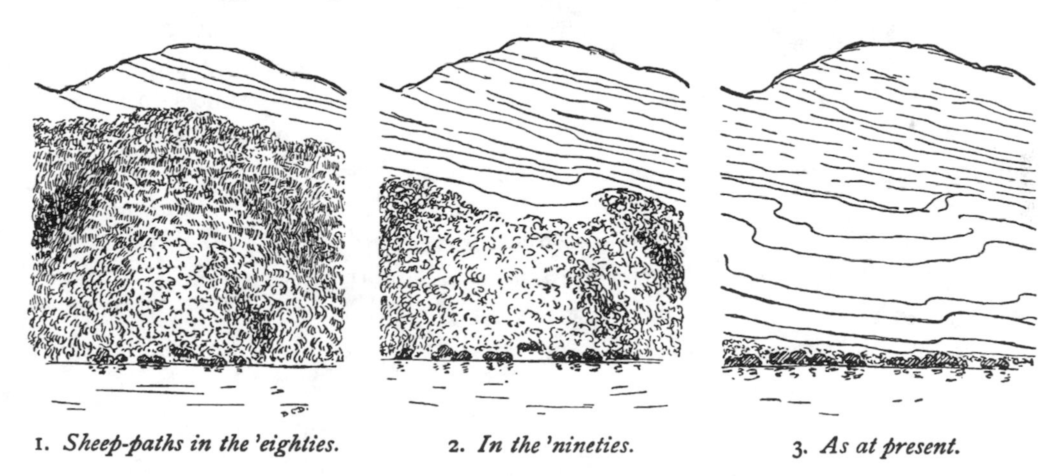

1. *Sheep-paths in the 'eighties.* 2. *In the 'nineties.* 3. *As at present.*

Following in the wake of his charge the shepherd's path too has declined in altitude. In the late 'seventies advantage was taken of the hill-top tracks. In the late 'eighties those of the higher sidlings were utilised. In the late 'nineties we followed sheep-paths along the low slopes. Every shepherd's beat on the run is a sheep-path broadened and trodden out. Now, after forty years, the latter number hundreds of thousands. From every camp they radiate like roads from a city. They are, in fact, roads from a city, for to sheep their camping-ground is as his town to man, at once a refuge and a resting-place.

There are on every station two types of path—the one, the line of morning dispersal and evening reunion, beautifully graded; the other, much more steep, called into being by the instinctive desire of frightened sheep immediately to climb to the tops. What may be termed the normal arterial system of each centre, of each sheep-camp, has also in modern times been affected by such arbitrary barriers as fencing-lines. By them sheep are forced to climb when they would prefer to wind, or in shepherd's phrase, to string to their camps on comfortable grades.

There are still on the largest remaining paddocks examples of these graded narrow paths, but the best, alas! have been ruined by the abominable utilitarian necessity for subdivision of land.

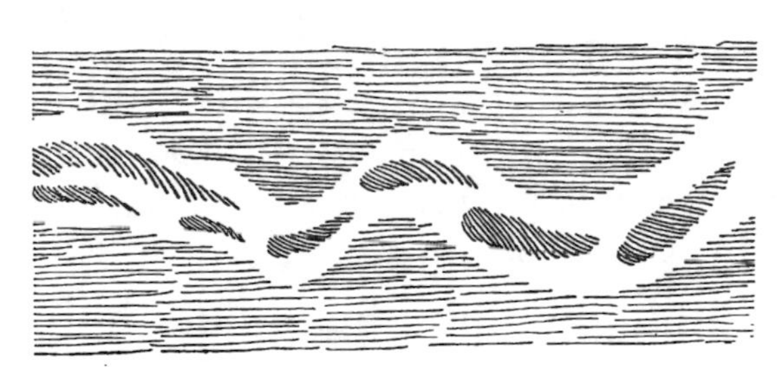

Showing single current.

Another kind of road is that made by driven sheep. Continual passage in one direction of any mass, animate or inanimate, stock, water, or blown sand, creates channels alike in their broad general features. As in the case of water, a living stream of sheep moulds and adapts itself to the lie of the land. Flowing through fern and scrub, the torrent is turned aside by the main obstacles—it passes over the less; a way is eaten, not through earth and rock, but through vegetation. Viewed from above and afar, a great travelling drove will, on the levels,

Stock route, showing double current.

break into countless shallow, rapid, irregular channels; it will pour itself in masses through choked defiles. On open land it will move slowly forward with broad blunt head, over declivities it will waterfall in cascades, over steep rocks drip drop by drop.

On closer inspection, too, the channel of a sheep-stream will show the typical scour of water action. There will be found just such islands and aits as separate the channels of a water-stream. On trails where, as sometimes happens, mobs are always driven out in one direction and always return by another route, these aits and islands of bracken permanently maintain a somewhat deltoid shape. On the main stock-route, along which an ovine current flows like a tide both toward the wool-shed and away from it, deltoids become lanceolate or sharply ovoid, like the raised "refuges" of city thoroughfares.

The final stage of a stock-route might well puzzle any man who has not followed each step in its strange metamorphosis. Nothing would appear more unlikely than that a drove-road through bracken should develop sometimes into a single sinuous line of shrubbery, sometimes

Sheep driven through dense fern.

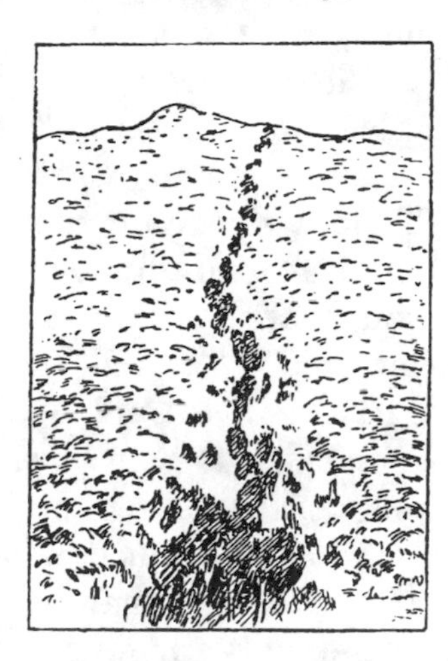

Single hedge line.

into a grassed space bounded by high hedges on either side. It has nevertheless happened,—single or double serpentine ribbons of scrub have moulded themselves to the shape and sinuosities of stock-tracks as the roots of a parasite grow to the form of their host.

The stages through which every paddock of Tutira has passed in its alteration from tutu and fern to grass are known to the reader. He will recollect that bracken and scrub, in early days, grew nearly everywhere tall and rank enough to shut out sunlight from the soil; that in the 'nineties the effect of stocking began to show itself in their weakened growth; he will remember, too, that from time to time the country used to be swept bare by extensive fires. Keeping these facts in mind, we shall be able to trace the change from an open track into a high single or double hedge. Like other minor phenomena of the station, a combination of special conditions has been essential to their creation. For the development of the former there has been particularly required a steep slope in soft soil, a more or less straight ascent and a heavy rainfall. The single hedge, moreover, belongs to very early times, when the flock was small, when vegetation—especially of the trough of the run—was practically unaffected by the few sheep carried, when after a fire, fern and tutu again choked the countryside. In those days, stock driven from one part of the run to another had to be jammed into this growth, a passage forced by the aggregate weight of the mob, as into an almost solid substance. There was but little spread in the movement of the driven animals,—the narrow spear-head of trampled bracken was flattened as if rolled by machinery; with repetition the trodden vegetation was destroyed; later again, in the centre of the wedge, a depression became worn by traffic. By the action of rain-storms and thunder-showers the little depression was gutted into a wide bare rut; on the sides of this rut manuka seed lodged and germinated. That is the first phase.

We have now to suppose that for some reason or another the route was abandoned. Perhaps a more convenient alternative track had been discovered, perhaps sheep were being exclusively used for breaking in another part of the run. Whatever the reason may have been, the line falls into disuse; the manuka which had germinated on the edges of the rut, the only open ground in the vicinity, grows undisturbed into tall plants.

The next factor we have to consider is fire. A little, an extremely little difference in surroundings will affect flames not running over thick growth; moisture emanating from a single fleecy cloud, green growth of plants that spring up alongside paths little used, sorrel, clover, capeweed, will damp down and extinguish them. Checked on either side, fires to a great degree die down in the vicinity of stock-routes; sections, at any rate, of such paths remain unburnt—sections so continuous that sometimes an old drove-track can be picked up by its tall manuka at a distance of miles, a line of hedge marking exactly where years before the wedge of stock had been driven into dense bracken. A particularly well-accentuated example of this single hedge development used to extend—ploughshare and axe have done their work—from the Maheawha crossing of the Tutira stream to the summit of the Image Hill.

The double hedge-line belongs to a later period, its growth synchronising with the increase of the flock. By the 'nineties, the numbers of sheep and lambs shorn on the station had trebled. In addition, ten, twelve, and fifteen thousand sheep used yearly to be borrowed for crushing purposes during the spring months. Huge hungry mobs driven in and out from the wool-shed during the shearing season passed along the stock-routes. About their centres—the current of a drove of animals is like that of a watercourse, strongest in mid-stream—vegetation was worn away, crushed and destroyed. On the flanks and wings, attrition, though less marked, was yet sufficiently strong perceptibly to check growth. The consequence was that on the edges of each stock-track seeds hitherto unable to germinate for want of light, plants hitherto unable to breathe for want of air, took possession, succulent green stuff such as white clover, suckling, cape-weed, and sorrel seized on the open soil.

Double hedge line.

Fire still ran over great areas from time to time, though not with the same sweeping violence as of yore. The bracken had become stunted and sparse, there was less of it everywhere to carry a fire. The heat, therefore, and height of a conflagration already weakened, was still further diminished as it reached the wings of the stock-track. As in the miracle of Gideon's fleece, what happened on the neighbouring lands did not happen on the stock-routes,—they remained green when the ground alongside was black.

The spread of Leptospermum scoparium — manuka — has been described; it appeared during the 'nineties on every scrap of land open to the sun, and germinated thickly over the breadth of the stock-routes. In the middle of each, however, where the current was most violent, its delicate seedlings were trampled and trodden into the ground, and were thus destroyed. On the other hand, plants on each of the edges or wings of the track survived and grew into tall shrubs.

There came a time at last when travelling mobs, driven to and fro over the run, moved between hedges moulded exactly on the windings and meanderings of the great stock-routes. It must not, however, be supposed that these hedges were anywhere continuous for more than a chain or two at a stretch. Fires, though less frequent and less sweeping, did sometimes manage to reach one wall of hedge, even on occasion to cross the trail, destroying the tall growth on either side. The main stock-route of Tutira could nevertheless at one time be traced for miles by hedges of manuka and by great individual bushes, one of which has for long gone by the name of Harry Young's shaving-brush. They are survivors of heavier fires that here and there had managed to cross the hedge-lines.

Conditions essentially similar, though less interesting because less naturally reached, tend to produce scrub-hedges along the modern highway bisecting the run. As its width of twenty-two yards is fenced on either side, there is no room for mobs to spread. On either side of the crown of the road, therefore, bracken has been completely worn away and manuka taken its place. Were it not for the requirements of the local roadmen, who use the scrub for repairs, the Napier-Wairoa road would throughout the length of Tutira pass between solid hedges of manuka.

The initial stages of these single and double hedge-tracks have been described as primarily worn in the thick vegetation of olden days during a period when the range of sheep was circumscribed by surrounding fern and scrub. On the crests of certain spurs on eastern Tutira down which sheep similarly confined by bracken-growth used to troop, where manuka never grew and where the surface took grass readily, there are tracks of another sort deeply fretted into the soil itself. During forty seasons each twist and turn has been deepened by the action of rain. They stand out as relief work and promise to last for an indefinite period, too narrow and deep for present traffic, but scooped afresh each year by flood-water.

Nearly all shepherds' riding-tracks have begun life as sheep-paths; only an insignificant minority have been deliberately made—mostly cut by myself on wet Sunday afternoons when it had become impossible to refrain longer from exercise. During the early years of the run, as the

Harry Young's Shaving-Brush.

A survivor of one of the scrub hedges that mark the line of old stock routes.

reader knows, sheep-paths perforce followed the tops. At a later date the sheep began to pare off the steeper rises on their roads of ingress and egress; better grades became possible as the bracken-growth retreated downhill. It was an amelioration, however, less rapidly accomplished than might have seemed likely. Sheep dislike sinking their feet into loose fibrous ground even when perfectly dry; one of their most deeply-rooted instincts is to tread firm land. Long, therefore, after the vegetation itself admitted of a rectification, after the fern was worn down, the sheep still preferred to climb to a higher elevation on ground hardened by himself than progress at a better grade on unindurated land; the former was still the line of less resistance. Seasons passed before the first faint tracery of a mere feeding deviation became an established track. Shepherds then, following the sheeps' lead, would begin to experiment on the seemingly sound track. Often at first, though shorter and better graded, it would prove a snare; often at first it would be rejected on similar grounds to those upon which it had been formerly condemned by sheep; hard enough now to endure their weight, the hoofs of horses still sunk deeply into the soil. At last in dry weather it was rideable—at any rate could be floundered through. In wet weather it still bogged the horses to their knees; many were the hasty returns to the harder, steeper path, many the hurried dismountings before it became in all weathers fit.

Sheep-track fretted into hillside.

Horse-tracks moulded on sheep-paths, when once stamped hard, do not readily alter their curves,—once a bend always a bend is the general rule. Sinuosities do, however, change with change of pace in the animals using the track. Two such instances occur to me: one between the Conical Hill and Caccia's Crossing, the other on a line roughly parallel with the present coach-road. Season by season I have seen their twistings straighten, their bendings disappear, in the same manner and for the same reason as have those of the curves and corners of the great main road between Napier and Wairoa. Their alteration suggests in how large a degree every accomplishment of man is but a development of subconscious action.

The two paths instanced happened to pass over ground where for two or three miles at a stretch no natural declivities or obstacles occur; where, in fact, the walk, the normal pace of mounted shepherds, could be quickened into the trot or canter. Horses ridden slowly follow sharp sinuosities without trampling the salients of each corner; trotting,

Path formed by horses walking, trotting, cantering.

these salients are impinged upon; at a canter they are trodden out. Horses ridden at racing speed would, I believe, in time rule out paths almost perfectly straight. In this elimination of curves, what the shepherd's horse has accomplished automatically and subconsciously, has been in later days ordained deliberately and of conscious purpose by the Hawke's Bay County Council. Now that a road has been installed, now that the advent of cars make it possible to travel at greater speed, sharp corners are also being spaded off the Napier-Wairoa

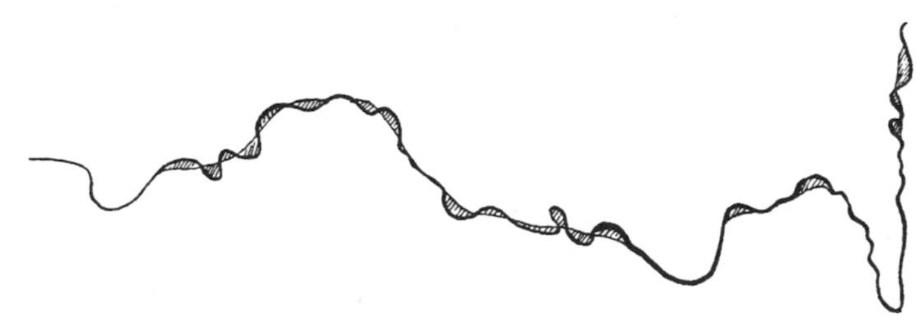

Napier-Wairoa road, showing curves straightened by H. B. C. C.

road. Man, who develops every hint provided in nature, has done on a grand scale what the shepherd's horse had accomplished without care and without forethought.

We have seen what has been done on the uncharted void of early Tutira by pig, cattle, sheep, and mounted shepherds. The pack-team's work, too, is written large on the surface of the run. It also, in olden days before the advent of a dray-road, played an important part in station activities. All material then was carried on horses' backs, wool deported coastwards, stores brought inland, fencing material and grass seed carried to the remotest corners of the run. Often the pack-horse trail was the development of the shepherd's riding-track, a farther stage of the sheep-path upon which both were based. Sometimes, however, it has happened that owing to the exigencies of station work a pack-trail has been suddenly and arbitrarily imprinted on virgin areas, an untrodden block has been invaded by a string of eighteen or twenty horses, the animals following in a general way the line of the leading packman, but settling details of the route each to its own satisfaction. At first, therefore, there is no single well-defined track; a multiplicity

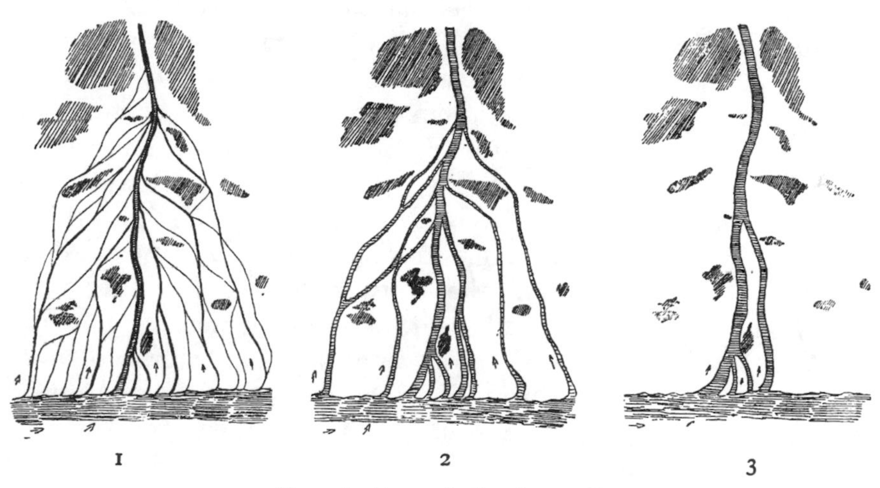

Horse-trails competing for traffic.

1. First day in use. 2. End of month. 3. End of year.

of temporary paths are set up, each of them, as if alive, appearing to be competing with the others for the new traffic. At first they are faint and ill-defined, the uneven ground merely brushed and bruised with hoofs. Each is exactly the width of a horse's body from the other, for when switched off an established track the units of the team jostle and jam together. Afterwards the character of the country determines the permanent nature of the track,—in open flattish land parallel paths are for long periods about equally patronised, in regions of high fern and tangled tutu they tend to converge; no horse wishing to waste energy in causelessly tearing through entanglements, the guiding line of the leading packman is thankfully followed. On dry spongy ground, too, the choice is quickly relegated to the lines which have most rapidly consolidated. In the vicinity of bogs, on the other hand, line after line is discarded as it becomes poached into a quagmire, dozens of tracks converging right and left.

The origins of the curves and windings of the original pack-trails of Tutira are known now to very few, some indeed only to myself, the surviving prehistoric packman of the station. I recollect the main obstacles just as some hind must, ages ago, have marked the trivial difficulties that account for the meanderings of an English footpath betwixt village and village. These, however, if consciously noted at all, have never been told. On Tutira such trails have been watched with interest from the beginning; their origins are now immortalised in ink. Where this sharp salient survives, flourished at one time three great tutu shrubs, whose projecting branches interfered with the pack-team's loads. I remember a fellow-packman felling them furious at the delay caused. This bend on the trail avoided a grove of manuka. I remember it tall and green. At this pronounced curve once lay the carcase of a horse, left where the poor beast had been dropped by natives on some hunting expedition. I have sniffed the reek of the beast, and recollect how the team day after day shied off to windward. This elbow marked the spot where at one time lay an immense totara log, afterwards sawn into strainers. The causes of the curves disappear, the stumps and roots of the offending tutu trees decay against which thirty years ago the loads of our pack-team used to strike. Fire passes through the manuka grove, its scorched poles fall to the ground; the stench of the dead horse passes away, its bones are scattered far and wide by pig; the great totara bole stiffens a fencing-line or supports a gate, yet still the curves themselves remain.

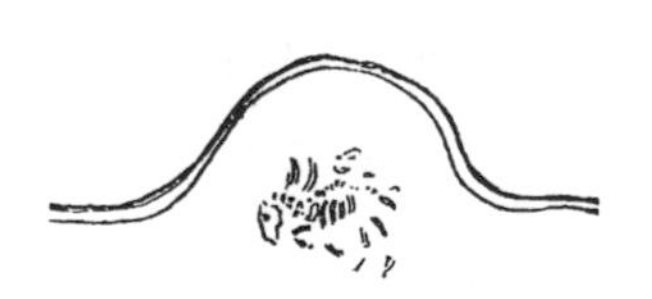

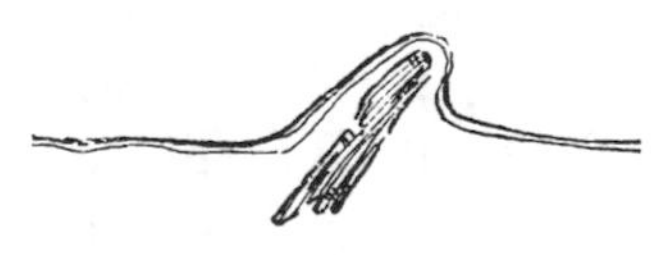

The lesser sinuosities of a pack-trail can only be generally accounted for. They result from a host of temporary insignificant local difficulties — little hollows and dips, dead brushwood cumbering the ground, projecting vegetation, loose spars of surface timber, spongy land, even thickets of thistles. Like the major impediments cited, they too pass away and are forgotten. Attrition by frost and wind wears down the little hillocks, rain fills the hollows with soil, the dead brushwood rots, its mould is blown abroad, strips of projecting vegetation are destroyed by stock, the surface timber is burnt, the soft ground hardens, with autumn rains the thistle stems fail. Each of these first causes, seemingly ephemeral as the reek from the dead pack-horse or the smoke from the scrub and thistle fires, is nevertheless still marked in the material world.

Another type of track created by the pack-team is worthy of note. Where horses follow one another horizontally in single file along a slope of clay hillside, a ditch-and-bank or ridge-and-furrow process is produced. Each animal taking the same length of stride plants his foot down where or whereabouts his leader has trodden: the consequence is an alternation of narrow bog-holes the size of a horse's hoof, apart

Pack-track on clay hillside.

from one another the width of a horse's stride, with ridges hard because untouched, but slippery with drippings of the coffee-coloured liquid spurted during traffic from the churned troughs. No track can give a worse fall to a shod horse. If the shoes of a front and hind hoof become locked in one of these greasy pockets, the animal is pitched sideways downhill, and must roll over without chance of recovery.

Besides the general opening up of the run by means of trails and tracks, the stocking of Tutira has produced phenomena which, though of minor importance, have been—and after all this is his book—of interest to the writer. One of these has been the metamorphosis of many of the hill-tops, a change which it is easy to imagine might in future confound and confuse the natural philosopher. The soils of Tutira are familiar to the reader, the uppermost layer, humus resting on a layer of pumice grit, this grit resting in its turn on a deposit of packed red sand. The original vegetation of the run has also been a dozen times described—fern rank and luxurious on the cold east and south aspects, less exuberant on the warmer, drier north and west hill-slopes, short and sparse on the tops.

The earliest sheep carried were merino, an easily scared breed, which upon the least alarm sought refuge on the heights. Roaming on the tops, they nipped off and trampled out the meagre covering of fern, their sharp hoofs broke through the dusty humus and pumice grit, allowing arid summer gales from the nor'-west to breach the hill-brow and blow away both humus and grit, leaving an absolutely naked surface of tightly-packed, slightly greasy, smooth red sand. In the 'eighties many prominent hill-tops were in this way blown bald and bare, such wind-blows being especially well-marked on the Rocky Staircase tops, on the crown of the Natural Paddock, on the Image Hill, on the Racecourse top, on Table Mountain, on the Second Range, on the Burnt Blanket. Each of these hill summits, where there had been originally a light loose covering of soil nourishing a sparse crop of

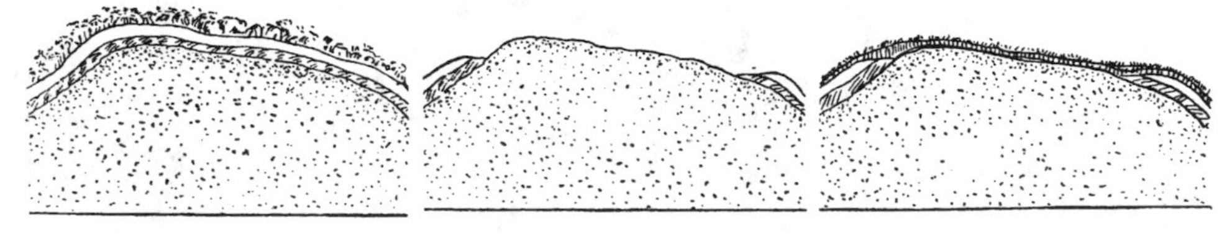

1. *Hill-top growing fern.* 2. *Hill-top blown bare.* 3. *Hill-top in grass.*

bracken, became, in the second place, a barren tract of red sand; in the third a deep-green luxuriant carpet of turf. Sheep which had caused the scar had also contrived the cure. Camping at night on the highest ground within reasonable reach, they gradually enriched the tops by their manure. On these bare naked wind-blows sheep lay, their numbers increasing as the run progressed. Their droppings and urine were washed by rain, or blown by wind, towards the edges of the scar. Grasses of creeping habit, especially Poa pratensis, certain members of the clover family, and certain weeds, Cotula asiatica, Geranium sessiliflorum, Oxalis corniculata, and others, crept over the bare space as the bark of a wounded bough envelopes the scar. The slightly greasy nature of the tightly-packed deposit of smooth red sand prevented direct absorption of the sheep manure; there was only, therefore, encroachment from the edges; inch by inch the turf crept upward until the wind-blow was completely carpeted, until the bare red sand had been transmuted to a glorious green. This change from indigenous vegetation, rooted in grit and humus, to absolute nakedness — from nakedness to a dense turf of alien fodder-plants — could only have occurred in the early years of the run, during a period when the station was but slenderly stocked. It is another example of what has been already noted, that many of the small phenomena registered in this volume have been possible only by a combination of many special factors, some of them, moreover, of brief duration. Hill-tops, for instance, heavily stocked, would have become rapidly enriched, the seeds of clover and grasses dropped in the animals' manure would have immediately germinated, and a matted turf been created within a few weeks that would have resisted alike the trampling of stock and the wear of summer gales.

Sheep viaduct.

He, therefore, who may in the future interest himself in natural phenomena, "my second self when I am gone," will know that although all tops may then be of a similarly luxuriant green, yet those sans pumice grit and humus have at one period been bare, naked windblows.

Another physical change on the surface of Tutira has been the growth of ovine viaducts. They also are dependent upon a multiplicity of special conditions. The ancient plateau system of the run, the varying elevation of its existing fragments, its attrition by earthslips and rain-storms, have been explained. In the course of time sections from which the cap of limestone has slid away have become mere ridges linking together narrow blocks of higher land. There are, moreover, here and there throughout the whole of Tutira, those narrow connections running east and west which have been called junctioning spurs. Topping their undulatory crests, sheep pass to and fro from their feeding-grounds; a well-marked path climbs to the rise and dips to the descent. It is at the bases of these loops that viaducts have been built, each year increasing their height, as if the little builders were deliberately taking thought of the morrow, scheming to save themselves toil. The tools of the sheep are his toes, his sharp hoofs act as gouges and chisels in the work; rain, sun, and wind, carrying down silt and dust from the heights on either side, supply building material; the centre of the path constantly scooped out is as constantly refilled. Though infinitely slow, there is no cessation in the raising of the little embankment; during storms the soft silt or liquid mud is squeezed out on either side; during droughts the trodden dust is hoofed to right or left. Whatever the weather may chance to be, fair or rain,

Sleeping-shelves.

dry or wet, the even top-dressing of the embankment by dust or liquid mud proceeds unceasingly. Its sides never slip or gap—they are bound together by a dense mat of Poa pratensis root. Of minute physical surface-changes, none have been more entertaining to watch than these viaducts. Some of them I have seen increase, inch by inch, until after forty years their height has risen to a yard and a yard and a half. The finished article, its close-nibbled verdant banks fed with rich dust and silt, is a beautiful bit of animal architecture.

Perhaps, however, of all surface modifications consequent on the stocking of land, the most curious is the formation of sleeping-shelves, ledges built by sheep themselves for their own convenience. Every day sheep from every camp on the run spread to feed—every evening they return to sleep. Their instinctive desire is at night to lie on a summit;

like other creatures, however, the sheep has to compromise with his ideal; economy of physical labour forbids too long a daily climb to camp, too great a daily descent to graze. The primitive instinct, therefore, that safety can only be attained on the highest possible top, becomes modified by custom. In practice, at any rate, the sheep does not always sleep on the summits or crowns of hills. On extensive stretches where there are no available natural camping sites, where hill-tops are distant, or where cliff formation makes their attainment difficult, camps are formed on the slopes. It is in such localities that sleeping-shelves have grown out from the hillsides, like the lip ornaments of the women of certain African tribes, or as fungus projects from dead timber. They are built with the same enormous patience as are the viaducts—decades going towards the construction of a perfect ledge. Each of them represents the labour of generations of sheep—the thousandfold repetition of a natural habit. Before a sheep lies down, his custom is to turn round twice or thrice like a dog. He then rests with his feet beneath his body. On an even slope, such as we have imagined, his slipping downhill is only prevented by the resistance of hoof and knee. His weight presses the turf upon which he lies downwards in an immeasurably minute degree, and inwards also to an extent almost equally intangible. For long these two opposite pressures, inward and downward, are the only factors that count. Later, however, the mere bulge in the hillside becomes the incipient shelf. Its projection begins to arrest the almost infinitesimal amount of water-borne silt that percolates through the grass-blades during wet weather. As the sheep, turning and scraping, settles himself for the night, in dry weather a small quantity of dust is likewise shuffled towards the outer edge of the shelf. At a more advanced stage—a sheep does not always rise to relieve himself—his droppings begin, instead of rolling down the slope, to rest on the lip of the now rapidly-growing ledge. In summer they blend with the dust of his nightly circlings and preliminaries of rest. With slight rains and dews they are trodden into a compost that nourishes the grass-roots. The sleeping-ledge in time becomes perfect, thrusting itself out at right angles to the slope, like a swallow's nest gummed to a wall. The tendency of the most highly-finished of these sleeping-shelves is to become in their last stage very slightly concave, their edges upheld chiefly by a mass of meadow-poa roots. They must then, on a well-drained slope, be most dry and comfortable couches.

Throughout the pumiceous area, where the soil is gritty and friable and where erosion is easy, there can be noted in these shelves a certain troglodytic tendency, their backs slightly concave, or, at any rate, perfectly upright and bare of grass. On stiff lands, in fact, the shelves tend to work outwards, on friable soils to work inwards.

Earth-bubble.

Another minor physical change, also attributable to stock, is the earth-bubble. These surface swellings are the result of a blocked soakage system; they commonly appear on the higher side of tracks, where the upper crust has been hardened by traffic, where free escape of water during heavy storms has been impeded. The subterranean creeks, or under-runners, on the upper side become gorged with water that cannot immediately drain away; reaching ground beneath which it cannot pass—ground toughened and kneaded into some sort of consistency by the tread of stock and comminglement of mud and manure,—it raises soft tumours which, when lanced by a sharp stick, eject con-

During flood.

siderable jets of muddy water. These minute phenomena are evanescent and soon subside. Though of considerable depth, the biggest do not exceed a couple of feet in diameter and a few inches in height; of the many surprises awaiting shepherds after heavy rains, not the least sudden is a blunder into an unnoticed earth-bubble.

Other curious though unimportant physical creations are mud-banks, built on the margin of the lake by passing floods. The waters of the lake, feet above the normal, blown violently from the direction of the gale, are piled up in big waves against the mass of flood-water pouring off the valleys. At the junction of these contending forces, a narrow width of calm, or at any rate currentless, water is produced. Beneath its line of quiescence mud and silt are quickly and copiously precipitated, until, with cessation of the storm, and a rapid drop in the level of the lake, a submerged mass or mound of mud is revealed, which, becoming grass-bound and solid,

After flood.

may exist for years, a long hog-backed monument to some mighty rainstorm.

Not one of these surface changes, directly or indirectly brought about by stock, can be considered other than insignificant; yet their aggregate has sufficed to alter the surface of Tutira in an almost incredible degree. Although, maybe, that change has been of greater interest to the writer than to his readers, at any rate it will have enabled them to realise the cumulative result of trivialities.

CHAPTER XXI.

STOCKING AND SCOUR.

THE difference between Tutira of '82 and Tutira of 1920 is the difference between youth and age: the face of the one smooth, that of the other wrinkled and lined. In the early days of the station its surface was unmarked by paths; now it is seamed with tracks. Before the arrival of the European with his domesticated breeds of animals, save for a few Maori footpaths the station was an untrodden wild: it was without path or track—in the language of Scripture, void; its surface is now a network of lines; it is reticulated, like the rind of a Cantaloupe melon.

In a previous chapter surface alterations of a minor kind, consequent on the stocking of land, have been considered; in this we can explain briefly certain larger effects. A single sentence—one is sufficient—will make clear to the reader what has occurred: the countryside has been transformed from a sponge to a slate. In this vast change the sheep, modifying the run with subconscious care to his peculiar requirements, has been the prime artificer. Nor, moreover, are these operations local; everywhere the flocks of the colony are transforming it with teeth and toes, crumbling it towards the sea. To visualise the magnitude of the general effect, the reader has but to compare the size of Tutira with that of New Zealand, the numbers of the Tutira flock with the number of sheep in the Dominion. He can then in part picture the alterations consequent on the importation of stock. This, however, is the story, not of New Zealand, but of Tutira; except to follow to the sea one of the rivers that rises in its hinterland—otherwise I could not illustrate ultimate results of scour—I shall confine myself as always to facts from the run. A previous chapter dealing with fern-crushing has shown in detail the destruction of the ancient vegetation of the run by sheep. The reader will remember how, during the first year of heavy stocking, the vast thickets of tutu which used to grow on the southern and eastern slopes were annihilated, how the bracken lost firstly its exuberance of growth, then became stunted in height, and ultimately in many parts perished. With the annihilation of these tall growths and others that succeeded them, the ground was no longer shaded from sun, no longer sheltered from the elements. Its naked surface was in summer-time burnt into dry dust, in winter beaten upon by torrents of rain.

The slipping of the marl hills on eastern and the subcutaneous erosion of central and western Tutira have also been described. These processes yet continue, and will continue; but in addition there has been established by the tread of sheep a new kind of wear and tear. The surface of the central station, once as absorbent as a sponge, now supports tens of thousands of sheep-paths, each of them acting as a shallow open drain. They are, moreover, so toughened and puddled with pulverised sheep-droppings that even after the heaviest of deluges they continue to rush off flood-water. I have seen them washed so perfectly clean of dirt and dust that the blanched, flaccid, fibrous roots of grasses and weeds showed up white like thread. Down these miniature gutters, grit, dust, and gravel are carried wholesale. Except on the flattest portions of central Tutira, the bed of every creek, rivulet, and rill has been metalled by the shingles and sands crumbled into it by sheeps' feet, or blown and washed down by wind and rain. Creek crossings, at one time barely capable of supporting a man on horseback, can now be negotiated by a loaded dray. Immensely increased quantities of material have found their way to the main streams too. The fluctuations of the streams themselves are more marked: there is a higher rise in flood, a lower fall in drought. Instead of a permanent percolation from the whole body of the countryside, there is a violent brief surface scour. Even locally the results of stocking have not yet been fully recognised. Loss of bridges and culverts, recurring flood after flood, cannot be due to miscalculation of catchment area or to lack of information in regard to rainfall; it is owing to insufficient allowance for the hardening of the countryside, for the enormous multiplication of open sheep-paths that rush off the surface water.[1]

Descending now to lower levels, the effects of stocking and scour are equally noticeable: the estuary of the Waikoau will serve as an example. Down this little river, in the 'sixties, my neighbour—the late John Mackinnon of Arapawanui—conveyed his clip from wool-shed to steamer in the offing—a distance of a mile and a half. It was carried down-stream by means of a punt capable of holding several tons. The Waikoau flowed then serene and smooth between

Estuary of Waikoau—past.

Estuary of Waikoau—present.

banks of exuberant greenery, growths top-dressed with liquefied leaf-mould and highly comminuted marl mud. Of old, however high the flood, except in the open course of the stream, its overlapping waters

[1] Explain it as philosophers may, the country settler soon comes to plume himself on any adverse peculiarity in his environment. To belittle it is to belittle a trouble which in his heart of hearts he believes only he himself is capable of enduring. A landholder in Hawke's Bay, through whose property flows a river, has no need to search for trouble ; there it is at his door, so much in mind that it becomes a part of himself until by some strange perverse mental process he becomes proud of its unruly ways, pleased when from time to time a wandering weed inspector or trades union official falls a victim, or a bridge is carried away. No aspersion is more readily resented than one cast on the dangerous depths of a ford, or the flooding powers of a river. After completion of the first bridge over the Waikoau, the work was viewed by a pair who knew Tutira and its weather ways. "How long do you give it?" says one. "The first decent flood," replies the other. He was not, however, perfectly correct. Only the earthwork on both banks was washed away, the bridge itself remaining an island, intact. The river's hint, however, was taken and a new span added ; yet the whole structure was wrecked in 1917. Good old Waikoau !

were calmed and tranquillised by resistance of stems, leaves, and blades which caught the alluvium, raised higher the banks, and further stimulated the rank jungle on either side of the waterway. The river in its lower reaches ran then like a canal, curving but little, and passing slow and deep between dense containing walls of luxuriant—almost tropical—vegetation. These conditions have been revolutionised by stocking and scour. Stock have destroyed the growth of the old banks: the accumulated silt of centuries, no longer bound by matted root-growth and protected from violent currents, has been carried oceanwards wholesale in the enormously larger floods of modern times. With edges stripped of their plexus of roots, with current no longer confined, the Waikoau changes its course in every flood; a score of wasteful channels trickle over a wide, stony, shallow bed. Nowhere now in the mile and a half between wool-shed and sea could wool be taken by boat or barge for fifty yards.

The cumulative effect of the work of sheep is in truth nowhere more apparent than on the alluvial lands of the river-mouths. They suffer a twofold deterioration—positive and negative; firstly, deposition of grosser grit and coarser sand—stuff in former times trapped by the riverside vegetation of the upper reaches now destroyed by fires and stock; secondly, loss of the finest forest mould held in suspension during flood. It is no longer allowed to settle in comparative calm amongst rank vegetation standing knee- or neck-deep in flood water. It is wasted now—carried direct to the ocean. In addition to cessation of income from the interior, there is an extravagant expenditure of capital banked, of alluvium deposited in former centuries. Nor does the harm done by scour cease even then; cross-winds blowing on the wider river-mouth raise wavelets of considerable size, whose lapping still further devours the banks.

Whatever may be the fate of large alluvial areas, smaller valleys run serious risks from breaching of ancient banks and from superposition of valueless sand, grit, and rubble.

We have now to consider, not indeed a minor, but a much less conspicuous aspect of stocking and scour. It is the permanent hardening of the crust of the ground and its effects on grasses. The surface, no longer subject to any natural process of mulching, has become less friable, the pounding and stamping of stock has affected the quality of the turf. There is no question but that the best fattening grasses are disappearing, or have disappeared. Though it is true that the constituents necessary for rye and clover are, after a couple of seasons, exhausted on the hungry pumiceous lands of which the trough of the run is largely composed, such is not the case in regard to the best marl surfaces of parts of eastern Tutira. Far less does it hold of the magnificent soils of Poverty Bay and of great parts of southern Hawke's Bay. In those districts, certainly, the gradual displacement of ryegrass and clover cannot be ascribed to exhaustion of the land. It is due to changes of the surface whereby certain natives are benefited at the expense of their alien rivals.

On an iron surface, however rich, germination is less easy; a sufficiency of moisture, moreover, in drought is unobtainable. I have described elsewhere how the sites of old Maori workings were in early days marked by grasses such as Microelena stipoides and Danthonia semiannularis. They grew where the surface had been stamped hard by man; now they grow where the surface has been pounded and trodden by sheep—that is, over nine-tenths of the province. A general deterioration in the turf has begun, entailing in its turn readjustment of the type of animals bred thereon. I do not say this is the sole reason responsible for the general change throughout Hawke's Bay from the Lincoln to the hardier Romney Marsh sheep; undoubtedly it is one, I believe the chief, reason. Fodder-plants such as rye, white clover, even cock's-foot, die out or flourish with less exuberance, inferior aliens and comparatively valueless natives taking their place; the flockmaster, adapting himself to the changed environment, breeds a hardier race of sheep.

It would be easy to stretch the links of cause and effect: the hills become like stone; the settler growls as, tipping his correspondence from mail-bag on to verandah floor, he opens an epistle demanding an increase in rates owing to the destruction of bridges. Stock trample hard a countryside 12,000 miles from the great cities of Europe; carpets are softer to the tread,—the coarser Lincoln fleece has been supplanted by the finer wool of the Romney Marsh.

CHAPTER XXII.

THE FUTURE OF NATIVE AVIFAUNA.

PERUSING this chapter, instances of British species which have struggled in vain, or are struggling against untoward environment, will doubtless suggest themselves to the reader. As in ancient England, so in New Zealand, so on Tutira, the axe, the fire-stick, the spade, the inroads of domesticated stock, have each of them played a part in the grand transformation scene. What has happened, or is about to happen, to the wild creatures of New Zealand, is in fact but a re-enactment of what has occurred to the fauna and avifauna of civilised Britain. The bear, the wolf, the beaver have disappeared; the places of the great auk and the bustard know them no more. The substitution of the olive and fig for the thorn and thistle has left no room in either country for animals regarded as undesirable, or breeds unable to fend for themselves.

Were Tutira an ordinary run, which like large tracts of southern Hawke's Bay could be flattened into a mere roll of turf, a mere carpet of grass, half a dozen, perhaps, of its bird species might survive. It is not; on its surface will probably continue to exist a greater number of species than on any other run in New Zealand. Although this, however, may be so, it is beyond all question that the numbers of each of these breeds will be lessened in the immediate future. It is impossible for those unacquainted with the past to realise the exuberance of bird life in the woodlands of the 'eighties. Bush-falling had barely been started in any part of the province, the North Island had been too much disturbed by war for anything approaching close settlement. Forest, wood, and water birds still existed in undiminished multitudes. It was then possible for Maoris to shoot on Tutira half a hundred brace of pigeon in a few hours. I have heard an observer describe how, as a boy at Norsewood, in the "Seventy-mile bush," he recollects pigeons so plentiful that, on certain favourite perching trees, their weight was sufficient to break down the smaller boughs. In the Tutira woods there were, besides the larger birds, thousands of wax-eye, warblers, and fantails. The rivers and lakes were as plentifully stocked: the cormorant tribe had not been then mercilessly persecuted. Tutira lake bore on its bosom a fleet of eight or nine hundred papango, widgeon (*Fuligula Novæ Zealandiæ*).

The fauna of Tutira will not detain us long. It was in the 'eighties represented by one species of bat.

Within the hollow boles of certain dead pines several small colonies existed at that time; later, when this timber was felled for firewood, bats became very scarce. The last I remember to have noticed used to fly at dusk about Harry Young's cottage, built in the early 'nineties.[1]

Proceeding now to the avifauna, I shall hazard a sketch of its future. Not very many species will fail to survive, though only in a countryside so broken by cliff and bog could so pleasant a prophecy be risked. The run has been so planned by Providence that the utmost industry of man cannot completely mar it. No farming, happily, can plane away cliffs or fill up gorges. So broken and so rugged must the surface of the station always remain, that twenty-six or twenty-eight out of thirty breeding species will continue to propagate their kind. Nor will species that disappear do so for the reasons so often assigned; they are not less vigorous than their acclimatised rivals, they will neither be ousted by imported species or annihilated by imported vermin. Much has been written about the inability of New Zealand birds to withstand the competition of the new-comers; their disappearance has been predicted on account of defective morphology. If this be indeed the case, then other qualities more than atone for such structural deficiency.[2]

Another reason assigned for the disappearance of the natives is inability to compete with alien breeds in regard to food-supply. The

[1] In more recent times I have come across bats at Waikaremoana about 1908, in the forests of the Motu in 1912, in certain islands south of New Zealand in 1913, in Little Barrier in 1919.

[2] I have seen the frail-looking fantail hawking nonchalantly for insects in a deluge that was killing the homestead sparrows, quail, pheasants, and other aliens wholesale. The fact is, that our imported birds are not bred to stand from thirty to seventy hours of tropical downpour, driven before a violent, sometimes an icy gale. In these storms species whose forebears have not been accustomed to face seven, fourteen, and twenty inches in three consecutive ceaseless days' rainfall, perish in great numbers.

Nesting-Hole and Egg of North Island Kiwi.

bird life of the forest reserves of Tutira does not support this theory. Natives are neither debarred from their fair share of food, nor intimidated by the presence of the new-comers. On these reserves I find the sparrow, starling, minah, yellow-hammer, chaffinch, greenfinch, blackbird, and thrush, the fantail, wax-eye, warbler, pied-tit, kingfisher, tui, and pigeon living together amicably. Native species, with perhaps the exception of the pigeon, lay the same number of eggs, breed as frequently per season, and rear as many nestlings as in the 'eighties, when few aliens had reached the station, when none were abundant. Food has proved ample for both stranger and native. The source of such beliefs is partly, I suppose, man's predilection for paradox and antithesis. The Maori's adumbration of his fate, doubtful at best in regard to himself, certainly false in regard to the indigenous grasses of his country, has passed current too long as an established truth. The white man's self-esteem has been flattered unduly in the belief that "as the *pakeha* rat has destroyed the native rat, as the *pakehà* grass has destroyed the native grass, so will the European destroy the Maori."

The real cause of diminution of native birds is easy to give: no creature can live without food and breed without covert; woodland species cannot exist without woodland, jungle and swamp-haunting breeds cannot survive without jungle and swamp, they cannot feed on clover and breed on turf. At one time there were on Tutira many hundreds of acres alive with forest birds; not one single individual now exists on many of these localities, because not one single tree remains. That is the simple explanation of the great decrease of natives. On the coastal portion of Tutira, where the country is grassed, comparatively few survive. On the ranges of the interior, where the forest is untouched, native birds far outnumber the aliens. So much for the immediate past. When in the future every acre of the run shall have become grassed, when everywhere the flocks and herds of the settler shall have subdued the remaining scrub and fern, a still more severe and searching test awaits the native avifauna. As has happened to other monsters of the prime, the easy-going sloth will have been succeeded by beasts lesser in bulk but more active, greedy, and fierce,—the squatter's room will have been occupied by the farmer. Under the sway of the yeoman class every yard of ground will be utilised; there will exist no longer unconsidered trifles of wild land, an acre here, an acre there, not snapped up, not ploughed, not grassed. Each homestead will support a colony of cats and dogs, each will be a nucleus for a settlement of rats. Wild covert will have altogether gone from the hills, the kowhai and fuchsia and hinahina which, either as single trees or in open clumps, have hitherto withstood fires, will have died out from lapse of time. Because of nibbling sheep and increase of danthonia—a grass easily fired—no seedling successors will have replaced the originals.

If, in fact, the squatter has chastised the ancient vegetation with rods, the yeoman will chastise it with scorpions. In the last, fullest, most energetic development of land for agriculture and stock-farming, shreds and patches of ancient Tutira will remain only in the deep gorges, the sinuous bogs, the cliffs of the run. There will nevertheless, as I have said, subsist on the station, though in sadly lessened numbers, nearly every native species that has bred on its 20,000 acres in my day. They will disappear indeed from the surface, they will sink, with the streams that are to prove their salvation, deep into the bowels of the earth, they will survive in the gorges.

Previous chapters have shown the effects of trampling of animals, drainage of swamps, destruction of water herbage, in general the disappearance of covert, the substitution for jungle and scrub of land open to the sun. The country under my régime has been shorn of its fleece; in the time to come it will be flayed of its very skin; yet in spite of himself, perhaps against his wishes, the settler of the future must on land of this type help perforce in the preservation of wild life. To a sheep-farmer producing sheep, wool, mutton, and beef on a great scale, the loss of one or two thousand sheep a year is accepted with comparative equanimity. It is unavoidable on land held as leasehold without compensation for improvements. The yeoman, however, will be a freeholder. He will have purchased his land like Mary her ointment of spikenard—at a great cost. Loss above the normal 2 or 2½ per cent will, on a small flock, be considered a serious matter—an evil to be remedied. The only sure and certain cure of that evil is fencing. If the smallholder is to prosper and to thrive, cliff and bog alike must be secured from trespass of stock. There will be conserved, therefore, on either side of each gorge and boggy creek on Tutira, a strip of covert. Above the rims of the cliffs will flourish manuka, bracken, certain heaths and dry-country plants; along the sides of the boggy creeks will grow flax, nigger-head, raupo, and rank sedges. Without gorge or bog such narrow belts of wild land would be of little avail; as additions to natural refuge-grounds they will prove invaluable. These fenced-off strips, moreover, will be kept inviolate from fire on account of the fencing material,—strainers, posts, and battens of wood.

Farming, moreover, in such lands as those of central Tutira will entail feeding of the ground. To obtain a return, marl, artificials, and lime will have to be supplied to its hungry though responsive soils. Of these manures a proportion will percolate beyond the limits of the fencing. Growth will be stimulated in the narrow strips of waste land as in the fields without. Reeds and water herbage will shoot up more tall and luxurious, manuka and heaths will put forth a stronger growth. Stimulation to plant life also means in the long-run stimulation to animal life, a bigger hatch of insects, an enhanced crop of land snails, grubs, and caterpillars.[1] This involuntary assistance to the avifauna of the station, though less striking and conspicuous, will prove of more importance than the planting of shelter-belts, orchards, and shrubberies about the homesteads to be; such cover is too open, too much overrun by cats, by dogs, by rats.

In the light of such vital factors in race maintenance as food-supply and breeding accommodation, we can proceed to consider the future of an avifauna, whose vicissitudes and disabilities may chance to suggest analogies to other readers in other lands.

Species that have during my day nested on the run may be divided into four lots—those that have actually been attracted to the place by novel conditions; those that have more or less adapted themselves to changed environment; those to whom change would have been fatal but for the broken nature of the run; and lastly, those to whom changed conditions have been fatal.

Nest and Eggs of Banded Dottrel.

The three species that have been attracted to the station by changed conditions are the Banded Dottrel (*Charadrius bicinctus*),

[1] Much in the same way as fish are attracted to the vicinity of shaggeries where seaweeds are stimulated by guano, and where, consequently, the marine life upon which fish feed is most abundant.

the Pied Stilt (*Himantopus leucocephalus*), and the Paradise duck (*Casarca variegata*).

Three or four pairs of the first named were induced to settle during a windy spring when several hundred acres of ploughed ground in the Waterfall paddock lay bare as sand-dunes beneath continued nor'-westers. Since that date the Dottrel has regularly reappeared each spring, nesting sometimes on tilled ground, sometimes on short grass. It may therefore be that by the chance combination of ploughed land and a windy spring a permanent change in the habits of the species has been induced—a change which may save it from local extinction, for it cannot be doubted that in the crowded future the coasts of Hawke's Bay will become less safe; apparently, however, on dry pumiceous lands there is sufficient food-supply of the sort desired; the Banded Dottrel following the plough and harrow may become a common species perhaps where before it was unknown.

We owe the appearance of the Pied Stilt to another station "improvement"—the great drain, to wit, cut in the 'nineties through the big swamp. The northern bay of Tutira lake has from that time begun to silt up; there has appeared a strip of muddy beach, a bank of sand, an area of shoal water. These conditions, though on the smallest scale, have on several occasions attracted pairs of Pied Stilt, eggs have been laid, and nestlings reared.

The Paradise duck bred with us for the first time after the great flood of 1917, when enormous deposits of silt covered Kahikanui Flat: of the fifty or sixty which remained during that winter, several pairs reared a brood. They have continued to breed on the station ever since. Given the chance of extension of range, it has been taken: Dottrel passing overhead have spied out naked soil; Pied Stilt—sand; Paradise duck—fresh feeding-grounds.

A dozen breeds have more or less successfully adapted themselves to change of environment. They have at any rate not lost everything by the alterations of the last forty years. Often, though not always, the harm done has been greater than the benefit gained, yet the species to be named have survived, and I believe will continue to do so even under the more severe ordeal of intensive land culture.

Of all the birds on the run, the Native Pipit or Ground lark (*Anthus Novæ Zealandiæ*) has been the greatest gainer by change.

It has lost nothing and won much. Its increase has been commensurate with increase of open ground. No longer is it limited to such oases in a desert of fern as landslips, wind-blows, bases of sun-dried cliffs, sand and shingle spits of open river-bed, pig-rootings, bare rocks, and the scanty cultivation-grounds of the old-time Maoris. The bird is a frequenter of neither forest nor marsh, so that operations which have almost annihilated certain species have but enlarged its domain. The area of land open to grasshoppers, daddy-long-legs, and caterpillars is a thousand times greater than of yore. Plough and spade are to this amenable species godsends, disinterring in multitudes the white soft larvæ of the green beetle. The Pipit is, moreover, very partial to the alien blow-fly attracted by every carrion on the place. Although truly a bird of the wilderness, it will haunt the garden too on occasion, watching the worker almost as an English robin does, and accepting tit-bits from the hand of a friend.

The Pukeko or Swamp-hen (*Porphyrio melanotus*) has, by its gregarious habits, productivity and general adaptability, proved able to thrive better on dry ground than wet, on grass and clover than on raupo and sedge. In the 'eighties the range of this fine bird was limited to portions of marshland where excessive moisture kept the water-logged flax yellow and starved, to quaking peat-bogs, to debris deposited by little streams emptying themselves into the lake, to narrow margins of soft ooze between the border of tall flax and the lake itself, to raupo beds in the shallows. The Pukeko has gained by every step in the development of the station, by the destruction of fern, by the felling of forest, by the drainage of marsh, by the increase, in fact, of treadable surface. Hundreds run in swamps now drained dry, hundreds explore the hills, breaking up the dead patches of grass in search of grubs; cropping itself, the anathema of many species, is a boon to the breed. Tender oats are sweeter than grass, they serve also to conceal the careless nest; amongst the ripening grain the bird weaves platforms upon which to rest. Nor does he willingly renounce the novel food-supply even when under thatch. The oat straws are carefully and deliberately drawn out one by one, the frugal birds devouring every single grain from one head ere beginning another. The Pukeko is, moreover, in the happy position of being able to regard with equanimity a further contraction of its feeding-grounds. It might even be an advantage to a breed where polygamy is largely practised, where combined clutches of ten and twelve are not rare.[1]

The social instincts of the species are already highly developed; at threat of danger the birds flock together for defence; foundlings, too, are welcomed and protected by individuals other than the true parents. The Pukeko, in fact, is within measureable distance of complete gregariousness, yet in spite of what has been said, in spite of its adaptability and great increase during my day, I fear for the species. He offers too large a mark, he loves the open and the sun; the gorge, the cliff, the inaccessible river-bed will be of no avail in his hour of need, for what are now condoned as peccadillos may in the days to come be classed as crimes. On a large run the damage Pukeko can do is trifling. Circumstances, however, alter cases; perhaps if even I myself possessed but a few acres I should feel annoyance at rape of oats, theft of straw, ravages amongst green maize, wholesale cropping of clover and grass. As man is constituted, the intelligence and high ethical standard of the Pukeko may not atone for mischief even on this petty scale. Still, it is hard to believe that amongst the future owners of Tutira one or two will not be found to protect in a semi-domesticated state a few of these splendid water-hens.

Young Harrier hawks.

The Harrier (*Circus Gouldi*), and in lesser degree the falcon, have also gained by the advent of settlement, by the admittance of light to the earth's surface, by the increase of creatures which live their lives in the open. The diet of the former must have in ancient days been scarce and precarious. It would then consist of lizards, insects, fledgling birds, and eggs. Not only must the quantity of small animals have been scantier over the whole run, but to a clumsy slow-flying species the prevalence of covert must have been particularly baffling.

[1] I have seen a nest containing seventeen eggs, the property, probably, of four or five hens.

Nowadays not only has the supply of birds, a percentage of which are perpetually falling out of the ranks from natural causes, increased, but the spread of open land has revealed to the harrier insect and reptile life formerly unknown in the land—frogs by the edges of water-holes, crickets below the dry cattle-droppings. Lastly, the stocking of the run has supplied to the Harrier mutton on a great scale. In a well-managed, well-fed, and carefully-culled flock running on perfectly safe country, 2 or 2½ per cent is about the normal death-rate. It is the unavoidable loss incurred through diseases more or less akin to those causing death in the human race. On Tutira, however, a minimum loss of 5 per cent has never been quite reached; it is the toll paid by the station to cliff and bog, and works out at the rate of from three to five sheep a day. Some of these are found and the skin at least saved; some are totally submerged in quaking morass, buried in holes, trapped in under-runners, or smashed by falls from cliffs. The balance, say half, is the perquisite of the Harrier; he is fed, therefore, at the rate of something not far short of 100 lb. of meat a day; for the sake of caution, say 50 lb. a day. Divide that again and say 25 lb. a day. Many, moreover, of these sheep are fat, so that it is not surprising that Harriers are sometimes killed on the ground by sheep-dogs, the birds so gorged as to be unable to rise. The breeding-quarters of the species will be in the future, fields of oats and clover, raupo swamps, and belts of low manuka.

Young Falcons.

The gallant little Falcon, too (*Hieracidea Novæ Zealandiæ*), has been more than compensated for loss of native prey by increase of ground larks, by the introduction and spread of pheasant, quail, thrush, blackbird, minah, starling, and lark. Especially is the Sparrow-hawk more plentiful about the centre and uplands of the run. These were districts in olden days barren of bird life, except of ground birds and sedentary species easily able to escape into the all-pervading fern. As to breeding-quarters, the Falcon has the choice of the whole dry-cliff system.

The Kingfisher, too (*Halcyon vagans*), has gained, or at any rate not lost, by the opening up of the land. His range of vision has been greatly lengthened; probably the increase of surface from which worms can be gathered has more than made up for the partial loss of cicada and dragon-fly, both of which are nowadays taken by the minah, and probably also by other acclimatised species. At any rate, the clutches of eggs are as large and the Kingfishers themselves as plentiful as in the 'eighties. Although the type of nesting-site preferred of old, rotten timber, has been destroyed by fire and falling of forest lands, sandbanks of a proper consistency remain in ample quantity along the open reaches of the Waikoau.

Kingfisher.

The Morepork (*Athene Novæ Zealandiæ*) has conformed to the requirements of civilisation, has become, indeed, a semi-domesticated bird, one or two pair living permanently in the vicinity of the homestead. Such residenters are attracted, especially during winter-time, by the influx of sparrows, rats, and mice. Indeed, from a merely utilitarian point of view, the Morepork is a useful ally

Morepork—male.

Female Morepork at nesting-hole.

to the settler, and when better known is likely to be of set purpose protected and encouraged. Though in great degree his ancient nesting-quarters, holes in trees, have been destroyed, yet like the Kingfisher he has adapted himself to novel conditions; on Tutira this small owl now chiefly breeds in dry dark cliff crannies. With habitations in the everlasting hills, and with an enormously increased food-supply, the Morepork is safe.

Grey Warbler.

Three small species, the Grey Warbler (*Gerygone flaviventris*), the Wax-eye (*Acanthisitta chloris*), the Fantail (*Rhipidura flabellifera*), though immensely reduced in number through the clearing of bush and scrub, will nevertheless always survive in the gorges and cliffs. The Wax-eye and Fantail, moreover, already breed about homesteads, the Wax-eye regaling himself on fig, cape gooseberries, box-thorn and other foreign dainties, the Fantail not infrequently carrying his friendly intimacy so far as to enter open windows in the pursuit of house-flies. Neither is the Grey Warbler, though rather less domesticated, quite proof against the superabundant supply of blights, caterpillars, and insect life generally, that infest every unsprayed New Zealand orchard and garden.

Fantail.

About the Shining Cuckoo (*Chrysococcyx lucidus*) I know but little at first hand; I have never found a nest containing the careless migrant's egg or chick. That the Cuckoo, too, is in some degree adapting himself to changed conditions there is, however, considerable proof, instances having occurred, I am told, where the chick has been reared in nests of imported species. On the tree-feathered gorges and about the strips fenced to keep stock from danger, there will always be sufficient food and cover for a few of these birds.

Of water birds, the Grey Duck (*Anas superciliosa*) is not suited by the conditions of Tutira and has never been plentiful; 20,000 acres, however, is a big bit of ground, and the surface of the lake will always attract flights, of which a few pair will remain to breed, chiefly about the open reaches of the Waikoau.

Young Grey Duck.

The future of the Widgeon or Scaup (*Fuligula Novæ Zealandiæ*) is less easy to forecast. Of the two vital factors in race maintenance,—food-supply and nesting accommodation,—the first at any rate is secure, for the feeding-grounds of the Scaup are the lake bottoms. Whatever other surfaces have been tampered with, that one at any rate has remained, and will always remain, intact; but whilst there will continue to be a superabundance of food, and whilst female birds will continue to be capable of laying large numbers of eggs, another danger threatens the Scaup—the loss of nesting-sites. No ordinary covert will suffice this pernickety species,—the nest must be hidden beneath many seasons' accumulation of rotting flax-blades. The Scaup, furthermore, never breeds except by the water's edge. Such particularity militates against the species, and is likely to do so in an increasing degree, since flax fibre in a dry season is excessively inflammable, and the plant is being fast destroyed by cattle. The case is interesting as an example of how a breed with ample feeding-grounds may decay in numbers solely and entirely from want of the particular cover required for nesting purposes; his prejudices in regard to housing accommodation will be his undoing if indeed he disappears.

Tui on nest on tree-top.

Another lake bird to whom the future is secure is the Little Grebe (*Podiceps rufipectus*), whose nest is practically undiscoverable, and whose ample food-supply rests secure, like that of the Scaup, on the lake bottom.

There are other breeds certain to have disappeared from the station but for its innumerable gorges; in them all pastoral changes cease, sheep cannot tread their banks nor cattle follow up their narrow beds. These scores of miles of gorge-bottom will never be trodden by man; they are unaffected by the alterations of plain and hill above, they are only to be reached by rope-work. Into a few I have myself from time to time attempted invasion, wading the shallows, climbing the barriers of piled flood debris, working hand over hand along the cliff scrub, swimming the cold, clear, unsunned pools, but always after a few hundred yards finding myself blocked by smooth-sided inaccessible cliffs and waterfalls. On the beds of these canyons there are shreds and patches of habitable

Tui nestlings hand-reared.

slope, deposits of deep soft flood soil rich with flakings of marl, vermiculations of sandstone, and leaf-mould from the fern-feathered precipices. Into these rifts during the course of ages the Kiwi (*Apteryx mantelli*) and Weka (*Ocydromus earli*) have sunk with the sinking of the streams themselves. Æons ago they were surface birds; now, perhaps, in the deepest ravines they may have almost differentiated as island races do from their fellows of the plains and hills above. At any rate, in these solitudes the species named are plentiful. There they will remain undisturbed till the day of judgment. These two highly interesting ground birds are safe on Tutira should the whole of the rest of New Zealand be turned into cabbage-gardens cultivated by Chinese.

Young Tuis tamed.

There also on creek beds unchanged with time, and rich with water insects and fly, will the Blue Duck (*Hymenolæmus malacorhynchus*) forever maintain himself. Into every ravine I have attempted to enter, signs of his presence are plentiful; where penetration is impossible and the gorge not too deep for sound, his delightful call may be heard far below.

Another recluse, the Pied Tit (*Petrœca toitoi*), will also survive about the shrub-fringed edges of the gorge.

Brown Duck.

The Pigeon (*Carpophaga Novæ Zealandiæ*) and Tui or Parson Bird (*Prosthemadera Novæ Zealandiæ*) are certain also to become rare birds. Elsewhere on the run food-supply and breeding accommodation alike will have been swept clear. A few pair of each will nevertheless maintain themselves in the gorges. The Tui will then as now haunt the homestead and shelter-belts when in mid-winter the eucalypts break into flower. At other times of the year kowhai and hill-flax (*Phormiùm Cookianum*) will provide nectar, wild fuchsia, poroporo (*Solanùm aviculare*), and other native plants, seeds and berries. The

Pigeon too will survive, though reduced to a few brace. This hardy bird can, I believe, digest almost anything green. I have known them devour immature male Pinus insignis flowers; I have watched them nibbling the dry fronds of Asplenium flaccidum, and stripping one by one the leaves of laburnums. They also freely feed on white clover leaves; on fallen forest, newly sown, I have known them grow excessively fat on rape and turnip shaws. Local survivors in the future are likely to obtain a portion of their food during at least a portion of the year on the surface of the run. They will take toll of the settlers' white clover, rape, swedes, and probably oats, to the value of a few pence per annum.

Fern-birds—male and female.

On shrubs growing at right angles to the cliff face, the silly platform of sticks which serves for a nest, though transparent from above and below, will be safe from rats, weasels, and prowling cats. The Pigeon in his gorge will be secure from man also, for without descent by rope the gunner could not shoot from below; shooting from above would be mere useless murder, as the birds could not be retrieved.

A few pair of Black Shag (*Phalacrocorax Novæ hollandiæ*) will also maintain themselves in certain very high cliffs. The Brown Duck (*Anas chlorotis*), the Fern-bird (*Sphenœacus punctatus*), the Philippine

Rail (*Rallus philippensis*), the Swamp-Crake (*Ortygometra affinis*) and Water-Crake (*Ortygometra tabuensis*) will find salvation in undrainable marshlands and boggy creeks. The Brown Duck cares nothing for the sunny river-reach, for the deep gorge, for the open width of the lake. His quarters are the slow-flowing streams with blind isolated pools growing sub-aqueous weeds, here and there stirring a reed-bed in their torpid course, here and there passing over an expanse of muddy shallow. There during the hours of light he hides in dense covert and in deep shade, only at night-time venturing out, but then showing himself strangely tame and fearless. Of late years the breed has become more scarce; banks have been trodden down by cattle, water herbage has been devoured. With the advent, however, of the yeoman freeholder, the bogs and marshes most dangerous to stock must perforce be fenced. It is not improbable that with more covert and better feed the numbers of the Brown Duck may again revive. He will be visible at least to those who care to watch him at dusk, and to note his utter unconcern in the presence of man.

Nest of Philippine Rail.

Another species that will gain, at any rate not lose, by settlement is the Fern-bird. It is very small, its habits are furtive, it breeds twice a year, it is adaptable in its choice of nesting-sites, its young are fed on alien as well as indigenous insects.

Young Kaka Parrots.

The Philippine Rail, the Swamp-Crake and the Water-Crake, will manage to maintain themselves. They too, like the Fern-bird, will be gainers by the larger crop of insect life resulting from the fertilisation and the intensive working of the pumiceous area. Certainly about the new homesteads there will also occur an increase of cats—tame, half-wild, and wild,—but strange as it may appear, this fact will not necessarily be fatal. On the contrary, in certain Hawke's Bay swamps I have found by far the larger number of rails' nests containing unspoiled eggs close to cottages. Injurious as the cat may be to the rail, its presence is still more baneful to the rat; relatively that is, the cat has actually come to exert a protective influence. No one of these species is common on Tutira, but their habits of concealment make them appear more rare than is truly the case. Of native birds that have bred in my day in Tutira, there remain two only to be considered—the Parrot (*Nestor meridionalis*) and the Parrakeet (*Platycercus Novæ Zealandiæ*). Neither species can survive locally in the absence of considerable areas of forest.

Such, under the conditions foreshadowed, is the future of the Tutira avifauna. In my sketch nothing whatever has been allowed for sentiment, sense of beauty, even for intelligent appreciation of usefulness. Yet surely in the future we may anticipate that a sufficiency of leisure will be the birthright of every man. If that be so, one step—and a lengthy step—towards rational content and wholesome happiness will lie in nature study. I do not mean the ability to systematise and classify—I mean the watchfulness that will awake in the student, fellowship, humour, and sympathy.

So far the best that has been done towards the conservation of species is but negative. At the best, man has here and there been content not to destroy utterly. What is required is positive work, the elimination of vermin and parasites, the study and augmentation of special supplies of food, the careful reservation of nesting accommodation.

At the worst, the species named will as species sparsely survive on Tutira; with a little trouble, a little watchfulness, a little expenditure of time and money, the numbers of individuals could be vastly amplified.

Baby Pukeko.

CHAPTER XXIII.

THE PARTNERSHIP OF H. G.-S. AND T. J. S.

REVERTING to the natural history of *homo sapiens* and the efforts of the earliest specimens of the breed to acclimatise themselves on Tutira, it will be recollected that at the end of Chapter XVIII. we had left the station in direst need; its life-blood had dried up in consequence of a fall in the price of wool. There existed no longer the wherewithal to pay interest on the station overdraft, let alone rent and working expenses. One of the partners had released himself from liability to the National Mortgage and Agency Company by forfeit of £600 into the credit of the station. The other had taken over the derelict half-share for 5s.; the martyrdom of man, in fact, had been consummated; Newton, Toogood, Charles Stuart, Thomas Stuart, William Stuart, Kiernan, Mackenzie, Cuningham, pass before me in sad procession, like the ghostly kings in "Richard the Third." They had perished in time or cash.

The miserable outcome of eight years' labour on Tutira was the writer of this volume. He stood, so to speak, on tiptoe, insecurely balanced on the piled carcases of his predecessors, up to his lips in debt. Because he was young and foolish, and because he had not then lived as he has since done—to see wool at bed-rock three times in thirty years and three times recover—he was filled with the gloomiest forebodings for the future, not only of himself and of Tutira, but of New Zealand; in his mature opinion the Dominion was doomed. His relatives, however, were wiser; after again demonstrating the lesser evil of drunkenness compared with the fatuous perusal of Henry George and the perpetration of verse, they proved, and this he readily credited, that things could not possibly be worse. The National Mortgage and Agency Company, moreover, did not feel inclined to release another owner at any price. The writer, in spite of his misfortunes, had already become attached to the place; he elected to hold on. The Company behaved decently in so far as it lieth in a loan company to do so; they had no malevolent desire to "bust" him; certainly they cannot have wanted such a station as Tutira then was on their books. They agreed to a small reduction in the rate of interest; what was more valuable, the station was afforded the chances of time and tide. Not only did the National Mortgage and Agency Company voluntarily forgo full interest, but they interfered on the writer's behalf when the local firm with whom he dealt in Napier attempted to charge 13 per cent on his current account.[1]

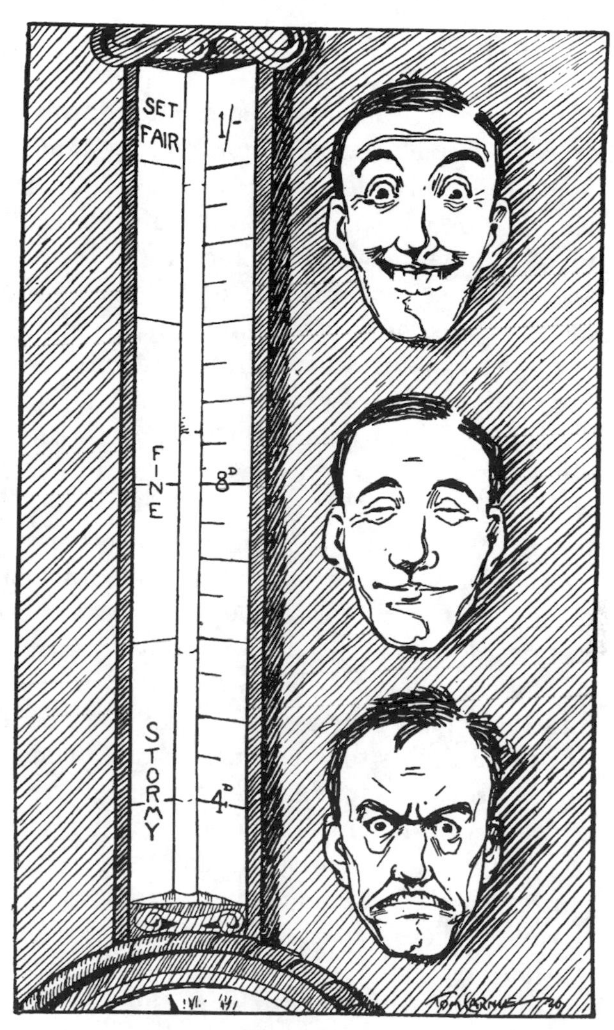

"*Truculent in the superlative degree.*"

For a couple of seasons the fortunes of the run hung in the balance. It was saved as many another station has been saved in New Zealand—by the process of sitting tight, by strict cessation of expenditure. Items such as interest, rent

[1] Years afterwards, during the great war, the writer found himself thus interrogated by a fellow-worker in a certain hospital: Did he know New Zealand? "Yes"—heartily—"the best place going!" Did he happen to know a town called Napier? "Rather"—very enthusiastically—"the prettiest spot in the world!" Then you must remember my brother, who used to manage for ——? . . . It was the man who had charged me 13 per cent!! Will the reader deem me implacable, ungenerous, unforgiving, will he blame me for lack of magnanimity when I confess that the acquaintanceship with my new-found friend cooled? Honestly, I could not effervesce over that brother.

and shearing expenses, were irreducible, but stores were cut down to a minimum; barring necessities, not a penny went out of the place; then and for years afterwards the joint personal expenses of myself and partner were under £60 a year. Napier we seldom visited—it was not safe; at any rate we did not consider it safe to be too often seen in the City of Destruction; out of sight was out of mind—to meet your banker was to recall to his mind your calamitous account, and in those days with wool at bed-rock prices our banker's facial expression was truculent in the superlative degree

The faithful Stuart, like the lover who has lost his mistress and cannot tear himself away from scenes of former happiness, remained on Tutira. I forget details now, but believe it was agreed that he should allow a part of his screw as shepherd to remain on credit, "to be paid when able." At any rate he and I together worked the place. What, however, we could do positively or otherwise was but little; the fate of the station was to be decided in the great markets of the world thousands of miles distant. For a couple of seasons the place was kept going, as I have said, by the negative process of sitting tight, helped too by a small rise in wool and by dry seasons, during which the death-rate dropped and the lambing increased. This period of watching and waiting was not unprofitable to our souls. Ample time was afforded for the reviewing of previous blunders, Stuart repenting him of his excesses in timber and grass seed, I myself of my orgy in two-tooth rams. Then, again, lacking coin and credit, our annual purchasings of outside stock ceased perforce, with the immediate result of a lessened mortality. We recognised then, for the first time, the full, the incalculable, value of station-bred stock, inured to a wet climate and poor soil; we began even to think that the disabilities of the run might be amended. The blind fury of labour for labour's sake had in myself given place to some faint glimmerings of sense. Errors of the past were duly considered; during winter evenings plans of future work, of future improvements, were talked over—supposing, that is, we should ever again possess the wherewithal.

In the meantime we were stuck for capital. It was out of the question again to ask, like Oliver, for more. As to the wool market, we had no faith in it, yet the event about which we had despaired did nevertheless happen. In June of '86, whilst the National Mortgage and Agency Company were still swithering as to what was wisest for us and for themselves, wool rose 2d. That rise saved the writer; the advance on the previous wool-clip had been what is technically known as a "safe" advance, perhaps 1d. under the price to be reasonably anticipated. The particular sale day, moreover, on which our clip was auctioned in London must have been one of maximum prices in a buoyant market. We obtained, at any rate, more nearly 4d. than 2d. per lb. over the prices of the previous year; when the account sales were rendered there was a surplus of something like £500 to the credit of the station. Stuart, who had been hesitating, now decided to risk what he had saved and been able to rescue from his particular debacle, and to take over the half-share. The sum he was able to command was as fruitful as the famous one million kept by Baring for the immediate development of the lands of Egypt. With it in hand, once again we began to improve. They were improvements done in a very different manner to the reckless, haphazard system of the past. Every penny—I had a most excellent, thrifty partner—had to be considered; nor were we content with the first plan that promised success; the scheme finally adopted was the best of many fully thought out, all of which promised success.

For the following twenty years Stuart and I worked together; we had each of us been through the mill. I believe that during our long partnership no considerable blunder was perpetrated.

The abiding difficulty of the run, the simultaneous making of the country and the proper feeding of stock, has already been touched upon. It was insoluble then, and remained so for twenty years. It was likewise impossible to change light lands into good lands; what, however, we could do was done.

We utilised this disability; we even made it produce results not very different in pounds, shillings, and pence from the famous stations of southern Hawke's Bay. The value of the trough of the run lay preeminently in its suitability for the rearing of young sheep. On its dry porous soil, suckling, the best feed in the world for hoggets, flourished amazingly; it was impossible to grow a heavy wool-clip, but this we determined to remedy by a large output of surplus stock. The limestone range of eastern Tutira enabled us to carry a sufficiency of ewes. The whole of the rest of the station was given up to hoggets; ewes only and hoggets were run.

In regard to feeding, the problem has been stated before: we could not afford to treat our flock properly, ideally; as long as our ewes were sufficiently nourished to bear and rear a fair lamb, it was more important to have numbers than condition; as long as our hoggets lived to tread the shearing-floor we hardly cared how many broken fleeces were thrown on to the wool-table; the carcase of every sheep that passed through the shed was worth more to the run than its wool. For a quarter of a century the choice before us every day of the year was the better treatment of a smaller flock, resulting in a less efficient crushing of fern, and therefore detrimental to the making of the run; and, on the other hand, an over-drastic grinding of fern with its accompanying future benefit to the run, but with its accompanying present damage to stock. We learnt to balance the rival claims of land and sheep to a nicety. It was found possible in practice to combine a heavy lambing, a small mortality, and a substantial increase in numbers, with a clip light enough to excite surprise in the bosoms of our bankers.[1]

Compromise between an unwise parsimony in regard to feed and the wintering of the largest possible number of sheep was in fact so evenly balanced, that during a few weeks of each winter, for perhaps ten seasons, eatable mutton was unknown.

We lived on wild pig, wild beef shot on Kaiwaka, and the fat wild sheep and double-fleecers that could be raked in from river cliffs; these were shot, dressed on the spot, and packed into the homestead. There was in those spartan days as little spare fat on the station as in Berlin during the last winter of the great war.

Well, then, with a certain amount of experience, a certain amount of local knowledge, and a certain amount of caution jammed into our heads, Stuart and I made a new start. Although no formal partnership was made out until later, the fresh capital put into the run was used from this time forward for the development of Tutira.

In spite of previous failure we were eager to be again engrossed in the most fascinating pastime in the world—land reclamation. The small sum now available would have been scarcely noted by a soulless company which knew not Tutira except by name, which would have swallowed this precious fund prededicated to improvement without

[1] I recollect an inquiry from the National Mortgage and Agency Company, about 1888 or 1889, asking if the number of wool pockets, so far shipped, was the whole clip. The average weight of fleece, including locks and pieces, had barely reached 5 lb., yet that year we had increased our flock by over 1000, and had docked nearly 95 per cent. of lambs, which were afterwards successfully reared.

so much as a preliminary grace. We decided to spend it without interference, as we thought proper. Fencing was our first care; compelled by conditions to remain perennially overstocked, it was important that there should be smaller paddocks. Into them weak hoggets and ewes hanging about corners could be shifted as into convalescent camps; by subdivision of the run we could at least average the feeding of the flock more evenly. Fencing was accordingly run from the Waikoau river along the top of the Newton Range into impenetrable scrub and fern behind Table Mountain. Further subdivisions erected at right angles from the new fence to the lake and elsewhere gave us nine paddocks in lieu of two. Certainly the posts, like angels' visits, were few and far between—the battens likewise were reduced to a minimum; still these unsubstantial fences did hold sheep, and that, after all, is the prime object of fencing.

The station had hitherto possessed but one set of yards, those in front of the wool-shed. Heretofore we had at docking-time driven our lambs from each of the two huge paddocks down steep slopes to lake level. Lambs, like bees, hate to run downhill; segregating to the rear, regardless of their deserted dams, they had broken back and scattered in hundreds over the paddocks, where, although the merino ewe is an excellent parent, a proportion had been mismothered and lost. As, therefore, we could not move Mahomet to the mountain, we tried the alternative and built on the hill-tops several new sets of yards. They were constructed strictly of posts and wire, not rails, for Stuart was as shy of timber as I was of yearling rams; not on unnecessary split and sawn stuff was our precious patrimony to be squandered.

Looking back now I am confident that, as a preliminary, we could have made no better move. We were able at once to increase our flock, and to carry them, if not better, at least more evenly. We were able, moreover, each year to nurse through the winter several hundred sheep that would otherwise have probably perished.

In the meanwhile time had not stood still; half of the term of the original lease had expired. No native lease, however, was on the east coast in those days ever allowed to expire completely. It was to the interest alike of native owner, squatter, and banker to see that when about half the period of a lease had expired, another should be substituted. A fair average of length tenure was thus secured in practice, its maximum twenty-one, its minimum about ten years. A fresh lease was accordingly drawn up, the rent as ever after on such occasions being doubled—an immediate gain to the native; the length of time during which the European was to remain in possession also doubled, twenty-one years instead of ten—a gain to the tenant and banker. Conditions otherwise were similar to those of the original agreement.

The natives whom we knew best signed, I do sincerely believe, largely to do us a good turn. The immense majority, at any rate, once more appended their shaky crosses, or signed their names in cramps and schoolboy scrawls. A percentage, however, resolutely refused to sign. They had no objection to our occupation of the land, they merely believed that the policy of *taihoa*—by-and-by, wait and see—was the proper policy in regard to all east coast native land transactions whatsoever. By this time, too, the mere mechanical obtainment of signatures had become a difficulty,—the number of our landlords had increased with the subdivision of shares; they lived in every province of the colony.

There must have been well over a couple of hundred of them. Although it is anticipating matters, an actual instance will show that even in the simple life—that Arcadian existence which was to preclude everything disagreeable—cares will arise, troubles will intrude. It was the old, old story revived; the reader will remember that terrible entry in an early diary, "30 per cent death-rate between 1st April and 31st March." Well, deaths in the flock were still what worried us.

For example, Raiha Pohutu, one of the original thirty-six owners of the Heru-o-Turea block, dies; we regretted his demise, of course, but the greater grief was his eleven successors, three of whom inherited one-sixth each of the original share, three one-twelfth each, three one-eighteenth each, two one-twenty-fourth each. Well, we were barely out of mourning for this sad event when bang would go another landlord—Karaitiana—who had inherited one of these eighteenths of a share. Again we were sorry, of course, but the bitterest pang of all was the fourteen successors of the deceased. They lay very heavy on our hearts, for three of them got five times as much rent as the remaining eleven. In money matters one cannot be too careful; by the native land courts Ranapia Taungakore was appointed trustee on behalf of six of the shareholders, whilst Te Huki Taingakore watched the interests of the three favoured minors who had inherited the big shares; still extra mystification and trouble was caused by the habit of many of the natives appearing in legal documents by other names than those borne in private life, their spoor being lost then, even to the bloodhounds of the law. If we did not personally love our hundred or so brace of landlords, at least we hated that they should die. "There's poor old So-and-so gone," one of us would exclaim sentimentally. "Yes, and a score more blessed grantees to deal with," would rejoin the other, who did the native work. Stuart did the native work, though, of course, I was there as with Cuningham over the station books to see that everything was done in a sound and proper business way.

Nevertheless, such as it was, our new lease gave us time; bankers may have looked askance at the imperfect document, lawyers wagged their solemn heads over it. It was what was known in legal phraseology as "a good holding title"; that is, it offered to the holder if blackmailed—and blackmail was always a possibility—a fighting chance. With a little ill-will on the part of a single Maori, with a little backing from an unscrupulous white man, each native nonconformist might have stocked the run with sheep up to the value of his share—in itself an unsettled problem, for no share was individualised,—each one of them could have put sheep on to the station with the right to shift them at any time anywhere. The station, it is true, might have called on these men to fence off their shares; it might have called, like the prophets of Baal, on their false gods, and with equally negative results. Between procrastination of native law-courts, dilatory habits of natives, rascality of low-class whites, sheep-farming on native lands—which, after all, is not a crime against the community—could have been made impossible. As a matter of fact none of these problematic disasters did happen. I will not say that throughout the period during which such conditions prevailed, no cases occurred of attempted personation and fraud, but I can truly say they could be counted on the fingers of one hand. As in warfare so in business, the Ngai-Tatara were a fine straight lot of men.

The landlord and tenant system is antiquated, absurd, and unsound, for the man who tills the land should own the land; yet as it did actually exist on Tutira, owners and occupiers met on fairly even terms, the former, indeed, getting his lands improved without having to pay compensation, but the latter, the lease once signed, becoming in his turn master for a time. It would be difficult, for instance, to imagine an English landlord writing to his tenant—say the Duke of Westminster, imploring an advance of rent from a tenant in Upper Tooting—in the following strain :—

"DEAR SIR,—Wroting you these few lines to let you know that —— told me there is a mistake about my money of the lease, the first time he drow my money the time when —— get married, it was the year in 1899 that the year —— —— get it since that time you have look at the book you see how many money it comes to now, there is some money for me there, about three pound ten shilling if nothing all the money I get from you £15 15-4 well you think from that year 1899 and how many money comes to this year, well you take out all the year money you add this money you gave me you will see the mistake in it if I am not sick I come there myself, well I am not well Dear Sir, that why I write to you give me some money I heard that you came to pay all the maoris of the lease will you kindly give me some money for I am very sick I am not well I was trying to come to Tutira to get me some money, well I cannot get on my horse I was very weak some maori told me you wont give me some money I told them I always asked you for some money you always gave me some money Please dont for get to send me some money if you send just £3-10 that is good enough that will do me for I cant get any money to buy me some food to eat Please dont for get to send me some money I want it very bad if you send me some money you give it to Riria Watene I have been bad all the Maoris got some money only myself Please dont for get Dear Sir be kind to the poor thing. Please dont for get to give to the one bring my letter to you I remain yours ——."

Doubtless we were "kind to the poor thing." Here is another example of the pathetic appeal :—

"To Mr Garthrie.

"DEAR SIR,—Will you kindly give my money to Joe that my letter what I gave to Watene Please don't you forget to give it to Joe Watene told me he never see you when he came back Please don't you forget to give some money if you can give two years I want £7 Please don't forget I am very hard up I get no money and I am not well Please don't forget.—I remain yours truly ——."

Then there is the business type of epistle :—

"DEAR SIR,—I have the Greatest Opportunity of writting you a few lines hoping to find you in the best of health And I may also have the pleasure of asking if you can Obelige me with the money of the lease for this year, etc., etc."

"This year" really meant next year, for always after signature a year's rent was advanced. Or—

"DEAR SIR,—I have the Pleasure of writting you these few lines hoping to find you in the best of health and asking if you don't mine giving the bearer £10, etc., etc."

Like master, like man—our landlords were perennially impecunious ; rents were spent always before they fell due, the station was expected to furnish perpetual advances, to replenish their landlords' pockets with sums varying from hundreds of pounds to shillings; marriages, births, and deaths were equally excellent reasons for demanding cash. All these loans and advances were quite irregular; in the absence of J.P.'s and licensed interpreters and stamps they need never have been repaid, yet it was rarely indeed that a Maori went back on his word. If a man must needs be burdened with a brood of a hundred couple of landlords, let him pray Heaven on his bended knees, I say, for Maori landlords.

Often I have wondered if any work at all done on the station was legally done, for if I am to credit the local natives, the original lease was signed by many who had no sort of claim on the Tutira lands; no proper supervision seems to have been exercised, many of the signatures were forgeries, or if that is too strong a word, one native signed for another; then again, was it clearly defined that Newton and the succeeding tenants of Tutira were permitted to destroy the ancient vegetation of the run, to cover it with clover and grass, to drain its swamps? Rumblings of distant thunder, that might have at any time broken over our heads, reached us now and again in the shape of legal remonstrances. To this day I remember one which threw Cuningham and myself into the utmost consternation,—we had not become calloused by custom to the sword of Damocles. This particular epistle was written, I recollect, by a Minister of the Crown, at the request, doubtless, of some good old crusted Tory, forbidding, under the most horrible penalties, the destruction of bracken. Another heathen reactionist cn another occasion forbade drainage, on the ground that it might affect the welfare of the eels in the lake.

To our solicitor, for comfort and sustenance, these communications were taken,—like Evangelist he guided us; he dried our tears; we clung to him like shipwrecked mariners to a plank; he stood between the station and eternity.

The rise in wool, and the new lease, were positive benefits; negative boons were the cessation of the purchase of ewe drafts and lines of young stud rams. During the reign of H. G.-S. and A. M. C. rams had been imported — heaven only knows why — from stud flocks in the South Island. They had been two-tooths, their price five or six guineas; we now bought aged rams at a fifth of the cost. The saving of money was great; the older sheep with set constitutions did their work better; they lived and throve where the others had died. These local sheep were at least as well bred, they had been used in famous Hawke's Bay flocks, and had only been culled on account of age. We believed rightly that if they had been considered good enough for southern Hawke's Bay, we could scarcely go wrong in using them on Tutira; we believed that by the time a line of rams had reached the age at which we purchased them—their fourth tupping season—they had proved themselves by the best of all proofs, good condition and sound feet, to be sterling stuff. The percentage of weak and feeble had died off as two, four, and six, and full-mouth sheep. I have always maintained that by our procedure in this matter the Tutira flock was sired by the most healthy sheep in the Province. In practice, at any rate, the change from young rams to old increased our lambing from fifty and sixty to over ninety per cent.

Our next step was to save our lambs. It will be remembered that about '82, immediately after the purchase of the station by H. G.-S. and A. M. C., "lung-worm" had broken out on the east coast; everywhere it had decimated the flocks of Hawke's Bay; on Tutira at least three-quarters of our weaners had perished during several successive seasons. Already we had learnt a little about the nature of stock. In strong reaction to "dirty" country the heroic device was attempted of weaning our lambs on newly-burnt bracken, forcing them to crop the springing shoots of fern; there could be, of course, on such ground no vestige of disease. That was all to the good, but the fronds had not sufficient nutriment for young stock. Our weaners, though perfectly healthy, had not condition enough to withstand the cold of the coming winter and the scour of young grass during the following spring. It was then that the newly-created fencing proved its worth. The smaller paddocks were "spelled," that is, kept empty of stock for five or six weeks previous to weaning, so that we should be able to turn our lambs into feed well matured, full of white clover and suckling We took endless trouble with the "tail" of the mob, which was specially nursed. The result was satisfactory: only two or three score out of nearly a couple of thousand failed to come in at shearing-time, and as nothing succeeds like success, especially success after preliminary failure, the result was a stimulus to further effort. We worked endlessly long hours; we had our reward in ewes that produced a big percentage of lambs, in weaners that survived the winter.

The station was still, however, selling only wool; our surplus stock was still an abomination. Once or twice we slew and skinned the brutes ourselves, once they were boiled down for soap, once again Merritt took them for his long-suffering swine. Dispose of them as we might, the surplus stock of Tutira for four or five years realised but 1s. or so per head.

A change was now, however, about to take place. During the tupping season of '88 six thousand ewes were put to Lincoln rams. From them five months later we got a lambing of over ninety-five per cent; of these we reared and sheared over five thousand. We had for sale, therefore, in '89, some two thousand five hundred half-bred wether two-tooths. Would anybody buy them? The station had so bad a name we were by no means sure. One day, however, a red-letter day for the station, amongst the monthly correspondence tipped on to the verandah out of the canvas mail-bag, arrived an offer of 4s.—four shining silver shillings—for the two-tooth wethers. My partner's countenance of solemn joy rises before me as I write. Lord! how delighted we were!

This first sale of sound and young stock marked a stage in the annals of the run,—it had definitely turned the corner: sheep were no longer bought, imported to Tutira; they were sold, exported from Tutira.

We now clipped 10,000 sheep. This increase from '82, when rather more than 7000 half-starved brutes, 3000 of whom, moreover, lived entirely on tutu, had passed through the shed, was due to natural expansion of feeding-grounds caused by dry weather and fire, to fencing, and to the ploughing of lands round the edge of the lake.

These swamps or flats—to this day they pass under the former designation—had been drained in the days of Kiernan and Stuart. Prior to the operation they had supported a stunted growth of water-sodden flax and starved spindly raupo; as the land dried and hardened, as its superfluous water disappeared, these native plants shot up in enormous luxuriance, so that the work had not proved immediately remunerative. The benefit of the draining was now, however, to make itself felt.

In one particularly dry summer Stuart managed to run a fire through the largest of them—Kahikanui swamp. The great blaze consumed everything dry; only the two or three latest crops of leaf lay on the ground or drooped from the crown of the plant, brown, flaccid, and parboiled. This jungle of dead and dying leaves was later trampled into fibre by sheep attracted by the huge succulent sow-thistles which germinated after the fire. During the summer succeeding the first conflagration a second fire was run through large tracts of this shredded flax, completely cleaning the land and destroying the crowns of such plants as had again begun to sprout. In thinner areas, where fire could not obtain a hold, the dead crowns were chipped with spade and adze, heaped up in piles, and burnt. The ground was then ploughed and harrowed, and the season proving propitious, the soil responded as virgin land does at its first working. There was no turnip-fly, save sow-thistles there were no weeds; the turnip seed, hand-sown by myself with great care, germinated evenly. We used to ride over and grovel in search of the earliest cotyledons. Returning from long days of labour, we refreshed ourselves with the sight of the growing crop—the development of the third leaf, the rise of the deep green shaws, the preliminary thickening of the roots, the bloom as of nectarine or plum on the bulbs, the immense leaves, the giant globes of these purple-top Aberdeens,—each change in the crop was a fresh delight. It was a sacrilege to ruffle their green luxuriance, a liberty to pull one or two for human use, a festival to think of them. Except to readers who care for the brown earth and all matters that appertain thereto, I despair of picturing our satisfaction at the success of this our first agricultural work. Our crop was the healthiest and heaviest I have ever known anywhere at any time. A second fine crop was grown the following season, a third was mediocre, a fourth no crop at all; weeds, chiefly docks, had taken possession of the soil.

Up to this date all improvements had been put on to the eastern corner of the station. Over it all fencing had been run, over it all grass-seed sown. On the other hand, nothing had been done to the great residue of the property—the fifteen or sixteen thousand acres then known as the "back" country. It remained still as it had been a hundred years before.

Our new step in development was the utilisation of this desert, which we handled bit by bit. The bare dry story of the breaking-in of a typical block of fern-land has been given, but nothing has been told of the personal equation, the human element, the hopes and fears, the ups and downs, the disillusionments and triumphs of the process.

Though the work done never fulfilled our expectations, yet these years were years of considerable realisation, and of still more pleasurable anticipation. The process from start to finish was of absorbing interest: it was pleasant exploring the details of the block, discovering how far cliff and gorge were available in lieu of post and wire; pleasant searching for surface timber, quantities of which then existed — timber which must have lain seasoning for centuries on the dry pumice; pleasant flagging out the exact line for the native fencers; pleasant taking delivery of the posts stacked

Packing-posts.

in piles of a hundred each, most of them split and shaped by the axe, although sometimes enormous boles unspoiled by fires necessitated the saw and blasting charge. Then came their packing on to the selected line. Timber is an abominable load, no two posts weighing the same nor fitting into one another; the jig-jog of the horses over hill and dale loosening the loads, untoward incidents occurred in each day's work—yet, in spite of all, jogging home in the dark, another half mile of fencing laid was something over which to ruminate.

During the whole of one winter we were thus occupied — one day preparing the forty or fifty loads, strapping them into evenly-balanced lots; the next, running in the teams at daybreak, saddling up, and, after the hastiest of meals, trotting our string into the heart of the run, loading up, and driving them with jangle of strap-rings on hooks, and groaning and creaking and straining of leather. Each day saw something accomplished, something done, until at length, flat on the ground for miles, lay the material of the future fence. How delightful then to note its erection, strain after strain, mile after mile, over gorge and slope, straight as a Roman road to its appointed end; to test the deep-sunk, hard-rammed strainers; to feel the adamantine fixity of the footed and rise-posts; to observe the neat pattern of the stapling, the trim-cut knots, the final result, six wires evenly parallel and taut as fiddle-strings. A fence-line can be erected to the glory of the Lord as truly as a cathedral pile.

The block first handled consisted of about 1600 acres of fairly good conglomerate land. The preliminary step was to get it burnt—not always an easy job, for it has always been impossible to be sure of a prolonged spell of drought on Tutira. There has been anxiety always lest wet weather should supervene, lest the bracken should not dry sufficiently to ensure a clean burn. With what trepidation, as autumn approached, have we not watched the skies! for not only had the bracken to be dry, but for a perfect burning day an atmosphere of scorching aridity was also required—a cloudless sky. On the particular March morning when we thus burnt out Stuart's paddock for the first time, all went well. A fierce sun blazed uninterruptedly from a sky of deepest blue; thin wisps of cloud, signs of the coming gale, lay high over the Maungaharuru Range. By eleven o'clock—be sure we were on the spot promptly—we were waiting, one eye on our watches the other on the sky, feeling for preliminary puffs of air, handling lovingly the lucifers that would give us black ground, a sward of English grass, increase of healthy stock, and supply a long train of benefits to the beloved station. There we waited in the fern barely restraining ourselves, "calming ourselves to the long-wished-for end," reflecting that every hour, every half hour, every minute of patience, was drying more and more thoroughly the layers of brake piled one on top of another.

What anxieties have I not known during the last hour or so of such a vigil! Supposing the wished-for breeze should fail? Supposing white fleecy clouds should diffuse a deplorable damp?—forebodings dark as those conjured up in a banker's parlour arise in the mind. Supposing—I have known it happen—the sky should become overcast, yet not actually forbid a fire? Suppose there should be the tragic choice of delay—perhaps for a week, perhaps for a month, perhaps for the season? or, on the other hand, a "burn" disfigured by patches of green, marred by strips and tongues of unburnt stuff, areas of thin fern unconsumed, breaks in the black at every trivial creek and sheep-track, a crestfallen return clouded with misgivings as to the wisdom of having attempted a fire, at not having waited for a better chance?

Upon that March day, however, though, like Elijah, we scanned the sky, no cloud even like a man's hand appeared. Although all went well, and although it might be sufficient to leave it at that, some readers may care to hear the details of such a day—at any rate the writer wishes to remind himself of pleasures past and gone. Towards noon, then, the fateful match is struck, the smoke curls upwards blue and thin, the clear flame, steady at first but soon lengthening and stretching itself, arises like a snake from its cold coils. Then, as often seems to happen, the draught of the fire summons at once the waiting wind; out of the hot calm bursts forth the new-born storm; the circle of flame lengthens into a streak which, widening at every edge, is pounced upon, flattened to the ground, and furiously fanned this way and that, as if in attempted extinction. A few minutes later a line of commingled flame and smoke, moving ahead with steady roar, sweeps the hillsides.

Few sights are more engrossing, more enthralling, than the play of wind and flame. Wind in the hills, like water in its course, never for an instant remains even in its force, but ceaselessly swells and fails, waxes and wanes. In the very height of a gale the rushing charge of fire will in an instant check, the flames previously pinned down will erect their forked tongues like a crop, or lift as if drawn upwards from the earth in the very consummation of their burning embrace; the smoke, a moment previously flattened into the suffocated fern, will rise thin like steam through the winged fronds. Upon slopes exposed to greater weight of wind the pace of the conflagration quickens, forked sheets of flame that singe and scorch the shrivelling upper growth reach far ahead; forward the conflagration rolls—sometimes grey, sometimes glowing, sometimes incandescent, according to the changeful gusts. As a lover wraps his mistress in his arms, so the flames wrap the stately cabbage-trees, stripping them naked of their matted mantles of brown, devouring their tall stems with kisses of fire, crackling like musketry amongst the spluttering flax, hissing and spitting in the tutu groves, pouring in black smoke from thickets of scrub. On the tops pressed forward by the full force of the gale the roaring conflagration passes upwards and over in low-blown whirlwinds of smoke darkened with dust of flying charcoal and lit with showers of fiery sparks and airy handfuls of incandescent and blazing brake. To leeward fire is no less wonderful to watch as it slowly recedes downhill, devouring in leisurely fashion first the driest material, then sapping the stems of the later, greener, still upright fronds, so that they too bow like Dagon and fall to earth, perpetually replenishing the flames. A fire thus fed, burning against the wind or downhill, presents at night-time a peculiar twinkling, winking appearance from the perpetually recurring fall of the green fronds into the blaze, and the consequent alternations of darkness and light.

In windless hollows yet another mood may be noted: there the flames, burning slowly, stretch and dip and curtsey and sway to the draw of the gale above; in the mazes of a magic dance they take their time and measure from the wind, veering now to one point of the compass now to another, sliding and gliding in accompaniment to the unheard harps of the air. So, on that afternoon of March, like the waters of Lodore, the fire passed over Stuart's paddock, roaring and pouring, and howling and growling, and flashing and dashing and crashing, and fuming and consuming not only the block so named, but hundreds of acres besides of the Rocky Range—then included in the Moeangiangi run,—the whole of the Black Stag, and nearly the whole of the Tutu Faces.

At nightfall, over every acre unswept by the wind lay a delicate grey veil—a light ash of shrivelled fronds still retaining their shape. A tang of salt, as from the ocean, scented the air, whilst here and there on the driest flats rose thin lines of blue from smouldering totara logs. Everywhere the contours of the countryside lay dim; the sard sun, low in the dun horizon, glowed a burning, blood-red ball; like the fog of a great city, a pall of smoke hung over the land. Oh! the ride home, salt with dry sweat and black with dust, not a hair left unsinged on hands and arms, but rejoicing, triumphant. Oh! the dive into the cool lake, the slow swim in limpid water past the snag Karuwaitahi, over the shoal Tarata; alas! that the run cannot once more be broken in; alas! indeed, that the past years cannot be relived; a fire on a dry day in a dry season was worth a ride of a thousand miles.

Our paddock thus fenced and burnt, the next operation was the crushing of fern and sowing of grass seed. Like the man in the public-school Latin primer, "Rich in flocks, yet who lacked coin," we could only afford a cheap mixture. The stores of Port Ahuriri were raked for fog (*Holcus lanatus*), for cock's-foot double-heads (*Dactylis glomerata*), for seconds of ryegrass (*Lolium perenne*), for goose-grass (*Bromus mollis*), for rat's-tail fescue (*Festuca myuros*). Of hulled fog at 1d. a lb. we bought, I remember, one hundred pounds' worth. Amongst tailings and sweepings acquired were included seeds of Poa pratensis, a certain amount of white clover (*Trifolium repens*), and a handful or so of crested dog's-tail (*Cynosurus cristatus*), which then first made its appearance on Tutira, together with foxglove (*Digitalis purpurea*) and vetch (*Vicia sativa*). Many hundred bags of goose-grass we got at little more than the cost of the sacks; an immense quantity of suckling, at about one farthing per pound, was also secured.

Goose-grass, rat's-tail fescue, and fog are, I am aware, damned in every respectable volume on British grasses. In the 'eighties they were, for all that, useful to the station; any plant that sheep would eat was an improvement on bracken. As for suckling clover, it is no exaggeration to say that more than once it has saved the situation. To this day, indeed, I believe that if one fodder-plant had to be eliminated, Tutira could least well afford to lose this so-called insignificant weed. Yorkshire fog, goose-grass, and rat-tailed fescue have long since ceased to give any appreciable benefit. In the 'eighties, however, their rapid germination and growth, their ability to thrive on light land, and their heavy seeding qualities, made them relatively valuable. Luxuriance and exuberance are not the words to describe growth of any sort on conglomerate and sandstone country such as that of central Tutira; yet each of the plants mentioned contributed its quotum to the service of the run. All of them had in the first place reached the run by chance. I had noticed them thriving, and adopted the hint dropped by nature.

These grasses, together with tailings and sweepings, the winnowed dust of a dozen stores, were, with a leaven of sound seed, poured in deep heaps along the wool-shed floor. There good and bad were mixed, bagged, and stacked, a dust, like the smoke of Tophet, arising from the work. The bags were next packed out and spaced about the paddock; there they stood in pairs leaning against one another, the seeds within meditating, we may imagine, on their coming opportunity of service, on their duty of germinating where no grass seed had germinated before. Sowers were engaged, camp sites chosen, and presently ten or a dozen natives were surface-sowing ground which a week before had supported a jungle of fern and scrub.

An immense interest was attached to the first appearance of the green needles of such fast-growing species as rye and goose-grass. Scarcely behind them came the flat cotyledons of white clover and suckling. Germination was earliest visible about the lines of pack-trails where seed had been spilt from accidental rents in the sacks, on damp localities, and on fertile outcrops of marl.

The reader knows that English grass did ultimately fail on the central run. At this date, however, the owners of the station were still in ignorance of that tragic future. Happily, yet undisillusioned, they saw in their mind's eye a spreading sward of velvet green. There was excuse for such a belief; in those days there was a whiff of virtue in the soil; seed germinated during first sowings as it never did again.

To return, however, to our paddock: each ride revealed a change—first, hillsides bristling with numberless needles of green and flat clover cotyledons, then plants in their second leaf, then plants tall enough to offer a bite to stock, then hillsides faintly green in favourable lights, until lastly, a green hue overspread the entire paddock. To persons careless to the reclamation of land, the delight afforded by the bringing in of the wilderness will perhaps appear a species of lunacy. It did not then seem so to us; our paddock was a long-drawn variety entertainment, more enthralling in the development of its plot than any novel. To paraphrase Hamlet, the land's the thing. I was twenty-five then; I am more than twice that now; some interests pass with passing years; one never palls—the development of land.

It is needless to follow the history of this paddock beyond its first autumn and spring, for the contractions and expansions of a typical block have been elsewhere related. Suffice it to say that, within the year, 1500 sheep were carried where not a hoof had trod before. I acknowledge we had not done the work well or properly, but may I again beg the reader to recollect our tenure and our lack of capital. We had cut our coat according to our cloth. We had used cheap wire, we had used cheap seed; nevertheless, after thirty years' consideration of the matter, I am confident we worked not only Stuart's paddock, but the balance of the trough of the run, on the right lines.

The Rocky Staircase was the second block crushed and sown; then came in order The Image, Tutu Faces, The Educational, Pompey's, and the Sand Hills. For many seasons we managed to increase our flock at the rate of about twelve or fourteen hundred sheep a year. Prices of surplus stock, too, rose from four and six to eight or ten shillings.

In the early 'nineties we felled, block by block, most of the light bush of eastern Tutira, obtaining on the ashes of the fallen timber great crops of rape and turnip, and afterwards fine swards of rye, cock's-foot, and clover. Year by year the flock increased, until in the middle 'nineties the station passed from the 'teens to the 'ties, eventually reaching the high-water mark of 21,300 old sheep shorn, and over 9000 lambs. Tutira clipped that season a total of a little over 30,000 sheep and lambs.

After that date commenced the ebb. We began to find it difficult to keep up the numbers; the soil was everywhere losing its first exuberance of fertility, white clover was largely disappearing, rye and cock's-foot flourishing with less than their pristine vigour. The felling of the light bush of the eastern run postponed, however, for some time, any very noticeable diminution in numbers. Besides, if the flock did diminish in size, we consoled ourselves that the sheep were better grown, better fed, and better woolled than formerly.

During this increase of flock and increase of area under grass, time had not stood still; though we had enormously improved the value of the station to its Maori landlords, our own interest in it, apart from increase of stock, had been annually lessening.

We now for the second time entered into negotiations with the natives in regard to a renewal. From a business point of view this third lease was an even more unsatisfactory document than its predecessors. For reasons elsewhere perhaps excellent, but of which the wisdom was, I think, doubtful when applied to areas which could not pass out of the hands of the state, we were debarred from the obtainment of a lease of more than half the run. The rent was again doubled, with the proviso, however, that, should the western moiety be taken at the termination of the old lease, still having nine years to run, the rent of the later lease should be halved. On the part of the station there was a certain degree of security in this, for the western portion of the run was by itself not worth the rent paid for it. Its occupant would have held it under an even more precarious title than ourselves. Anyway, the new tenure was not such as encouraged improvements except those that would give an immediate return.

Again, as before, a considerable minority of native owners did not choose to sign. Looking back, I am surprised that any native owner signed at all; all of them knew that we would have consented to almost any terms sooner than be dispossessed. For the right to remain in possession they might have blackmailed us to almost any extent. They did not do so; advances of rent were asked for, perhaps, rather more often; certainly they were not requested in the abject, hang-dog, heart-broken strain I am accustomed to toady my banker; perhaps, approaching his point, the aged petitioner would allow himself to speculate as to when the station would be asking for yet another lease: "I tink some a day you rike te new rease, Te Mite. I sign all righte." Well, well, if one impecunious gentleman won't help another impecunious gentleman, what is the world coming to? Then, too, having passed all its own life in debt, the station had a fellow-feeling for its landlords, the more so as normal payments of rent to them were snapped up by creditors—frittered away in paying debts; half the storekeepers in the district knew when the Tutira rents were due—where the carcase was there were the eagles gathered together.

Advances of rent were made, however, only to the older men, and then only for serious objects.

For several years there was a rage for tombstones: one pious native had erected a monument to his progenitor; the custom caught on, others followed suit, until I was hail-fellow-well-met with every undertaker in town, and an adept in appraising the degrees of grief represented by broken columns, mourning doves, crosses, and obelisks. Then after a few seasons tombstones were "off," the station was required to advance for houses built on the European model—houses very often not used, the older folk especially, when pride of possession had palled, reverting to the warm smokiness of the raupo *whare*. A third fashion was in wedding raiment; the daughters of the tribe had to be attired for the altar. In those days Stuart and I dreaded to see a model clad in bridal attire in a Napier draper's window, lest it should attract the eyes of one of our people. We hated the advent of a wandering parson, God forgive us, lest he should rope in the morganatically unmarried couples and splice them, and the station be forced to advance for unlimited material of virginal white.

Barring the trouble given in these ways, and it was great because of the numbers of landlords, they could hardly have behaved better to us than they did behave. The Maoris are a fine race; I have lived with them for forty years, and can find almost nothing but good to say of them. The system, however, was wrong—one party in the transaction doing the work, the other holding power to evict without compensation for improvements. Even from the owners' point of view, the imperfect title and lease of only twenty-one years—too short considering the conditions—was an error. Improvements that should have been poured into the place were either withheld or dribbled out in niggardly fashion.

Little more remains to be said of this period. In the late 'nineties, the station being free of debt, the adjoining run of Putorino—twelve thousand sheep—was bought from Mr George Bee; later the Heru-o-Turea block was added.

In the summer of 1903, the station being for the second time free of debt, the writer purchased his partner's share in the three properties, together with thirty-two thousand sheep.

Toetoe Grass.

CHAPTER XXIV.

THE NATURALISED ALIEN FLORA OF TUTIRA.

We think of the colonisation by England of the temperate regions of the globe as for the benefit of her citizens alone, their domesticated animals, their domesticated plants. The scores of tribes of smaller living things are overlooked whose desire to multiply, whose lust for land, is quite as keen as that of man himself. It is of them that the following pages treat.

In the wake of our sailors, explorers, soldiers, and pioneers, they steal unnoticed, unobserved. The proverbial sun that never sets on the flag never sets on the chickweed, groundsel, dandelion, and veronicas that grow in every British garden and on every British garden-path.

Elsewhere the ancient vegetation of the run has been described. Following its destruction through man's agency by fire and stock a huge area of virgin soil was, to use a New Zealand political term, thrown open to selection. Upon the decline of the tyranny of omniscient fern, a host of ancient and eager rivals rushed upon the soil. With the assistance and assent of stock the ground was seized, not only by indigenous plants, whom we may imagine to have been for centuries eagerly waiting for expansion and jealous of their hungry foe, but by aliens brought from thousands of miles—from Europe, Asia, Australia, and America, from, in fact, the four quarters of the globe.

Each of these plants had in one way or another to reckon with the sheep, for Tutira, like other parts of pastoral New Zealand, lacks the shady lanes of the Old Country, its roadside banks and hedges, its strictly preserved game coverts into which no stock can stray.

Wanting these natural sanctuaries, perhaps not a few weeds that reach the colony may never manage to spread beyond the precincts of the sea-coast towns; on the arterial roads it is possible they may be each year eaten out by the great mobs of travelling stock that pass down country seawards, towards the freezing-works. Nevertheless, though a few may thus be exterminated, or at any rate retarded, in their up-country progress, the partiality of the sheep for others is an aid in the struggle of life. Each of them, at any rate, has to take him into account. The grasses and clovers, for instance, bargain with him for the right to live; whilst providing him with food and raiment, they utilise his body as a distributing agent. Others elude destruction by enormous seed production, or nauseate him by their taste, or escape him on cliffs and rocks, or quietly withstand him and endure his perpetual crop and nibble. One plant alone—the blackberry (*Rubus fruticosus*)—is his master, seducing him to destruction with the bait of its black fruit, openly trailing great runners for his entanglement, and finally feeding on his carcase.

Hooker's 'Handbook of the New Zealand Flora,' published in '67, gives a list of 130 foreign species then naturalised in the colony. Cheeseman's 'Manual of the New Zealand Flora,' printed some forty years later, enumerates over 500 plants thoroughly well established in the colony. Not far from half of that number are now acclimatised on Tutira alone.

Under the designation of naturalised aliens, plants have been included which, surviving the accidents incidental to germination and early growth, have reached the run "by themselves." With the exception of fodder-plants brought up and scattered wholesale on the run, species have likewise been included which, although originally carried up by man purposely, have, after arrival, proved able to spread abroad and propagate themselves under the normal disabilities of variable seasons, trampling and grazing of stock, ravages of slug and snail, competition of other members of the vegetable kingdom, not infrequently also the active hostility of man.

Scientific procedure, according to order and species, would quite fail to show the true interest of this invasion. The plants have been segregated, therefore, into groups, according to date, method, and manner of arrival. The date of arrival is in many cases certain, in all approximately correct, but it may well happen that error has occurred in regard to method and manner of travel. Some species, for instance, which I have enumerated as having probably reached Tutira in one way, may also have reached it in another, possibly by two or more routes simultaneously. Only those who have spent a lifetime noting the establishment of an alien flora can fully appreciate the multitudinous channels by which seeds can be carried, the impossibility of precise verification of their journeyings. The writer does not dare to make positive pronouncements, he has watched too long and seen too much to dogmatise. A notice, therefore, more or less detailed, has been given of each plant named, and the reader allowed to draw his own conclusions. The appearance of new species from time to time has been carefully noted. Since the afternoon of the 4th September '82, when I spotted the blossoming dandelions, golden in the turf of the Twenty Acre paddock, the rise, decline, and, in almost every case, the fall of every weed has been watched. I may say, indeed, with a fair degree of confidence, that not one has been overlooked. To begin with, probably change of every sort has been more clearly watched on Tutira than on most sheep-stations. I myself, for thirty-six years, have been on the prowl, seeking, like St Paul's men of Athens, for something new. Harry Young has been an admirable second; the shepherds of the place, cognisant of my crotchets,—I hope no more sinister word has been applied,—have also been more or less on the *qui vive.* Twenty thousand acres, nevertheless, will sound at first hearing a big patch of ground about which to make so confident an assertion. Readers will, however, recollect that in early times four-fifths of the station still lay in deep fern, amongst which no other growths could live. They will, furthermore, have learnt from the chapter on fern-crushing that outlying corners and poorer portions of a paddock almost at once reverted to bracken. In practice a comparatively small portion of the station supplied the weeds enumerated. Then, again, in this residue of the run 90 per cent of aliens have appeared about the homestead, the gardens, orchards, garden-paths, and roads. Lastly, as nearly all aliens watched on Tutira have increased in the ratio of unit, hundred, and hundred thousand, many species have been impossible to miss. It may have been easy to pass by the unit, the second year's spread could hardly have been so overlooked. The attention of the most unobservant could not but have been drawn to the hundred thousand stage.

Guns who have shot rabbits in cover not yet withered by frost and rain, know the gain of a cubit or so added to a man's stature; that advantage, too, has been mine—the advantage of height. An immense proportion of my life has been spent on horseback; from the saddle it has been possible to spy out the land from an elevation, not of six feet, but of eight or nine. Again, a great proportion of shepherd's work is work done in the early hours of the morning. Plants, grasses perhaps especially, then stand forth from one another with extraordinary clearness; discrimination, difficult at noon, is simple at dawn, when dew emphasises the most minute dissimilarities.

Tutira plants have been marked not for a day and never again, but year after year, each in its own self-selected spot. There has occurred the rarely vouchsafed opportunity of watching aliens on one particular bit of land, not as a stranger passing by, who views a particular species temporarily rampant, and who continues on his way with that most misleading fact stamped on his mind, but season after season.[1] Thus have been watched the appearance of the pioneer plant, its rapid increase, its vast multiplication as if about to overspread the whole district, the check in its numbers, its slow diminution till perhaps a weed viewed in turn as a nuisance, menace, and actual peril, dies back to the normal, specimens appearing so rarely that instead of being cursed as a foe it is welcomed as an old friend reviving an old interest.

The annals of Tutira can be read in its weeds. Each phase in the improvement, each stage in the development of the run, has been marked by the arrival and establishment of aliens particularly fitted for the particular condition. Each of the main periods in the history of the station has produced an especial flora.

In the 'sixties, when Maoris were still in occupation of the run, its acclimatised species consisted of a grass or two, such as rye (*Lolium perenne*), a few purposely planted edible fruits—Cape-gooseberry (*Physalis peruviana*), peach and potato,—a few pot-herbs, such as mint (*Nepeta cataria*) and thyme (*Thymus vulgaris*).

In the 'seventies stocking was attempted. That period was marked by the establishment of plants carried up in the body of man, as blackberry; in the stomachs of stock, as members of the clover family; in the wool of sheep, as Australian burr (*Acæna ovina*).

During the 'eighties the house and wool-shed were built and a permanent homestead established. As if by magic, there appeared those plants which seem to be almost parasites to mankind—plantain

[1] I cannot but think that the great botanist Hooker may have been misled by reports of some such temporary multiplications of aliens into the fear expressed by him as to the "actual displacement and possible extinction of a portion of the native flora by the introduced."

(*Plantago major*), shepherd's purse (*Capsella bursa-pastoris*), annual poa grass (*Poa annua*), chickweed (*Stellaria media*), groundsel (*Senecio vulgaris*), and others.

Later again, when surface-sowing commenced on a great scale, in the train of valuable fodder-plants purposely scattered abroad, numbers of weeds and inferior grasses, stowaways such as foxglove (*Digitalis purpurea*), vetch (*Vicia sativa*), hop-trefoil (*Trifolium procumbens*), hair-grass (*Aira caryophyllea*), and many more made their appearance.

When a greater degree of leisure had made possible the care of a flower-garden, there reached Tutira one by one a multitude of those plants that seem habitually to consort with their more lovely relatives—white dead-nettle (*Lamium album*), common fumitory (*Fumaria officinalis*), couch-grass (*Agropyrum repens*), pimpernel (*Anagallis arvensis*), and a host of others.

Sheep's-bit.

With tillage of the rich swamps round the edge of the lake appeared species that follow the plough—charlock (*Brassica sinapistrum*), common erodium (*Erodium cicutarium*), and others.

With the obtainment in 1908 of a satisfactory lease,—an event yet to be chronicled,—agricultural operations began in the long-neglected trough of the run. In order to discover what plants were likely to thrive on its dry soils, experimental plots were sown with grasses and fodder-plants of sorts other than those hitherto purchased. Themselves hailing from dry downs, deserts, and highland pastures, they too brought in their company weeds to correspond, weeds of a type quite new to Tutira. Thus with burnet, milfoil, crested dog's-tail, fescues of sorts, species of lotus, all purposely sown, arrived stowaways such as sheep's-bit (*Jasione montana*), bladder campion (*Silene inflata*), and other upland weeds.

Nay, even such a trivial factor as the private taste of a maid for caged canaries has enriched the station by three aliens.

In 1901 the first sods of the Napier-Wairoa road were cut. As in other cases cited, the labour of man was the opportunity of plants. In large numbers aliens such as vervain (*Verbena officinalis*), Mayweed (*Anthemis cotula*), strawberry-clover (*Trifolium fragiferum*) and many others, reached the station by pedestrianism—on their own legs, so to speak.

I verily believe were a menagerie to be established or a musical festival ordained on Tutira, plants corresponding to these forms of human activity and ingenuity would be forthcoming. Species possessing tastes in accord with the dust of cages, heaps of mixed dung, horse-flesh, and monkey-nuts would follow the menagerie. Top-hats, violins, ground resin, old catgut, long hair, and broken piano-wire would doubtless likewise produce its specialised flora.

Nor is this correspondence between certain lines of human enterprise and a certain type of plant altogether local. Not only does an alien vegetation spread in the wake of man; still more curiously it responds to the pace set by him. Since motor traffic has been possible on the Tutira road, weeds are reaching the station more rapidly from greater distances; seed that used to travel per diem twenty or thirty miles in mud adhering to a buggy, now clings to a motor-car for twice or thrice that distance. In the 'eighties and 'nineties I was well acquainted with almost every single travelling weed, long before it actually reached the station; it was a perennial interest to mark its modest movements run-wards; I knew of these strangers miles north and miles south of Tutira long before they actually attained their goal. Nowadays they come from beyond my ken, though that—for I, too, move with the times—has also been extended three or four-fold. Plants indeed have sometimes appeared as though attracted to the run merely by thought, a magic procedure which can, however, to serious readers be prosaically explained. It is resolved, say, that yarrow (*Achillæa millifolia*), of which several million seeds are required to weigh a pound, is to be largely sown in certain blocks; a second step in our chain of cause and effect is inspection of samples in the Napier stores. In one of them, the seedsman inadvertently brushes against the purchaser's coat; in another, a draught from the dusty floor overhead is blown down the ladder by which he mounts; his hand touches the tiny seeds, or his sleeves; dusting himself with his handkerchief, they are transferred from pocket to pocket; they hide themselves beneath his finger-nails, they fall into his shoes; departing, he carries away seed lifted unbeknown in a dozen ways. Reaching the station, they are shed on floors and carpets, they are swept out dry in dust; they are carried abroad glued to wet boots; they adhere to gaiters and saddle-gear. What, in fact, had been visualised but a short time ago as silvery seed, appears as if by miracle growing and green. Had Romeo to such matters seriously inclined—and there is not a word in Shakespeare to suggest that this was not the case—I am confident that, after his interview with Juliet, her nightie and hair must have been plastered with seeds.[1]

Lastly, I would fain forestall criticism as to the groupings of Tutira aliens in chapters to come; I have, at any rate, found it impossible quite to satisfy myself. The claims and qualifications of a species are not infrequently so evenly balanced that it may with nearly equal propriety be classed in several categories. To give but a single instance: I have transplanted the weeping-willow (*Salix babylonica*) from group to group till the unfortunate plant must be dazed. It has figured at one time as a garden escape, at another as a pedestrian, before becoming firmly rooted and grounded amongst the missioners. Before segregating my aliens into groups according to their manner of arrival, I propose enumerating:—

A. Species in possession of the run prior to 1882.
B. Plants reaching Tutira between 1882 and 1892.
C. Plants reaching Tutira between 1892 and 1902.
D. Plants that have appeared between 1902 and 1920.

A.

List of Plants Naturalised on Tutira prior to 1882.

Peach (Prunus persica).
Dwarf Cherry (Prunus Cerasus).
Apple (Pyrus malus) (var.).
Blackberry (Rubus fruticosus).
Strawberry (Fragaria elatior).
Vine (Vitex vinifera).
Cape Gooseberry (Physalis peruviana).
Watercress (Nasturtium officinale).

[1] Seeds in their life-history strangely resemble jests—both are distributed every day in millions over the surface of the globe, both depend on a sympathetic soil for their germination, perpetuation, and increase. Such "faceetiousness," for instance, as the pronouncement that it is "easier for a coo to climb a larrik tree tail foremost to harry a craw's nest, than for a mooderate to enter the Kingdom of Heaven," could strike root only in minds humorously cognisant of the bitterness of religious sects during a particular period in a particular country. Transported to Polynesia, such a jest would lie as dead as a cocoanut planted in a Stirlingshire garden. Punch's delightful bit of humour—Tyro at shooting party to keeper at termination of drive: "Are all the beaters out?" "Yes, sir." "Are you sure?" "Yes, sir." "Have you counted them?" "Yes, sir." "Then I've shot a roedeer"—spread in Scotland after its perpetration over forest, moor, and covert as I have noticed in Tutira the spread of a new plant perfectly suited to its environment. In the minds of guns, keepers, and beaters it had found a congenial nidus, yet such a jest broached, say, at a conference of Seventh-day Adventists could never have reproduced itself; it would have perished in an unresponsive soil.

Gorse (Ulex europeus).
Broom (Cytisus scoparius).
Weeping Willow (Salix babylonica).
Nonsuch (Medicago lupulina).
Toothed Medick (Medicago denticulata).
Spotted Medick (Medicago maculata).
Field Melilot (Melilotus arvensis).
White Clover (Trifolium repens).
Suckling (Trifolium dubium).
Red Clover (Trifolium pratense).
Yorkshire Fog (Holcus lanatus).
Lesser Quake-grass (Briza minor).
Ratstail Fescue (Festuca myuros).
Fescue (Festuca bromoides).
Sheep's Fescue (Festuca ovina).
Red Fescue (Festuca rubra).
Field Brome (Bromus mollis).
Brome (Bromus racemosus).
Meadow Poa (Poa pratensis).
Cocksfoot (Dactylis glomerata).
Rye (Lolium perenne).
Mint (Mentha viridis).
Cat's Mint (Nepeta cataria).
Garden Thyme (Thymus vulgaris).
Horehound (Marrubium vulgare).
Potato (Solanum tuberosum).
Tobacco (Nicotiana tabaccum).
Small-flowered Silene (Silene gallica).
Yarrow (Achillæa millifolia).
Great Mullein (Verbascum thapsus).
Australian Burr (Acæna ovina).
Prickly Thistle (Cnicus lanciolatus).
Cape-weed (Hypochæris radicata).
Dock (Rumex obtusifolia).
Sorrel (Rumex acetosella).
Dandelion (Taraxicum officinale).
Sweet Vernal Grass (Anthroxanthum odoratum).

B.

List of Plants Naturalised on Tutira between 1882-1892, given in Approximate Order of Arrival.

Raspberry (Rubus idœus).
Horse Radish (Cochlearia armoracia).
Parsnip (Peucedanum sativum).
Oats (Avena sativa).
Wild Oats (Avena fatua).
Rape (Brassica rapa).
Mouse-ear Chickweed (Cerastium glomeratum).
Vetch (Vicia sativa).
Crested Dogstail (Cynosurus cristatus)
Foxglove (Digitalis purpurea).
Austrian Pine (Pinus austriaca).
Insignis (Pinus insignis).
Blue Gum (Eucalyptus globulus).
Bermuda Grass (Cynodon dactylon).
Sweet Briar (Rosa rubiginosa).
Daisy (Bellis perennis).
Shepherd's Purse (Capsella Bursa-pastoris).
Narrow-leaved Cress (Lepidium ruderale).
Chickweed (Stellaria media).
Groundsel (Senecio vulgaris).
Procumbent Speedwell (Veronica agrestis).
Thyme-leaved Speedwell (Veronica serpyllifolia).
Wall Speedwell (Veronica arvensis).
Annal Poa (Poa annua).
Centaury (Erythræa centaurium).
Greater Plaintain (Plantago major)
Buckshorn Plantain (Plantago coronopus).
Ribwort (Plantago lanciolata).
Flax (Linum marginale).
Milk Thistle (Silybum marianum).
Self-heal (Prunella vulgaris).
Boxthorn (Lycium horridum).
Canadian Groundsel (Erigeron canadensis).
Hair Grass (Aira caryophyllea).
Persicaria (Polygonum Persicaria).

C.

List of Aliens Naturalised between 1892-1902, in Approximate Order of Arrival.

White Goosefoot (Chenopodium album).
Small-flowered Buttercup (Ranunculus parviflorus).
Wheat (Triticum sativum).
Ox-tongue (Picris echioides).
Vervein (Verbena officinalis).
Mallow (Malva verticillata).
Small-flowered Mallow (Malva parviflora).
Haresfoot Clover (Trifolium arvense).
Alsike (Trifolium hybridum).
Spurry (Spergula arvensis).
Tansy (Tanacetum vulgare).
Borage (Borago officinalis).
Periwinkle (Vinca major).
Elder (Sambucus nigra).
Californian Thistle (Cnicus arvensis).
Meadow Foxtail (Alopecurus pratensis).
Giant Fescue (Festuca elatior).
Hedge Mustard (Sisymbrium officinale).
Hop Clover (Trifolium procumbens).
Ox-eye Daisy (Chrysanthemum leucanthemum).
Italian Rye (Lolium italicum).
Brome (Bromus unioloides).
Timothy (Phleum pratense).
White Lychnis (Lychnis vespertina).
Californian Stinkweed (Gillia squarrosa).
Field Madder (Sherardia arvensis).
Erodium (Erodium cicutarium).
Linseed (Linum usitatissimum).
Trailing Hypericum (Hypericum humifusum).
Four-leafed Polycarp (Polycarpon tetraphyllum).
Thyme-leafed Sandwort (Arenaria serpyllifolia).
Lesser Stitchwort (Stellaria graminea).
Petty Spurge (Euphorbia peplus).
Pimpernel (Anagallis arvensis).
Hawkweed Picris (Picris hieracioides).
Water Forget-me-not (Myosotis palustris).
Evening Primrose (Œnothera oderata).

D.

List of Species reaching Tutira between 1902–1920, in Approximate Order of Arrival

Knotweed (Polygonum aviculare).
Lesser Swine-cress (Senebiera didyma).
Pearl-wort (Sagina apetala).
Amaranth (Amaranthus (sp.)).
Nettle-leaved Goosefoot (Chenopodium murale).
Thorn-apple (Datura stramonium).
Modiola (Modiola multifida).
Couch-grass (Agropyrum repens).
Black Bindweed (Polygonum convolvulus).
White Dead-nettle (Lamium album).
Field Stachys (Stachys arvensis).
Fumitory (Fumaria officinalis).
Cockspur Panicum (Panicum Crus-galli).
Pennyroyal (Mentha pulegium).
Ragwort (Senecio Jacobæa).
Green Panicum (Setaria viridis).
Canary Grass (Phalaris canariensis).
Annual Beardgrass (Polypogon monspeliensis).
Mayweed (Anthemis cotula).
Hyssop Lythrum (Lythrum hyssopifolium).
Reversed Clover (Trifolium resupinatum).
Strawberry Clover (Trifolium fragiferum).

Clustered Clover (Trifolium glomeratum).
Suffocated Clover (Trifolium suffocatum).
Meadow Buttercup (Ranunculus acris).
Pearlwort (Sagina procumbens).
Wall Barley (Hordeum murinum).
Fennel (Fœniculum vulgare).
Viscid Bartsia (Bartsia viscosa).
Basil-Thyme (Calamintha acinos).
Fiorin (Agrostis alba).
Paspalum (Paspalum dilatatum).
Lesser Dodder (Cuscuta epithymum).
Selaginella (Selaginella kraussiana).
Celery-leaved Buttercup (Ranunculus sceleratus).
Fiddle Dock (Rumex pulchra).
Bedstraw (Gallium parisiense).
Lesser Broomrape (Orobanche minor).
Beardgrass (Polypogon fugax).
Bladder Campion (Silene inflata).
Corn Cockle (Lychnis githago).
Dwarf Mallow (Malva rotundifolia).
Sheep's-bit (Jasione montana).
Wood Poa (Poa nemoralis).
Burdock (Arctium lappa).
Melancholy Thistle (Carduus heterophyllus).
Musk Thistle (Carduus nutans).
White-leaved Senecio (Senecio cineraria).
Barberry (Berberis vulgaris).
Sunflower (Helianthus (var.)).
Chicory (Chichorium intybus).
Sand Brassica (Diplotaxis muralis)
Cleavers (Gallium aparine).
St John's-wort (Hypericum perforatum).
Flax-leafed Hypericum (Hypericum linarifolium).
Reflexed Poa (Poa distans).
Mugwort (Artemisia vulgaris).
Honeysuckle (Lonicera japonica).

CHAPTER XXV.

STOWAWAYS.

LIKE stowaways from a vessel's hold, many weeds have leaped on the new-found land to seize their share of its advantages. Such species have managed to secrete themselves in grass seed, oat seed, and turnip seed, amongst the grain itself, in the corners of the sacks, or in the interstices of their material. Stowaways have arrived also in packets of flower or vegetable seeds. Perhaps in these ways a particularly large number of strangers have smuggled themselves on to Tutira. Owing to the great extent of second- and third-class country sown, also to the parlous state of the finances of the run in early days, cheap seeds were largely purchased; hundreds of bags of "seconds," of Yorkshire fog and warehouse sweepings, have been at various times scattered broadcast on its pumiceous areas.[1]

Between '82 and '88, owing to financial reasons, but little progress was made in the development of Tutira. About the latter date, however, such improvements as sowing, which had altogether ceased, were recommenced. Season by season grass was scattered broadcast over many thousands of acres. There are but few of the larger blocks, therefore, that have not from time to time yielded new aliens.

The paddock known as "Stuart's" was the earliest of the large areas thus successively fenced, seeded, and crushed. It furnished us with four stowaways—hair-grass (*Aira caryophyllea*), crested dog's-tail (*Cynosurus cristatus*), foxglove (*Digitalis purpurea*), and common vetch (*Vicia sativa*). These four arrived with stuff supplied by a friend who, at that time, owned a business in flower, vegetable, and field

[1] When in the 'eighties on one occasion tailings were required for a neighbouring run,—except Arapawanui, all east-coast runs north of Napier during the 'eighties were on their last legs,—none were forthcoming. "For," said the store-man to the would-be purchaser, "a fellow called Guthrie-Smith bought them all a month ago; he should have got seven years for it."

seeds. The two grasses had secreted themselves amongst the purchased sacks. The round peas of the vetch had no doubt been spilt on the floor of the seed store, and been subsequently swept up. The minute seed of the foxglove, too, had doubtless thus also been shovelled into the sacks. For ten or fifteen years, at any rate, before crested dog's-tail was purposely sown on the run, a few plants of this grass were always to be seen in Stuart's Paddock; vetch and foxglove flourish to this day in the Black Stag country, then included in Stuart's Paddock.[1]

Basil Thyme.

The sowing of the "Second Range" gave us basil thyme (*Calamintha acinos*), hop-clover (*Trifolium procumbens*), and meadow or "giant" fescue (*Festuca elatior*). The first-named has remained always on the original spot of its appearance, but though thus stationary has managed to survive the smothering of bracken and subsequent fires, as described in the shrinkage and expansion of open land. Hop-clover has never appeared happy on Tutira. After a struggle for two or three seasons it disappeared from the original site, and though renewed from time to time by later arrivals the plant has never managed to hold out for long. Meadow or "giant" fescue, as it is often called in Hawke's Bay, was for long represented on the run by a single plant, near the crossing of "Smother" creek. This grass, both the typical form and a German sub-species, was at a later date purposely sown, but without success. The soils of central Tutira do not suit a plant which has become a curse to the alluvial lands of the Province.

White Lychnis.

The ploughing of forty acres of alluvial land on Kahikanui Swamp

[1] We had bought from Mr Fred Fulton, besides tailings and sweepings, £100 worth of hulled fog. The fact has always remained in my mind because of the visit paid to us on that occasion. No doubt he had been informed that we were just about bankrupt, which indeed was the case, and had ridden up to see about payment. His relief, I have often thought, must almost have equalled ours when he got the cash. These were the times of touch-and-go, when we were never quite sure that any considerable cheque would be honoured.

introduced four new aliens. With the crop itself appeared wild oats (*Avena fatua*), a single plant of white lychnis (*Lychnis vespertina*), and stork's-bill (*Erodium sicutarium*) in abundance. Later, when the paddock was shut up for a crop of ryegrass seed, there appearod a few plants of meadow foxtail (*Alopecurus pratensis*). At a later date this valuable fodder-plant, purposely sown on the same piece of land, failed utterly. It failed again when sown in 1902 on one of the trial plots of the experimental farm.

The sowing of Kahikanui Hill paddock was responsible for the appearance of ox-eye daisy (*Chrysanthemum leucanthemum*), a plant unpalatable to cattle only, and which has therefore never spread on lands devoted almost exclusively to sheep.

Corn thistle (*Cnicus arvensis*), in New Zealand rechristened Californian or Canadian thistle, was first detected by me on Putorino, having arrived either in badly dressed seed oats or in oaten chaff fed to the plough team. It appeared after the laying down of a small field near the homestead.

Cnicus arvenis is in New Zealand amongst the weeds prescribed by law, weeds for which the owner of the ground upon which they appear, even to the roadsides, is made responsible. As, however, in the early days of Canterbury settlement, the attempt to deal with another thistle (*Cnicus lanceolatus*) failed, so in later days the effort to cope with the corn thistle has broken down. A few Hawke's Bay landowners have been prosecuted, a certain amount of thistle-cutting has been done in a half-hearted and perfunctory manner—sufficient, in fact, to satisfy the local inspector of noxious weeds.

Country settlers in truth have a pretty good idea of what can and what cannot be done in practice. They knew in this instance that on Government and native lands, unstocked and untenanted, the thistle was spreading unchecked. They knew, moreover, that though here and there in arable areas inspection was stern and severe, yet that elsewhere its presence was winked at. The fact is that all the King's horses and all the King's men cannot catch up a weed that has obtained a start. No action is ever taken in time; to begin with, the new plant is not noticed in its unit stage; when it numbers hundreds a few of the more observant settlers become interested; when thousands appear it is talked of as a new-comer; only when the hundred thousand phase is past, when the plant has been carried or blown abroad to every corner of every province in New Zealand, is legislation attempted. Thus at a date when Hawke's Bay settlers were being compelled to cut Californian thistle, I noticed elsewhere, in a district from which large consignments of oats and oaten chaff were forwarded over the whole colony, three different oat-fields of from ten to fifteen acres each heavily infested with patches of corn thistle in full bloom. Small holdings, thorough tillage, not legislation, are the cure for undesirable aliens.

Viscid Bartsia.

The surface sowing of felled forest-land on the distant Maungaharuru block introduced viscid Bartsia (*Bartsia viscosa*), the grass seed amongst which this stowaway reached the run having been drayed to the station. Had it been packed, the chances are that the weed would have been first noticed between Tangoio and the Tutira wool-shed, whereas I did not pick up the line of the blossoming plant until near "The Dome." It told a story easily decipherable. Near that hill during the previous autumn a pack-load had evidently been ripped; between the spot where I first detected the plant and the boundary gate separating Putorino and Tutira, seed had been jogged out in considerable quantity. After the boundary gate the plant altogether ceased. There, where the horses had been stopped, the packman had doubtless noticed the rent, and stuffed or plugged it. No more specimens at any rate appeared, until three miles further on, when the fallen bush was reached; there once again I found the new-comer sown inadvertently, flourishing in profusion, sharing the soil with its rightful owners —purposely imported plants, like rye, clover, and cock's-foot.

Corn Cockle.

In 1910, when about to handle the arid centre of the run, four new stowaways appeared on my experimental plots for dry country fodder-plants—corn-cockle (*Lychnis githago*), sheep's-bit (*Jasione montana*), bladder-campion (*Silene inflata*), and woolly thistle (*Cnicus eriophorus*). Besides the new-comers named, several other weeds had also managed to secrete themselves among the fodder-plants sown on the experimental farm; as, however, they had already reached the run at earlier dates and in other ways, they need not again be specified.

Bladder Campion.

Undeniable as it is that too much dirty seed is each year placed on the market, yet the spread of weeds is inevitable. Greater care might at best stave off the evil hour of their arrival; no legislation can completely check their journeyings from spot to spot. Not only do they travel in the seed, they cling to the sacks themselves. In this way have arrived wheat (*Triticum sativum*), barley (*Hordeum vulgare*), turnip (*Brassica napus*), rape (*Brassica rapa*), white goose-foot (*Chenopodium album*), evening primrose (*Œnothera oderata*), small flowered buttercup (*Ranunculus parviflorus*), wood poa (*Poa nemoralis*), and parsnip (*Peucedanum sativum*) also, though it had already appeared and is mentioned elsewhere.

During the progress of contract work done at any considerable distance from the homestead—ploughing, fencing, grass-seed sowing, and draining—camps are established; about them rubbish accumulates in a marvellous way, the untidy premises soon becoming strewn with torn bags and littered with old filthy sacks, many of which conceal stowaways. Straggling plants of wheat, cape-barley, turnips and rape, are nearly always to be found on such spots. Wheat and cape-barley have never been sown or used as horse-feed on Tutira,—they have arrived jammed in corners of sacks or tangled in interstices of their rough material. Oats, too, many times I have found on seed-sowers' camps, where the plant could not have been carried in by machinery or borne in the bodies of hard-fed horses. If the species named have thus reached the run, no doubt other weeds, especially crop weeds, are passed over the colony by rail, coach, and dray in surprisingly brief periods.

The average life of a sack is, I daresay, about five years, each sack in its time playing many parts.

Starting at the Bluff, the southernmost port of the South Island, a sack may only become finally useless in the far north of the North Island, having spread blights and noxious weeds from one end of the colony to the other. It may commence its career with all sorts of high ideals, with the determination to carry only Timaru wheat, Hawke's Bay ryegrass, and Akaroa cock's-foot, but has in later life to abate the lofty pretensions of youth and ultimately to submit to the carriage of ordinary grain, ordinary ryegrass, and ordinary cock's-foot. Later, still whole and presentable, our bag will be considered fit for tailings and oaten chaff. It will now perhaps cross Cook's Strait and be passed about a farming district bearing perhaps in one short jolt apples, in another onions, becoming at each trip more stained with rain and marked with mud. It is now filled with potatoes—another downward stage—and forwarded to Auckland. By this time ragged, rent, disreputable, with senses blunted in regard to weed-carriage, it may reach some struggling settler's little home in the roadless north; there, with no pride left, it will cover a bee-hive, roof a leaky hen-coop, or in a buggy act as mat for dirty boots. Lastly, the poor creature takes to drink, and hangs in a besotted state about a native settlement. There, utterly degraded, it may serve as a saddle-cloth to some galled Maori hack, and ultimately dropped, hatch out some long-secreted weed, that like a wicked action comes to light at last. It is not very often that a stowaway is thus caught red-handed emerging from his hiding-place; yet white goose-foot (*Chenopodium album*) was seized by me in the very act, a magnificent specimen, his great roots embedded in a rotten sack, one of many strewn about the site of a Maori drainer's camp.[1]

White Goose-foot.

Evening primrose (*Œnothera odorata*) has always been a fairly common plant between Napier and Petane, and between Petane and the coastal hills; yet this Patagonian would, I believe, never have made unassisted the inland journey necessary to reach Tutira. It appeared about a plough camp, a plant here and a plant there, on the site of the tents

[1] Dr H. A. Gleason, of New York Botanical Gardens, Bronx Park, tells me that the seed of Chenopodium album has been found in the dwellings of prehistoric man in Europe. Then, apparently, as now, the plant was parasitic to man, since it only grows on tilled lands,—surely an extraordinarily interesting glimpse into the long life-history of a weed.

of contractors whose last work had been done on a coastal run. Away from the warmth of the coast this yellow evening primrose is unable to perpetuate his race, and after a season or two dies off.

Along the grassy margin of the lake in hot weather the native shearers often prefer to live under canvas rather than stay in the permanent accommodation provided. It was on the vacated site of one of these temporary camping-grounds that *Ranunculus parviflorus* first appeared. This little buttercup I had known extremely plentiful about the Tangoio *pa*, whence came the shearers, but it had never appeared on the road; and as masses of it suddenly took possession of the tent sites, I have no doubt it must have arrived in the old sacks so often used by Maoris for saddle-cloths, sleeping-sheets, and other purposes.

Wood poa (*Poa nemoralis*) also first reached the run in sacking; I found it, at any rate, flourishing amongst old bags on the site of a deserted splitter's camp.

There yet remain for mention stowaways which have arrived in penny or sixpenny packets of flower and vegetable seeds; goose-grass (*Galium aparine*) thus hidden reached Tutira in a packet of spinach seed. Wall mustard (*Diplotaxis muralis*) secreted itself amongst Virginian stock; yellow pimpernel (*Lysamachia nemorum*) smuggled itself on to the station in the company of verbena. A species of silene, of which I only got withered specimens impossible to identify, came up in Harry Young's garden with mignonette seed.

With the recrudescence in recent years of white clover owing to ploughing and manures, bee-keeping has been again revived on the station. In 1913, opening a box from Rouen, France, I found imbedded in the artificial wax a single plump sunflower seed. The stowaway was planted, guarded with special care from slugs and snails, and eventually matured into a magnificent ten-foot specimen with a head of great circumference. Though doubtless unable without assistance to have reproduced itself, and therefore, perhaps, not to be properly included amongst acclimatised aliens, it is yet a good example of the strange manner in which seeds pass from land to land. Its arrival at any rate corroborates what has been said before, that no new phase of station work can be undertaken without the appearance of a corresponding flora.

CHAPTER XXVI.

GARDEN ESCAPES.

INCLUDED under the above heading are chiefly trees, shrubs, and hedge plants; escapes from the garden proper occur in lesser numbers. Some of them would, I believe, survive for considerable periods were mankind, fires, and domesticated beasts banished from the land; a few perhaps would take a permanent place in a reconstituted flora of New Zealand.

On Tutira the groves of golden willow (*Salix alba*), as well as fine specimen trees, owe their origin to a stout riding-switch brought at my request in '83 or '84 by Mr T. J. Stuart from Meanee. The original stick placed in a corner of our first garden unfortunately had to be destroyed on account of its great growth, but not before its limbs had been widely distributed over the run. Although so free a grower no seedlings appear: increase has been by stake, pole, or less often, by branchlets carried down in slips and floods.

White poplar (*Populus alba*) is another species which, although not reproducing itself by means of seed, yet merits inclusion amongst garden escapes. Planted in '85, it has of late years in a small way become a nuisance by reason of inordinate suckering.

Two or three score of pines (*Pinus insignis* and *Pinus austriaca*) were planted on Taupunga peninsula by Stuart and Kiernan in '80. Both species have in their immediate neighbourhood spread towards the south-east, their light flat seeds having been blown from ripe high-placed cones during north-western gales. In the lee of the original plantation a younger generation of each has arisen. Pinus insignis, in addition, has managed to convey itself great distances north, south, east, and west from its orginal site.

Individual pines of this species have appeared in localities so barren and miserable that sheep seldom graze over them; sheep nevertheless have, I think, been the transporting agency which alone can account for these cases of distant germination. Unless seed had found lodgment in a fleece and been thus carried far afield, it is hard to account for specimens discovered miles distant from the parent plantation. The insignis, as it is universally known in New Zealand, makes a double growth each season, one in spring, another in autumn; it is not surprising, therefore, that several of the trees planted forty years ago have attained a diameter of more than five feet.

A score of eucalypts (*Eucalyptus globulus*) were also planted by Stuart and Kiernan; though, however, the species has reproduced itself, the blue gum compared to the pine is but a sedentary plant. Only in spots where fire has swept over the ground, and but at limited distances from the parent trees, has germination occurred.

Prickly acacia (*Robinia pseud-acacia*) and wattle (*Acacia dealbata*) perpetuate themselves not only by ample suckering, but also by seeds. I have found young plants of each on favourable sites several hundred yards from the homestead. Though I have never seen birds feeding on seeds of either of these plants, they oftenest appear in company of seedling gooseberries and blackberries, obviously dropped from roosting-boughs.

The original elderberry (*Sambucus nigra*) growing on Tutira had also been on my suggestion brought up as a riding-stick. Although the small thicket grown from this riding-switch has been destroyed, yet the plant has managed by means of seed to migrate a distance of a couple of miles.

Of broom (*Cytisus scoparius*), a single plant grew for over twenty years on the site of an old clearing on Putorino. Owing to ploughing, of late years it has largely increased.

Gorse (*Ulex europæus*) had also preceded me. On Tutira there have always been half a dozen patches, spread possibly by horses, possibly by pig, in very early days from a hedge which partly enclosed a native cultivation-ground near Lake Orakai. No new thicket has appeared in my time, although the original patches have increased in size through ill-advised attempts to destroy them by fire.

African box-thorn (*Lucium horridum*) and barberry (*Berberis vulgaris*) have each been used as hedge plants; each has spread not only about the policies but to the distance of several hundred yards by means of seedlings, all of which I may add have been destroyed.

In 1914 seed from a honeysuckle hedge (*Lonicera japonica*) had germinated beneath a favourite roosting-place of minahs. The young plants had found a suitable nidus in the shade and leaf-mould of a long-established cherry grove. When, after the war, I returned, the little thicket had been smothered, macrocarp and berberry hedges about the homestead had likewise been overrun; honeysuckle had also established itself everywhere in the plantations and manuka gullies reserved for native birds.

The strawberry (*Fragaria elatior*), planted during '78 or '79 in a temporary garden on the Taupunga peninsula, was nearly thirty years later still surviving. It had been growing in turf,—turf, moreover, except for a few weeks in each year, perpetually nibbled by sheep. Proximity of scrub may have prevented stock biting very closely herbage rather less palatable on that account, otherwise this highly valued garden plant had endured for over a quarter of a century the hardships of the commonest weed. Moved to a somewhat neglected corner of a later garden, the transplanted roots have taken a new lease of life and appear determined to maintain their grip on the station both by seed and runner; but though vigour has revived, flavour has altogether gone. Grown again with full exposure to the sun, the fruit of these rescued strawberries is extraordinarily tasteless—literally is not worth the labour of gathering.[1]

When it is considered how in the animal kingdom some long-domesticated species are unable, or hardly able, like the camel, to reproduce their kind and maintain themselves without the assistance of man, the sustained vitality of many of our domesticated plants appears remarkable. It would have seemed more probable that after centuries of rich feeding, culture, and care, the stamina of such plants would have become impaired, or at any rate modified to meet these artificial conditions. Yet the strawberry, gooseberry, parsnip, tansy, horse-radish, carrot, and potato appear after centuries of care to have retained the pristine virility of their wild progenitors.

Comment has already been made on the hardihood and virility of the garden strawberry; the long persistence of the potato (*Solanum tuberosum*) seems as remarkable; that such a plant, manured and care-

[1] It may be worth adding, in view of the open question as to the relative flavour of British and New Zealand small fruit, that the taste of another variety of strawberry grown in the Tutira garden seems to me to be excellent. I believe, too, that our station rasps, gooseberries, and currants are as full-flavoured as those grown in England.

fully tended by man for hundreds of years, should endure, as it has done on Tutira for more than half a century, an entire absence of tillage, the strangulation of matted turf, the trampling of stock, the competition of cherry-suckers and the shade of trees, has always been a matter of surprise to me. Yet in 1906 I gathered tubers, healthy though deteriorated in size to big peas, from native cultivation-grounds deserted for fifty years. Very carefully disentangled from turf and replanted with fibrous roots undamaged, these peas produced that same season potatoes as large as damsons. Next year I had a profusion of well-grown tubers, blue-skinned and blue also throughout the flesh. Though of no great size, they possessed the peculiar flavour of the plant in a marked degree; in taste they were superior to the more shapely field and garden varieties of modern times.

Carrot (*Daucus carota*) grows sparsely though vigorously in some of the homestead clover-paddocks. Parsnip (*Peucedanum sativum*) and horse-radish (*Cochlearia armoracia*) maintain themselves, in spite of weeding, chipping, and digging, on the site of the original garden of 1884. The former has also in light lands persisted for years on the site of a deserted drover's camp, having probably arrived there as seed in sacking.

Though here and there in the district the vine (*Vitis vinifera*) still flourishes, it cannot be considered truly a garden escape. As, however, a solitary specimen on Tutira has survived for half a century on the site of George Bee's deserted garden near the foot of the zigzag on the Heru-o-Tureia block, I include it in our list. Another grows on the banks of the Waikoau, midway between the eastern ford of that river and the coast. The vine, I believe, never in Hawke's Bay renews itself by seed, though raisin stones, accidentally reaching dry soils, germinate freely.

Tansy (*Tanacetum vulgare*), not even now to be found as a pot-herb in the station garden, grows but on one spot on Tutira. As its arrival illustrates what must occur in the way of combination of favourable chances before a new species can appear in a new locality, as also the plant is one about whose manner of travel there can be almost no doubt whatever, it deserves the distinction of a paragraph.

I found it in occupation of ground directly beneath an angle-post in the Tutira-Arapawanui boundary fence. The upkeep of a mutual march, renewal of wire, replacement of broken and rotting posts, is usually shared by adjoining run-holders; it is alternately kept by one or the other. It was a season or two after such an overhaul by Arapawanui that I first noticed tansy. As in the case of another alien (*Bartsia viscosa*), its story is particularly easy to piece together. To begin with, in the Arapawanui garden I knew amongst the pot-herbs that there existed a substantial tansy plot. With this fact in mind, it was not difficult to imagine the order issued as to repair of the boundary fence; to note the man shoulder his spade; to observe the soil adhering to the tool; to visualise the tiny seed wrapt in its coatings of clay. So far quite conceivably all may have happened as on former occasions—the order given as before, the spade as before taken from the garden, with also, as before, earth and seed adhering to it. Now, however, under more fortunate circumstances, the earth might not, during the strapping on to the saddle, during the brushing through scrub, during preliminary repair work, have become detached along a section of the fence-traversing bush where the seed would perish for want of light; it might not, as before, have been choked on dense sward or rotted by exposure, or bitten below the crown by stock, or perished by too deep burial, or been annihilated by slugs, or washed out by torrential rains, or crushed under foot, or mildewed by blight, or baked by drought. Yet in these ways, and a score besides, the appearance of tansy on Tutira may have been for years postponed; seed may have again and again been brought up on claggy spades from Arapawanui, only to perish. On former occasions there may have been an excellent tilth provided, but invalidated by too deep burial; the season of the year may have been propitious, but spoilt by abnormal weather. At last there had occurred a combination of favourable factors, resulting in the acclimatisation of a new alien. Probably, of seeds that reach New Zealand, not one in ten thousand succeeds in establishing itself. My tansy patch is in itself an example of the particularity of certain species as to conditions facilitating germination. Though every year tens of thousands of winged tansy seeds are launched into the air, not a single one

Tansy.

has taken root. The original patch increases exclusively by spread of roots.

As remnants of conquered races take refuge in the mountains, so amongst rocks do persecuted plants longest survive. Cape-gooseberry (*Physalis peruviana*) still manages to hold out precariously in the crannies of certain limestone cliffs. The only plant of tobacco (*Nicotiana tabaccum*) got by me on Tutira was procured as far back as '83, also beyond the reach of stock, on one of the huge limestone quadrilaterals of the Racecourse paddock. Each of these plants had doubtless escaped from native cultivations in very early times.

Of the species included in this chapter, few have spread beyond a couple of miles, whilst several survive only by suckering or the carriage of broken branchlets in floods and landslips. Six years ago it would have been correct to say that asparagus, elderberry, box-thorn, barberry, gooseberry, raspberry, red-currant, and honeysuckle had strayed but a score or so yards from garden and orchard. That is no longer the case. Alien birds are year by year proving more active agents in the dissemination of alien vegetation. The blackbird, thrush, and minah especially are becoming more and more parasitic to garden and orchard; by them seeds are being carried season by season further afield.

CHAPTER XXVII.

CHILDREN OF THE CHURCH.

ANOTHER lot of Tutira aliens has carried a message which assuredly no other group of plants has anywhere been privileged to bear. They have reached the station as heralds preparing the way, forerunners making the path straight for the coming of a King. I can never view a row of thyme or clump of mint on the long-deserted site of a far inland *pa* —gifts brought from afar of frankincense and myrrh—without seeming to hear their native carrier tell his tale of the mission garden whence the plant had sprung, of the white men from across the sea, of their strange new gospel of peace and goodwill. Assuredly not one of these mission garden aliens, these children of the church, has been handled, tasted, or smelled without discussion of the donor, the austere example of his life, his beliefs.

No white man in early days visited the district of which Tutira forms a part. The population was too insignificant, the locality too wild; rumour and report of Christianity was beyond doubt first carried up-country by the medium of plants. Prior to translation of the Bible into the Maori tongue, fertilising messages from holy writ, texts from scripture, had been scattered over heathendom in the form of drupe, rootlet, and seed. As in Antioch, the followers of the new faith were earliest known by the name of its founder, so during discussion of missionary plants were Christian precepts first ventilated on the wilds of Tutira.

As we shall see, some of the plants in this group have reached the run almost directly from mission stations, others by more circuitous peregrinations from the same source.

It is impossible in this volume and in this chapter to deal, however briefly, with the story of the introduction of Christianity into New Zealand. Marsden had already visited the country, but it was not until early in the nineteenth century that a Church of England Mission Station was permanently established at Paihia in the Bay of Islands. Internecine strife was then everywhere raging betwixt the Maori tribes. In the North Island over the whole wide land, it is hardly an exaggeration to say that Paihia was the single oasis of peace and culture, the one good deed in a naughty native world. It was from this missionary centre that influences radiated which in the beginning modified the rigour of strife, and which in the end terminated tribal warfare.[1]

Between this date and that more or less general combination of the Maori clans at a later period against the encroachment of white settlement, plants from mission gardens were widely distributed. The pot-herbs, for instance, still found growing on the sites of deserted hill *pas* of Maungaharuru, must have been taken there at a very early date, for these fastnesses had been abandoned long before the Mohaka massacre in the 'seventies, long before the battle of Omaranui, long before the stocking of Tutira. They survive there still as scraps of past history, as relics of the primordial introduction of Christianity.

Although, however, the plants of this group have probably all reached Tutira more or less directly from mission sources, I do not mean to say that several of them had not reached New Zealand prior to the advent of the church. The peach (*Prunus persica*), for example, is certain to have been imported from New South Wales at an early date; its stone is of just such a size and shape as would lend itself for transport, too big to lose readily, yet small enough for easy portage; peach-stones were habitually carried in early times as gifts to inland districts. We can leave the likelihoods at this, that though the peach and probably the tobacco plant (*Nicotiana tabaccum*) also had originally reached New Zealand from Australia and had skirted the coast in the trail of the sealing and whaling industries, yet neither had been carried far up-country or far from these industrial centres.

The few Europeans then in New Zealand, sailors and beach-combers,

[1] The old-time Maori's devotion to warfare, and the levity with which he engaged in it, would be incredible were it not attested on all hands. A typical instance related by Darwin in his 'Voyage of the Beagle' will suffice: "A missionary found a chief and his tribe in preparation for war, their muskets bright and clean and their ammunition ready. He reasoned long on the inutility of the war and the little provocation which had been given for it. The chief was much shaken in his resolution and seemed in doubt, but at length it occurred to him that a barrel of his gunpowder was in a bad state and that it would not keep much longer. This was brought forward as an unanswerable argument for the necessity of dictating immediate war—the idea of allowing so much good gunpowder to spoil was not to be thought of, and this settled the point."

were not the type of men that in trim gardens take their pleasure, not the type to encourage the native in paths of peace and plenty. Neither, moreover, was *muru* an institution likely in either race to foster foresight or assist in the accumulation of private wealth. Although, therefore, it is possible that a few seeds may have been carried short distances inland, away from trade centres, it is to the Mission Station that places at the back of beyond, such as the inland east coast, owe even the peach and the tobacco plant.

Except the potato, the former was probably the earliest alien to reach the station. Peach-groves, indeed, like the two native grasses already named, everywhere marked the sites of native villages, native cultivation-grounds, even the smallest homes of outlying Maoris.

The peach-trees of Tutira in the 'eighties appeared to be about forty or fifty years old, though after a certain time their boles increase but little in girth, making determination of their age difficult. Growth is rapid at first—fruit may be gathered from seedlings in their third or fourth season,—but the very few survivors now on the run seem to have hardly altered in girth during my residence. The trees composing the little groves dotted about Tutira, Puterino, and Maungaharuru varied in numbers from several dozen downwards. They were of two distinct types, the more common variety akin perhaps to the wild progenitors of the race, its fruit ovoid rather than round, smaller than garden varieties, slightly though pleasantly bitter, its stone easily detached, the skin downy, and when fully ripe, yellow, not red; the other type in all ways similar to the peach of commerce, except that I have never seen yellow-flesh varieties growing wild.

Beneath the laden branches of these old orchards, pigs could be stalked during moonlight nights; to sheep camped in hot weather in their shade the thud of a dropping peach was a signal to rise and feed; horses, too, were fond of the fruit and could neatly manipulate and eject the stone.

About '83 or '84 the groves and outlying trees on Tutira became diseased with die-back and curl-leaf, so that ten years later but few remained, even those, like the emblem of the stranger knight in Pericles, only "green on top."

The cherry too (*Prunus cerasus*), although it had never been established locally as was the peach on old native workings, has also, in all probability, sprung from the mission garden. Comparatively speaking, the plant is a new-comer. The grove planted by Craig from roots carried from Havelock North in the late 'sixties, and flourishing near the present homestead until overrun by honeysuckle, has been the source from which suckers have been taken to other spots on the run. The cherry on Tutira, unlike the peach and almond, which germinate readily, even if but partially covered with soil or rotting grasses, has never sprung up from seed. Of all the stones that have been scattered about the homestead, men's quarters, and Maori camps, thrown from verandahs on to dug soils, emptied on rubbish-heaps, dropped in the orchards, basketed for picnics, riding and rowing expeditions, or, lastly, fallen from the trees themselves, never a stone has germinated. Nor has the cherry been given only by man an excellent chance of spreading by means of seed. The fruit of the fine plantation near George Bee's old homestead at Maungaharuru has for fifty years been chiefly gathered by native pigeons, yet there too no seedlings have appeared either in the open bush or about its edges.[1]

One wretched gnarled specimen of an apple-tree grew in '82 on the site of a deserted clearing on Pera's Flat. It was literally fleeced with American blight, which throughout the province was then threatening the existence of the wattle (*Acacia dealbata*).

Whatever may be the degree of relationship of these aliens to mission gardens, the tie between the missionary and the pot-herb tribe is very close and very intimate.

Catmint (*Nepeta cataria*), spearmint (*Mentha viridis*), thyme (*Thymus vulgaris*), horehound (*Marrubium vulgare*), were in very truth born of the church. Doubtless all of them first reached New Zealand with drupes of stone fruit, pips of apple and pear, with grain, with grasses, with seeds of trees; they were imported of set purpose to multiply and replenish the earth, for the policy of the Church Missionary Society was from the beginning practical; the earliest laymen sent forth were persons "trained to useful arts." It was to this system, indeed, that years later Archdeacon Henry Williams attributed in great degree the lasting effect of the work done.

[1] True until after 1914. Since then seedling cherries have appeared not infrequently in the manuka-clothed ravines near the homestead. I am inclined to believe these seedlings now for the first time germinating on the station have been taken by birds from the so-called barren double-blossomed Japanese varieties, many of which mature a tiny fruit, or from the great cherry-trees of several garden varieties in Harry Young's garden, the whole of whose fruit has for years been stolen by birds.

That pot-herbs especially should have been so brought out from England seems the more natural, when it is considered how large a part the still-room played in the lives of gentlewomen of a century ago.

Each of the plants which in New Zealand we now look upon as a mere weed was then well known for its medicinal virtues. The warm sweet seeds of the fennel (*Fœniculum vulgare*) were valued as a carminative medicine for infants. Pennyroyal (*Mentha pulegium*) was held in high esteem for culinary and pharmaceutical preparations. From other mints was expressed an oil used in medicines as an excitant and stomachic for promoting digestion. Thyme was in high favour as a flavouring; the extract of horehound was a remedy for coughs and asthmatical complaints.

Only a generation without physicians, dependent on medicines from its own gardens, can fully appreciate these simples of an old-fashioned past. The spread of pot-herbs was very rapid, because their value was very great.[1]

Catmint.

Of catmint (*Nepeta cataria*) one clump grows, and has grown for years, on the site of a native clearing in the Maungahinahina, another on a Maori cultivation-patch on the Maungaharuru Range. Its seed never germinates on the surrounding turf; the plant never spreads. I am convinced these two clumps could only have reached their present sites by the intervention of man; doubtless they were brought direct as rootlets in very early times.

Spearmint (*Mentha viridis*) flourishes on the margin of streams, and often covers roods of marsh. Once introduced inland, its rootlets carried down in floods, its seeds attached to the plumage and feet of wild-fowl, the plant would rapidly overrun the travertine deposits where it specially luxuriates.

[1] Doubtless scores of other medicinal plants were imported, though comparatively few may have been able to propagate themselves. I have it from members of the Williams family that Charles Darwin, walking at an early hour in the Mission garden at Paihia, gathered sage-leaves for breakfast. Of his hosts and of their garden he was more appreciative than of their country. "I took leave of the missionaries with thankfulness for their kind welcome, and with feelings of high respect for their gentlemanlike, useful, and upright characters. I think it would be difficult to find a body of men better adapted for the high office which they fulfil." It is heart-rending to read his additional remark. "I believe we were all glad to leave New Zealand. It is not a pleasant place."

Thyme (*Thymus vulgaris*), another plant that could only have reached Tutira by deliberate carriage, manages, after half a century, still to retain its trim compact shape. On Maungaharuru and elsewhere the original rows were, until lately—and may yet be, if not grubbed by pig—distinct on the ancient garden-plots of long-deserted native villages. The seed of thyme, like the seed of tansy, never germinates in turf; no beast eats it. When met with, the plant is a sure and certain indication of bygone settlement.

Horehound (*Marrubium vulgare*) grows plentifully on many parts of the run, but especially prefers sheep-camps on conglomerate outcrops. Under certain conditions the plant acts as an aphrodisiac.

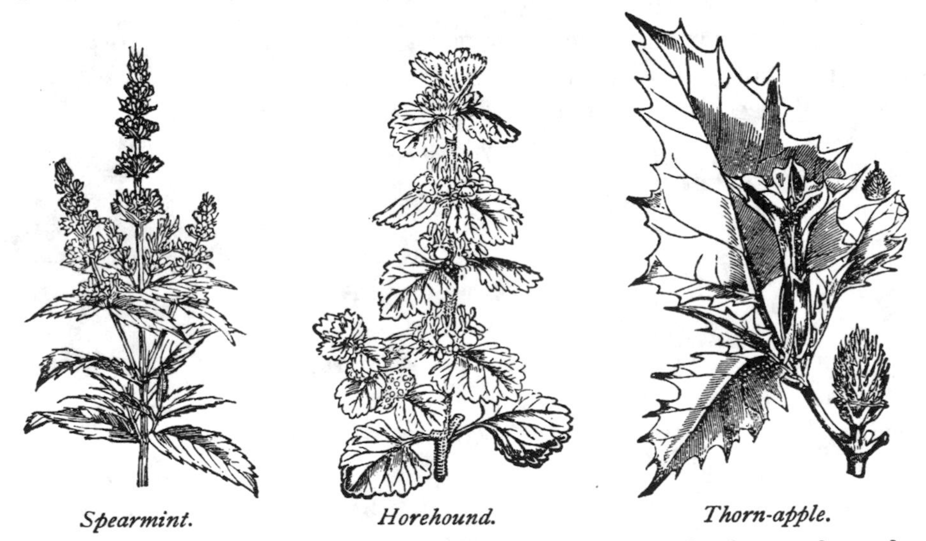

Spearmint. *Horehound.* *Thorn-apple.*

Twice I have noticed that rams pawing and nosing the leaves have been stimulated as by the proximity of a ewe eager to mate. On each occasion I believe the discovery was a chance discovery, but having been experienced, persistence in the bruising of the plant was prolonged with visible results.

Thorn-apple (*Datura stramonium*), or, as it is still called in Hawke's Bay, "Priests' Weed," has on two occasions appeared at Tutira. As a plant likely to be of use in pulmonary affection, it was distributed in early days throughout the *pas* of Hawke's Bay by the Rev. Father Regnier of the Meanee Mission Station. It offers yet another example of the esteem in which medicinal herbs were held in the early years of last century.

The universal spread of water-cress (*Nasturtium officinale*) points also to early importation. As it is not a plant the frolicsome sealer is likely to have burdened his memory with sailing for New Zealand waters, the chances are that it also is of a missionary origin.[1]

The manner of arrival of ryegrass (*Lolium perenne*) is still another instance of that appreciation by early converts of mission plants—a proof, too, of the immense distances seeds were carried by neophytes and scholars. Its collection, carriage, and subsequent neglect are also typical at once of the mingled intelligence and carelessness of the Maori character. Ripora, afterwards the mother of my friend 'Pera, was as a girl educated at the Bay of Islands Mission Station. She it was who first brought ryegrass to Tangoio about '34 or '35. It had been gathered at Paihia either from one of the newly-sown missionary fields, or saved from plants that had already spread about the native quarters. Stowed away safely during the overland march, guarded from sea-water during the long canoe-voyage south; on arrival neglected in the *whare* at Tangoio, the cloth of the containing-bag torn by mice, the once-treasured seed was finally flung out in forgetfulness. There, falling on fertile soil, it germinated like the barley and rice cast forth by Robinson Crusoe before the entrance of his cave. The exact manner of its ultimate arrival at Tutira can only be surmised, but probability points to the equine stomach. Horses ridden from the one place to the other scattered it along the trail and dropped it on the station. It is curious to think that this, the most valuable grass in the province, should have reached Tutira

[1] There are no snakes in New Zealand, but water-cress stories seem in early times to have been almost equally alluring. I hope—I trust—that none of those who wrote about the plant to their friends in England were missionaries; at any rate, the tradition of giant water-cress in New Zealand yet lingers in the Old Country. Not once, but several times, at home I have been commiserated with on the fact that our rivers in New Zealand were blocked by the plant—that inland navigation was hindered by its growth. On one occasion, when recovering from typhoid, and under the care of a doctor whose forebears had been connected with missionary enterprise, as I had foreseen and dreaded, condolences were offered about the plant and the misfortune to the colony of its importation. At first I struggled to state the facts; but finally—he was very insistent and positive, and I weak—I let him go away in the belief that its stems were larger than those of the British oak, and that if by chance a pair of moa still lived they would infallibly choose to nest in a water-cress jungle. The truth is, that on the Avon, and possibly elsewhere, small areas of water really were densely covered with a sud of rootlets; possibly, too, the navigation of small row-boats may have been in some degree hindered over insignificant distances.

long before the run was taken up, and years before a grain of grass-seed had been purposely sown on the station.[1]

We now come to the best known and most widely spread of all the missioners—the willow (*Salix babylonica*). A willow leaf must indeed in early times have been to the natives of New Zealand as the olive leaf to the inhabitants of the Ark—an emblem of hope, an indication that the deluge of bloodshed, strife, and rapine was abating from the face of the earth. It has been carried from the tree weeping over Napoleon's tomb at St Helena to the original Church of England Mission Station at Paihia in the Bay of Islands, to other Mission Stations of later date, and thence spread everywhere. About this tree—the ancestor of the willow-groves of the Dominion—Sir Henry Galway, at one time Governor of the island, has kindly given me the following information. He writes: "I have to-day received a reply from St Helena *re* the Napoleon willow, but there is nothing in that reply to show the country from which the original willow was imported. That being so, I am satisfied nobody in the island can give information on that particular point. My correspondent says that the willow, with other trees, was imported into the island by the East India Company, and that the Tomb Valley, then known as either Sane Valley or Geranium Valley, was one of the areas in which the willows were planted. The willow under discussion was growing before Napoleon arrived in St Helena, and the grave was dug quite close to it. I send you, under separate cover, a print of the Tomb, the original having been drawn after the exhumation in 1840. The original willows disappeared very many years ago, and those now growing are the great-great-grandchildren of the original trees."

Like the sago-palm to the Indian, the willow to the settler in New Zealand is useful in a score of ways: it can be pollarded for stock during drought, it can be planted for the drying-up of marsh and well-head; as no fencing is required, individual trees can be

[1] The Poverty Bay ryegrass, so famous throughout Australasia for germinating quality and weight, is also directly descended from missionary sources. Mr J. N. Williams has told me it was first noticed shortly after the shipment of a couple of cows from the Bay of Islands to the later-established headquarters of the Mission on the Waipaoa river. There the grass appeared, having either been carried in the animals' bodies, or amongst hay shipped as fodder for the voyage. Mr Williams' brother, the late Bishop of Waiapu, has informed me, too, how rapidly and thoroughly it killed out the native Microlœna stipoides then in possession of the whole of the Poverty Bay flats.

Napoleon's Tomb at St Helena.

Showing the original Weeping Willow from which have sprung the single trees and groves that beautify every part of New Zealand.

placed where wanted without cost or care. Slips as thick as a man's wrist, and of sufficient height to stand above reach of cattle, thrust into suitable soil, will in a few years provide a fine circumference of shade. It is the harbinger of spring, the verdure of its pendent trailers bearing promise to the struggling settler of warmth returned, grass-growth, and lambs on hillsides once more green. Bare of leaves for only six or seven weeks of the year, this exquisitely graceful tree has become—in the north at any rate—almost an evergreen. The growth of the weeping-willow, in fact, is so rapid, its vitality so exuberant, that, had it not perforce remained celibate, the waterways of the colony would have been seriously affected. The old-world origin of our New Zealand groves is, as stated, the celebrated tree growing over Napoleon's tomb at St Helena. From it cuttings were brought to the Bay of Islands Mission Station by two English ladies—Mrs Malcolm, wife of Admiral Malcolm, and Mrs Abel.[1] These ladies, reaching New Zealand by one of the sailing ships which in the early part of last century were accustomed to call at St Helena, presented a box full of small rooted twigs to Henry, afterwards Archdeacon Henry Williams. His daughter, Mrs Davies—now a venerable lady of ninety-nine, to whom I am indebted for the anecdote of Napoleon and for other information—recollects well the circumstance of her father's call on the newly-arrived ship, and his enthusiasm over his cuttings.[2] From the Bay of Islands the willow spread south.

Reaching the Meanee Mission Station in Hawke's Bay, a slip was taken by Colenso, who planted it at Tangoio, where in '85 a willow-tree grew measuring over seven feet in diameter. From Tangoio a slip was carried to Tutira and planted on the *pa* Te Rewa. That tree is dead, but its branches have populated the station. In the 'seventies the Stuart Brothers and Kiernan had begun that planting which has since so beautified the place. Trees were still, however, in the 'eighties, few and

[1] Mrs Abel had a small daughter on board. This child had been a favourite of Napoleon; with her at St Helena he used to play, even apparently to romp, if such a skittish term can be applied to the movements of the Man of Destiny. On one occasion, at any rate, the girl during their play managed to capture his sword, exclaiming in glee, as a child might, that she had done alone what the nations of Europe had leagued to accomplish. Napoleon never forgot this speech, or cared again to play with the child.

[2] There are probably willows of the same descent in England also. Major-General Smith, to whom this book is dedicated, when returning from service abroad, recollects brother officers having to make restitution of slips clandestinely gathered from St Helena; doubtless, however, all offenders were not detected by the sentinel on guard.

far between, one row growing on the western edge of Waikopiro lake, another on the south of the Taupunga peninsula; three trees stood on the wool-shed peninsula, three more on the little flat where the present homestead has been built; now there are hundreds round the lake itself, and along the edges of the alluvial flats, planted chiefly by Harry Young, Jack Young, George Whatley, T. J. Stuart, and myself.

Sweet-briar (*Rosa rubiginosa*), "Missionary" as it is still called, has been spread abroad by the horse. Though I could do so, it would be wearisome to the reader were each plant located. It is enough to say that in the 'eighties, between Petane, probably the local starting-point of the plant and Tutira, bushes grew scattered at long intervals. Many years later, long after the local extirpation of these pioneers, the station was again invaded, plants appearing plentifully on the Tutira–Heru-o-Tureia track. Grass seed packed from the station was being sown on that distant block, where there was then no holding paddock. The hungry horses fed about the old Maori briar-infested cultivations, devouring amongst other rubbish quantities of red ripe hips. Returning without loads and driven fast, their stomachs were emptied throughout the "Wild Horse Country," "Nobbies," "Educational," "Second Range," "Dome," "Image," and "Natural Paddock." Sweet-briar originally, therefore, reached the run from the south, but later from the north-west. It has never spread so dangerously as the bramble. The equine stomach is more expeditiously emptied; as no rider, moreover, during a journey would willingly allow his steed to gorge on hips, each outlying bush has not served, as in the case of the blackberry, for a fresh reservoir of replenishment. As its local name—"Missionary"—implies, sweet-briar too is a child of the Church of England.

CHAPTER XXVIII.

BURDENS OF SIN.

ANOTHER group of new-comers can quite well be pictured as having been dumped on Tutira—carried up, that is, and cast down as suddenly as Christian in Bunyan's allegory rids himself of his load. Many burdens of sin have thus on the station been dropped by many pilgrims, some of them living animals, some of them larger members of the vegetable kingdom itself, and some of them not living at all, insensate, inanimate, though endowed with motion. The larger have carried the lesser on their backs, in the manner that peasants in Egypt to this day account for the simultaneous yearly appearance of the stork and quail.

Aliens of this group caught in the very act, whose origins are known as positively as there can be certitude and finality in such matters at all, are given priority of place. Others there are about whose method of arrival the writer himself entertains no doubt, but regarding whose appearance innumerable details of likelihood, if given to prove the point, would too greatly cumber our story; others again there are which may have reached the run in one of many several ways. Sweet vernal-grass (*Anthroxanthum odoratum*), growing in '82 in close proximity to the cherry-grove near Craig's *whare*, was in all likelihood carried up in roots or earth adhering to them. Suckers had been taken by Craig from a long-established clump at Havelock North, a township within a short distance of light river-bed country scores of acres of which used to be densely covered with sweet vernal. For twenty years after my arrival, at any rate, sweet vernal was elsewhere unknown on the station.

Procumbent speedwell (*Veronica agrestis*) was carried up in soil attached to the roots of certain moss-roses planted immediately after completion of the original cottage in '83. To this day I think with pleasure of their blue blossoms flourishing splendidly on our first tiny ill-tended plot. The plant has never strayed far; it is a garden rather than a field species, and has for more than thirty years remained within a few dozen yards of the spot of its first appearance.

Seed of modiola (*Modiola multifida*), which I had long noted in Napier as a persistent garden weed, was brought up by Harry Young in a bundle of cabbages; at any rate, the plant appeared where they had been grown.

Pondweed (*Potamogeton polygonifolius*) was introduced from Pouriri with water-lily roots which were sunk in the lake. They died, but for a season or two the pondweed flourished with the same exuberance of vitality as is shown by land plants enjoying their first taste of fresh soil.

Daphne from a Napier nursery carried up white dead-nettle (*Lamium album*). During one of the many alterations in shape, size, and locality of the station gardens, it managed to extend its range into the rose-beds. Later again, taking advantage of my absence during the war, it has still further enlarged its domain. In spite, or perhaps because of, thorough autumn forking, it survives through the persistent rooting of each broken fragment.

Lilium candidum, which used to grow with me over seven feet in height until disease seized upon the plant, brought up black bindweed (*Polygonum convolvulus*).

With bamboos from Sir William Russell's garden at Flaxmere appeared fumitory (*Fumaria officinalis*).

Pot-moss (*Selaginella* sp.), though but a very doubtful acclimatisation, appeared and spread for a single season where a pot of nerine bulbs had been sunk into the garden soil.

Twitch-grass (*Agropyrum repens*) was introduced in clumps of kniphofia bought from a Hastings nurseryman.

Portulaca (*Portulaca oleracea*) came up attached to a bulb of Lilium giganteum given me by the late J. N. Williams from his Frimley garden. Petty spurge (*Euphorbia peplus*) and field stachys (*Stachys arvensis*) were almost certainly attached to earth adhering to roots. They appeared, at any rate, immediately after the importation of a number of herbaceous plants from England.

Amaranth (*Amarantus* sp.) and lesser swine-cress (*Senebiera didyma*), I believe, reached Tutira with a consignment of fruit-trees for a new orchard.

On the other hand, I have often been surprised at the non-arrival of plants. An introduction, for instance, of many score roses from Auckland, a source of weed-supply hitherto untapped, brought not a single new weed.

The earliest turf turned over in my day on Tutira, and devoted entirely to potatoes, produced none of the following plants: groundsel (*Senecio vulgaris*), chickweed (*Stellaria media*), pimpernel (*Anagallis arvensis*), shepherd's purse (*Capsella bursa-pastoris*), thyme-leaved speedwell (*Veronica serpyllifoliam*). They were not in the soil of Tutira in '82. When, however, we moved to the spot where the homestead and the garden were permanently established, when gooseberries, currants, strawberries, raspberries, stone-fruit, apples and pears were planted, the above-mentioned aliens appeared one by one. There are, in fact, certain species that live particularly about gardens, garden-walks, and garden-bed edgings. Their tiny seeds in a score of ways reach the pockets, clothes, and boots of every labourer employed. He manures them when he dungs the ground, he plants them with his cabbages, he sinks them in his celery-trench, he forks them with his asparagus. They cling to his tools, his pea-stakes, his matting, his garden line. He mixes them in his potting-shed with shredded turf, with sand and leaf-mould. In a hundred ways they are disseminated. It is impossible for the best and most careful firms to forward only the plants ordered. All sorts of things are added gratis. I have received a Somersetshire worm with Kelway's delphiniums; with Mariposa tulips, even Barr & Sons have forwarded weeds.

Other plants in this list have arrived by other modes of transit, though always, like those already named, dumped down suddenly. The private taste of a maid has, for instance,—it is another example of what has been already noticed, that every episode in station life, even the most trifling and ephemeral, has been marked in weeds,—the private taste of a maid, I say, has been responsible for *Setaria viridis, Panicum crusgalli*, and canary-grass (*Phalaris canariensis*). These three plants appeared within a season or two of the arrival of her canaries. Before that they had not been seen. The last-named has extended its range; I have noticed it on ploughed ground across the lake. Setaria viridis and Panicum crusgalli have not, to my knowledge, yet strayed beyond the precincts of orchard and garden.

Spurrey (*Spergula arvensis*) was dumped on Tutira by machinery. It first appeared directly beneath a second-hand hay-rake brought from Taradale and allowed to stand temporarily on newly-ploughed land. This particular settlement of spurrey, which must have reached its destination in clay glued to the machine, was destroyed. The plant has, nevertheless, elsewhere taken possession of soils suitable to its requirements. Probably in a second attempt it smuggled itself in as a stowaway.

Other pilgrims depositing burdens on the station have been animals. After the camping for a night in Whatley's Paddock of several hundred travelling cattle, *Senecio cineraria* appeared plentifully. Musk thistle (*Carduus nutans*), too, appeared where travelling cattle had stayed a night in another paddock, the "Twenty Acres." There is the further evidence of the arrival of this plant in this particular way, that several head of the mob had been "dropped" by careless droving in the Natural Paddock and had found their way to the main camp, where in due time several colonies of Carduus nutans also appeared. The idea, by the bye, that donkeys only eat thistles, is quite erroneous, horses, cattle, and sheep alike being partial to the purple flower-heads.

Knot-grass (*Polygonum aviculare*) was first found in front of the stable doors; as I have elsewhere seen horses freely cropping the plant, it may have, in the first instance, been carried up and dumped down by a horse.

Clustered clover (*Trifolium glomeratum*) was brought up in the stomachs of stock borrowed from Mr Bernard Chambers. The plant, at any rate, had not previously been noted on the run, and did grow thickly on Te Mata, whence came the sheep.

Suffocated clover (*Trifolium suffocatum*) has also reached Tutira in this way, the plant appearing thickly on certain camps where there had been none previously.

A magnificent plant, milk-thistle (*Silybum marianum*), has spread from Arapawanui, where, according to the late John Mackinnon, it appeared in the early 'seventies. I believe it has been carried to Tutira by pig; at any rate it has been dumped down on the run where cattle and sheep at that date never fed, and years prior to the sowing of grass seed. The plant, furthermore, in its leafy prime, is too prickly to have been willingly touched by stock. During winter, however, when the vast foliage fails, it is not improbable that seeds may have been picked up by rooting wild pig and thus carried from spot to spot. The plant has a remarkable local history. It grew on Tutira in a locality where we had occasion to erect sheep-yards. These were not only used many times every season for drafting, docking, &c., but were built, besides, on a flat-topped ridge over which one shepherd or another rode weekly, or oftener.

Milk Thistle.

Lastly, there was a spring in the immediate neighbourhood convenient at noon for boiling the billy which shepherds carry slung on their saddles. Traffic, in fact, on that ridge was continuous from one year's end to another. The original forty or fifty specimens of milk-thistle were spaded out—cut below the crown by myself; undoubtedly not a single plant was missed, for, apart from the fact that I would be careful in my own interest, the weed had elected to settle on a fertile sheep-camp, where the grass was closely nibbled, and where, because of the fertility of the soil, any specimen missed would have become in summer-time a plant five or six feet high, peculiarly apparent and conspicuous. Seedlings, nevertheless, appeared for twenty-five years on an area 30 feet by 60 feet—one season a rather less, another, a rather more, numerous germination taking place. Evidently the seeds possessed, like the units of egg-batches of certain moths, the property of hatching out at widely different intervals of time, thus ensuring a propitious period sooner or later.[1]

The daisy (*Bellis perennis*) merits mention not only on account of its manner of arrival, but because the plant has proved quite exceptional in its rate of spread. Unlike the majority of aliens, it has increased slowly, even on soils afterwards found to be entirely suitable. Locally

[1] It came during the great war, whilst the station was depleted of its men; specimens then not only seeded on the original site, but spread elsewhere; I notice the alien greenfinch seems now to be feeding on the seeds.

the daisy is still, indeed, almost unknown outside its original spot of deposition—the Home Paddock. Whilst still a rare weed elsewhere in the province,—I remember its absence in the 'eighties from the beautiful tennis lawns of southern Hawke's Bay,—it was plentiful near the original Tangoio wool-shed and drafting-yards. The daisy reached Tutira in the following way: During the 'eighties our station stores were drayed to the Tangoio wool-shed, and there deposited until such time as called for by the pack-team, which was penned in one of the yards, the horses being led forth and loaded one by one. Easily balanced cargo, such as flour and sugar, was disposed of first, odds and ends, small parcels, &c., were reserved for the "last" horse, an imperturbable beast treating shouts and stock-whip crackings alike with bovine indifference. These odds and ends were not infrequently placed in sacks, slung directly from the iron hooks of the pack-saddles, and therefore, if properly balanced and firmly fixed with surcingle, secure from the chances and changes of jostle, jog, trot and canter, over miles of execrable going. On the particular occasion to which I refer, the load of this "last" horse was badly balanced. To right the equilibrium I remember hastily spading up two or three divots from the turf of the paddock and flinging them as ballast into the mouth of the lighter sack. Upon arrival at Tutira its contents, parcels and earth alike, were tipped on to the ground nearly opposite the front gate of our newly-finished cottage, and there the daisy first appeared on Tutira. Though, however, thus established, the increase of the plant was extraordinarily, exceptionally slow. It seemed as though the daisy, almost alone among aliens, had been unable to devise methods of dissemination, or perhaps that the plants which ultimately flourished were variants from the type, more exactly suited to novel environment. Fully fifteen years passed before the few acres about the house were overrun; then the multitude of expanded blossoms was a marvel; nowhere else have I seen such an exuberance of bloom: in spring sunshine the paddock lay white as if under snow.

Prickly burr (*Acæna ovina*) was well established in the paddocks near the wool-shed prior to my time. Although sheep will crop the leaves and tender seed-stems, they are left untouched when tough and stiff. It is likely, therefore, that this very early arrival may have been carried up in the wool of early imported merino sheep.

Two plants yet remain which probably owe their transportation to a very intimate relationship with other members of their own kingdom. One is lesser dodder (*Cuscuta epithymum*), a single small specimen of which I first discovered growing on mint (*Mentha viridis*) in Peras Swamp. Afterwards the plant made, I believe, an independent second appearance in the homestead paddocks, where there are now established large circles of dodder amongst the red clover crops. The other parasite is broom-rape (*Orobanche minor*), a single specimen of which was first seen on the turf of the home paddock, where it appeared to be attached to the roots of cat's-ear (*Hypochœris radicata*). Several years later it also appeared plentifully amongst red clover.

Broom-rape.

CHAPTER XXIX.

FIRE AND FLOOD WEEDS.

PREVIOUS to the 'eighties the effects of burning out the indigenous vegetation of the run had been almost imperceptible; ground temporarily cleared had immediately lapsed into its former condition. As, however, the flocks and herds of the station increased, the ground consolidated and the bracken growth diminished; above all, as light penetrated to the surface, certain weeds one after another temporarily took possession of the fire-blackened tracts. On the better soils of eastern Tutira appeared such plants as Melilotus arvensis, Medicago lupulina, Medicago denticulata, and Sonchus oleraceous; on lands good and bad, Carduus lanceolatus and Briza minor; on grass lands over which in dry summers fires had run, Bromus mollis; on pumiceous lands, Hypochæris radicata, Silene gallica, Cerastium glomeratum, Trifolium dubium, and, at a later date, Erigeron Canadensis.

Taking these aliens in the order named, field melilot (*Melilotus arvensis*) has never spread beyond the alluvial lands around Tutira lake. Only after flax-fires great or small did the plant show itself; then on rich grounds left black and bare, it appeared, tall, rank, and luxuriant, for a single season. Toothed medick (*Medicago denticulata*) and nonsuch (*M. lupulina*), other fire weeds, throve only on fertile hills and flats; they never even germinated on the pumice lands of the trough of the run, though their seed was prominent in the numerous sacks of tailings scattered broadcast over that area. Neither of these members of the pea-flower family took possession on a great scale: I have never seen Melilotus arvensis spread over more than thirty acres as a dense crop, whilst the others never overran more than a few square yards outside of the garden; they were only to be found prominently on land over which fire had passed.

Sow-thistle (*Sonchus oleraceous*) was another most prominent fire plant temporarily possessing hundreds of acres of newly-burnt forest land. I have seen brairds of this weed so thick that whilst the plants were still young and flat on the ground, the surface seemed rather blue-green than green, owing to the young leaves' peculiar hue. On soil deep in leaf-mould and grey with ash, millions upon millions of seedlings germinated. As the cotyledons appeared immediately after the first rain, it was evident that seed had been already strewn on the forest floor; carried by the winds from scattered individual plants surviving on cliffs and natural escarpments, lightly buried in leaf-mould and debris, they had but awaited the call of the sun. The sow-thistle flourished with an equal exuberance after the destruction by fire of flax-swamps.

Prickly thistle (*Cnicus lanceolatus*) was, in the 'eighties, only to be seen in quantity on the seaward fertile portion of the run. Unlike the weeds already mentioned, each of which blossomed as annuals, thistle growth was dependent on the nature of the soil. On sound marl land after autumn fires the seedlings germinated, became great prickly stars during winter, and blossomed during the succeeding summer. On the other hand, throughout the pumice stretches of central Tutira, it became a biennial, reaching maturity only during the second season.

Another plant, not to be found except after fires had swept the land bare, was the little quake-grass (*Briza minor*). I have found this handsome species on almost every part of Tutira, but never in any single instance plentiful. It has indeed been discovered by me in such out-of-the-way spots at such early dates that I have sometimes wondered if after all it may not be an indigenous species; repeatedly, moreover, I have got it where before fire the countryside had been a sheet of dense fern, where alien grass had never been sown, where almost no human foot had trod. Guesses, at any rate, can be made at the methods by which most species transport themselves from spot to spot, also as to the agencies animate or inanimate employed, but the propagation and spread of this grass remains a puzzle.

In the early 'eighties occurred two dry seasons during which grass fires were run over the hot western and northern faces of certain portions on eastern Tutira; on the blackened ground then and once again seedlings of brome-grass (*Bromus mollis*) appeared in so great profusion that other grasses were temporarily submerged in hillsides of waving hay.

"Poor Pretences" was then the name by which this brome-grass was known on the east coast: it may be worth putting on record a suggestion as to how the plant came by such a curious designation. It was a "poor pretence" compared with ryegrass, cock's-foot, and white clover, the mixture sown then almost as a religious duty on Hawke's Bay runs, good and bad alike. So much for the sense; as regards the sound, "poor pretences" can, I think, only be a corruption of *Poa pratensis*; the Latin name of the one plant done into English has been fitted to another. The reader will recollect how this grass—more widely known as goose-grass—was sown wholesale over the trough of the run. Nowadays, like scores of other aliens once prominent, it has practically disappeared.

Small-flowered Silene.

Each of these seven weeds appeared after fire, though none of them overran, like Silene gallica, hundreds of acres; like Cerastium glomeratum, Hypochæris radicata, Trifolium dubium, and—at a later date—Erigeron Canadensis, thousands of acres. The extraordinary spread of small-flowered silene (*Silene gallica*), after allowing for its taste for soil of a loose light texture, was due to two especial factors,—one the viscidity of the plant's stalks and stems, the other the nature of the sheep then on Tutira; they were merino, not a bare-legged breed, but sheep, on the contrary, wooled to the toes. Fragments of silene adhering to their shanks were thus carried wherever sheep trod; in the vicinity of "Flower Hill" a single stretch of more than two hundred acres was in 1886 densely covered with silene. In another locality the plant came up at a later date in equal abundance. It is still a common weed alongside pumice paths stirred by sheep traffic, but has elsewhere almost ceased to appear.

Mouse-ear Chickweed.

Another fire weed that thrived prodigiously on newly-burnt land was mouse-ear chickweed (*Cerastium glomeratum*). Scattered as great healthy plants, though never forming anywhere anything approaching a matted growth, this alien in the height of its luxuriance grew not over hundreds but over thousands of acres. After fires on the "Staircase" and on the "Second Range" these paddocks assumed in certain lights a strange grey-green hue. It was caused by thistle-down blown from other blocks and lodged in the chickweeds' sticky trails. It also has nowadays become almost a rare plant.

Cat's-ear.

In '85, during a dry season, fire ran over the Image Paddock, destroying the only considerable groves of manuka then growing on the run,—groves perhaps in all some twenty or thirty acres. Throughout this block, especially on the edges of the dead groves, "capeweed" or cat's-ear (*Hypochœris radicata*) germinated in millions of millions of millions; its seed, blown from the adjoining grassed lands, had been caught by the manuka tops and fallen to earth as the seed of the sow-thistle had elsewhere been trapped by woodland boughs. Whilst the plant was at its zenith, from the hilltops across the lake I have watched, looking downwards, the phases of a marvellous colour scheme develop, a change from orange-brown, the hue of the tips of the closed blossoms, to the dazzling yellow of the fully unfolded flowers; as the plants began to expand their blossoms, I have even temporarily turned my back on them, the more fully to appreciate the change when viewed again. By nine on a hot dry morning over scores of acres a sheet of gold was spread, a dense bright carpet of colour only possible on land tenanted for the first time by an alien thoroughly appreciative of its environment.

Suckling.

Suckling (*Trifolium dubium*), though at an early date noted as a plant perfectly adapted to the pumice soils of central Tutira, was nevertheless comparatively slow in taking full possession. It reached success by no short cut, its little seeds were neither blown abroad as thistle, or sow-thistle, or capeweed, or Canadian groundsel, or glued to the legs of sheep like mouse-ear chickweed or silene; it was carried in the stomachs of sheep. Its spread in the centre and west was slow because in early days those parts of the station were lightly stocked or altogether unstocked; nevertheless, as time progressed, the range of the plant extended until after fires it has become the most important fodder-plant on the station. Unlike the majority of aliens on Tutira, suckling appears and reappears on the same ground more and more thickly. I never look on this insignificant weed without thankfulness: to it I owe my continued ownership of the station; it has produced more wool and saved the lives of more hoggets than any other single fodder-plant on the run. Now that a larger proportion of the pumice soils are open to light and air, its germination, early or late, profuse or sparse, according to meteorological conditions, decides the nature of the coming clip of wool.

After fires in the 'nineties, over central Tutira paddock after paddock was temporarily overrun by Canadian groundsel (*Erigeron Canadensis*). It also, like other aliens named, has almost completely disappeared.

Spread of plants after fire has, however, been by no means confined to aliens. Readers will recollect that it was as a fire weed that manuka attained its grip of the run; several of the terrestrial orchids, the common catch-fly (*Drosera rotundifolia*) and Pelargonium australe, have also sprung up and spread after fires, particularly after fires through manuka thickets.

Of late years, too, after fires on rich damp swamp-land, has appeared in profusion Polygonum serrulatum. It is a weed which follows the flax-mill, carried in men's boots, sacking, and machinery.

One weed only on Tutira owes its rapid spread to flood. In the late 'nineties Gillia squarrosa appeared thickly on a sheep-camp on the top of the Image Hill. During the following year the paddock containing that hill was crushed, immense mobs of sheep being run on it, and innumerable new paths stamped out. Then occurred one of the floods which pass at irregular intervals over the station; paths became runnels, runnels became brooks, the Papakiri, into which they poured themselves, rose feet above its banks; everywhere along its course sand was deposited from a couple of inches to a couple of feet. During the following season, on this flood-drift as on tilled soil, germinated masses of the evil-smelling plant, Californian stinkweed it is called, on account of its malodorous savour. Next year it was gone, the following season or two there was a sparse recrudescence of the plant, now it has become a rare weed, appearing only where by chance the surface has been broken and the soil stirred.

CHAPTER XXX.

PEDESTRIANS.

About forty plants have attained their goal by pedestrianism—not, of course, by unbroken continuity of root-stretch from beginning to end of the journey, but by repeated portages over short distances, re-establishments again and again for another and another step inland, up-country, Tutirawards. Neither do I mean to affirm that these wayfarers have been too proud to have accepted from time to time a short lift on a roadman's shovel, the warm shelter of a stomach, the grip of a mane or pastern, a brief trundle on the wheel of a dray or buggy, the hospitality of a friendly hoof or woolly shank, the assistance downhill of a brimming water-table. They have, nevertheless, to all intents and purposes reached Tutira on their feet. Dozens of times I have met or passed each of them on its trek towards the station; I have watched them drawing nearer and nearer to its sacred soil; I believe, in fact, that not a weed thus moving towards the run by road has been overlooked. A natural inclination, I suppose born in me, to note small changes in my environment had grown gradually into a habit of watchfulness. Each ride beyond the run contained the element of anticipation, of hope, the possibility of the discovery of a new wayfaring alien; for nearly forty years the fortunes of these wayside weeds have been an interest in my life.

The mode of approach of no two members of the group has been exactly similar. Each has gone forward in the manner best suited to its predilections and peculiarities. Some have advanced with celerity and confidence, by leaps and bounds as it were; others have progressed hesitatingly, slowly, step by step, feeling their way; others again I can recall, laggards, faint-hearts, that were on the road in September '82, but which have not even now attained the station. Mangel-wurzel, for example, doubtless in the first instance swept as seed from one of the Port Ahuriri stores, then grew plentifully as far as half-way to Petane. It has made no further progress. Perhaps it was not to be expected that a plant loving the salts of the coast should willingly forgo them by an inland journey—but why this stoppage before the Tangoio Bluff? Why, again, has Alyssum maritimum, which used to scent the whole beach road for hundreds of yards south of Napier, never ventured northwards farther than the street edgings of Port Ahuriri? Sown inland as a garden flower the plant thrives, the seeds germinate in profusion; during the construction of the Napier-Wairoa-Gisborne road ample space of spoil was open to settlement, depths of pulverised earth, yet for some reason or another no single plant of alyssum appeared.

Garden scabious is another species which in the 'eighties flourished thickly over roods of reclaimed lands between the western foot of Shakespeare Hill and Port Ahuriri, yet although in up-country gardens scabious thrives, and although roadside conditions for a period might have been considered eminently tempting, not a specimen has stirred abroad. In the early 'nineties the shingle road between Hastings and Roy's Hill was overrun during four or five seasons by Centaurea calcitrapa, yet again there occurred no movement Tutirawards. Had Red Valerian, firmly established on the road cuttings of Napier in '82, been a plant of any enterprise, it too might now have been happily domiciled at Tutira. Breaks of fissured limestone cliff obtrude at intervals for practically the whole distance, yet it has never budged from its original home. I have often thought that the passivity of certain aliens provides only less food for conjecture than the spread and progress of others.

In descriptions of the physiography of the station it has been explained that the lands of Hawke's Bay south of Napier are more fertile than those of the great pumiceous area immediately to the north and west of the Province. Before these wastes were thought of as fit for sheep-breeding, draining, planting, even ploughing had progressed in the south. North of Napier, on the other hand, the country was a wilderness: the pioneer had barely set his foot on it, there was absent from it that unfailing indication of man's presence, an alien self-settled flora. It produced no stock; mobs of sheep travelling northwards disappeared then as in story-books travellers vanish into an ogre's den, never to re-emerge; the surplus stock of the south were driven up to die on the

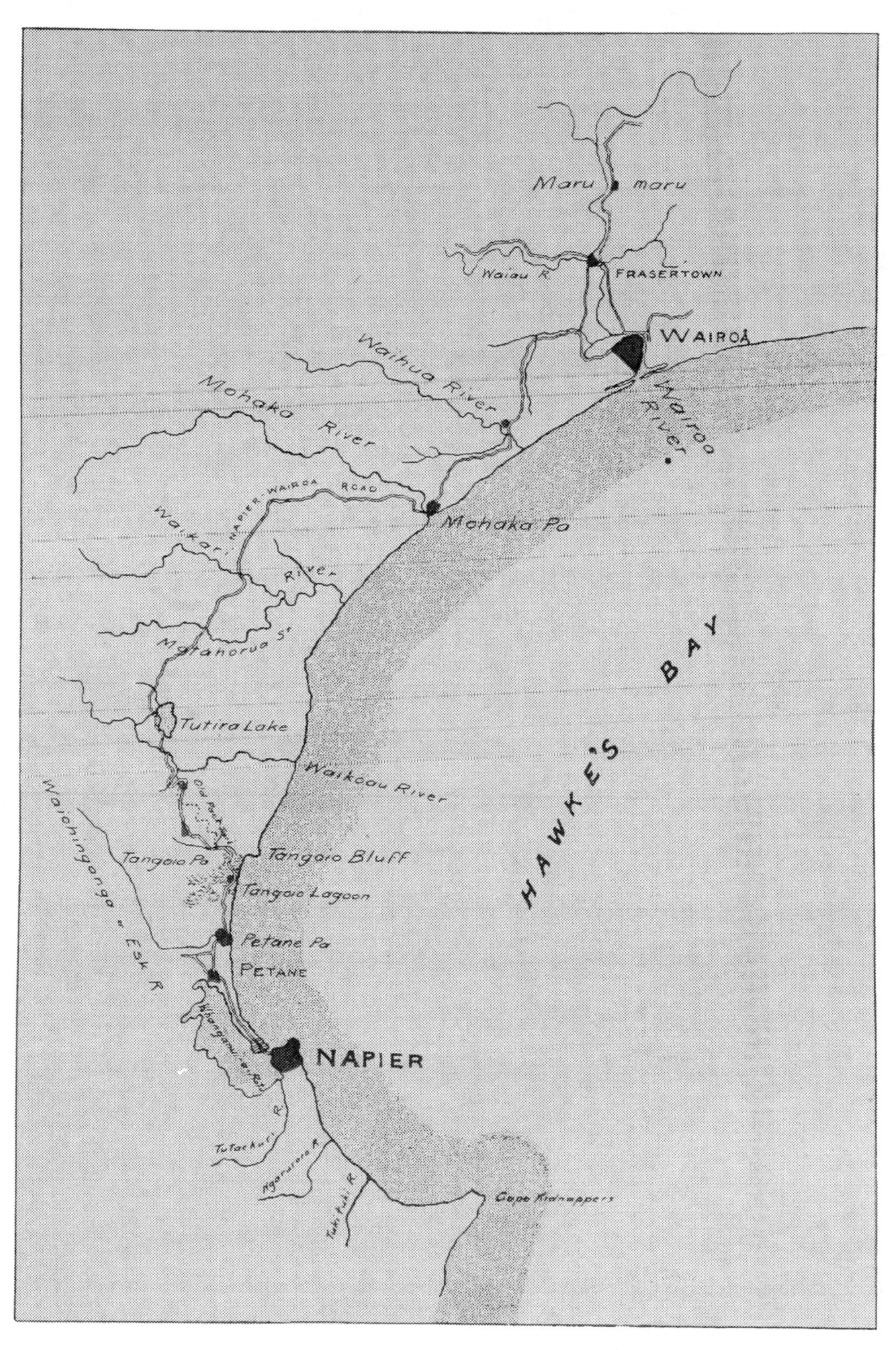

Recruiting Grounds and Multiplication Centres of Pedestrian Weeds.

"east coast" runs. The difference between the north and south, then so strongly marked, still holds, though in a lesser degree; the majority, consequently, of pedestrian plants have reached Tutira from the south. Almost all of these wayfarers, moreover, are comparatively recent arrivals, the construction of the coach-road, up which most of them have tramped, being itself a modern event.

A short description of the way by which alone, before and during the 'eighties, the run could have been reached, will show the almost insuperable difficulties pioneer pedestrians had then to surmount.

We can take Napier, the port of the province, as the main centre of weed liberation. Immediately after leaving its streets an estuary of several hundred yards in width lay athwart the route. Over it sheep could be ferried, though horses towed behind had to swim the distance. Thus at the outset of the journey any weeds lodged in hoofs, or about the mud of pasterns, were destroyed. The feet and legs of sheep, too, were saturated with salt water, both in the leaky punt and in landing operations. Then came several miles of barren shingle; furthermore, a considerable river had to be negotiated, sometimes by swimming, sometimes by deep wading, in either event the sheeps' legs, feet, and belly-wool being washed clean. Then again, the road followed the coast-line for several miles, sometimes over hard, sometimes over soft shingle and sand. The hoofs and legs of all manner of stock were in fact thoroughly well cleansed before reaching Tangoio; for this purpose, perhaps, the long, dry, barren stretches of loose shingle and sand proving as effective as water itself. Droving, a tardy process at the best of times, became under these untoward conditions even slower, stages even shorter; more ample time still was afforded for sheep, cattle, and horses alike to empty themselves. Moreover, in such going there was but little chance of picking up new weed supplies. By the time Tangoio was reached the likelihood was gone, thenceforward the way lay over closely-nibbled hill-tops.

A more unpropitious track, in fact, for the perambulation of weeds could scarcely have been selected. It would almost seem as if the road, like some great beast, had consciously attempted to free itself of parasites by washing and dust-baths.

Under these conditions it is not surprising that travelling plants should in early times have been few and far between. Only a fraction of the whole length of road had been properly formed; from Napier to Petane it consisted of a narrow line of clay blinding superposed on loose shingle. From Petane, still following the general line of the coast, drays could be taken to Tangoio over natural flats and strips of sand, except when after heavy weather the pressure of lagoon water had burst the beach and temporarily blocked all wheel traffic. From Tangoio for two miles a switch-back riding-track sprawled over the hills roughly parallel to the sea. At First Fence the way degenerated into a trail faintly marked by the Tutira pack-team.

About the early 'nineties, however, Government determined on a dray-road to connect Napier and Gisborne. Natural difficulties of the more formidable sort were attacked simultaneously in many sectors, swamps were drained, cuttings blasted and picked out of ravines and gorges. Within a short period the road-line became open after a fashion to riders and pack-teams, with the completion of six-feet cuttings aliens began to move inland in larger numbers, with cuttings widened into fully-formed sections of dray-road they reached us yet faster. Traffic increased enormously, contractors' camps sprang up along the route, pack-teams multiplied, a weekly mail-coach was subsidised by Government. Finally, almost the whole of the stock traffic that had formerly followed the coastal pack-trail was diverted; the invasion of the station by road weeds was facilitated and accelerated by vast mobs of cattle and sheep that poured themselves along the new road. A living stream flowed through the whole length of the run from both north and south.

There remain to be recorded the chief recruiting-grounds, convalescent camps, cities of refuge, and multiplication centres of our pedestrian weeds. To begin with, all of them have reached the Dominion through one port or another. Most of the Tutira settlers have of course disembarked at Napier, or, as it used to be called, "Scinde Island,"—within the memory of man an isolated block of several hundred acres connected with the mainland, north and south, by shingle spits. Long prior to my arrival Napier had been thickly set with gardens and orchards; there were broad spaces, too, of beach, dumping-grounds for ballast, waste lands along the railway track and along roadsides, and wildernesses of reclaimed ground. Every such spot maintained its wild alien.

Scinde Island then, twenty-eight miles distant, was the first and foremost of the spots where weeds have germinated and multiplied ere starting forth to stock Tutira. Another weed depository would comprise the lands about the township of Petane, and especially about the native village of the same name, twenty-two miles distant. Unlike the ancient cultivation-grounds of the Maori, which have been described by Colenso as models of neatness and careful culture, a modern native village is forlorn, unkempt, and untidy in the last degree. Every such settlement contains a superabundance of land, only half of it half-tilled; vacant corners, unsown headrigs, widths of mud road, offer ideal germinating ground for virile ambitious weeds.

Still approaching Tutira we reach the Coastal Hill, of which the larger seaward portion is called Te Uku, the smaller Puke-Mokimoki. It is a low bluff or promontory fenced off from the neighbouring run, and therefore a secure camping-ground for travelling stock. When first known to me it still maintained what was probably its original vegetation—coast grass (*Microlæna stipoides*) and sparse spray-swept bracken. As, however, pastoral interests developed and stock traffic increased, these aborigines were speedily ousted, the Coastal Hill became the resting-place, sometimes for a few hours, sometimes for the night, of considerable mobs of travelling stock. In later days, consequent on the progress of the east coast, hundreds of thousands of sheep yearly camped on, trod and manured, the little promontory.

Besides stragglers of many kinds resting on their way, I have seen this camp at different periods under a dense crop of prickly thistle (*Carduus lanceolatus*), of ox-tongue (*Picris echiodes*), of buckshorn plantain (*Plantago coronopus*), of Bathurst burr (*Zanthium spinosum*); for several years it then grew a sward of pure ryegrass (*Lolium perenne*); at present it carries a hirsute mat of Chili grass (*Sporobolus indicus*); doubtless when there has gathered on it a superabundance of manure, or when a severe drought may have exposed the dusty trampled ground to extra light, some new weed will take temporary possession.

Passing the County Boundary hill, Pane-Paoa, a fourth weed-centre exists at Tangoio, another unkempt briar and bramble-tangled native settlement. At eight miles distance from that station, where drovers customarily halt their mobs at midday, there is still another weed-centre of lesser account; and lastly, a sixth, where travelling stock, temporarily blocked by a gate, used to tread the ground into dust, or poach it into mud.

From northern ports, too, such as the Bay of Islands and Gisborne, pedestrian weeds have also reached the station; accommodation paddocks of roadside inns, drovers' camps, and Maori villages have, as in the south, proved their chief recruiting-grounds and multiplication centres.

Of these pedestrians, the blackberry (*Rubus fruticosus*) was, if not the earliest, one of the earliest to move inland. It stands forth—that fatal and perfidious plant, sown in the eclipse and dug with curses dark—as the single alien that is the master of the sheep, the one plant that makes a victim of him. Its normal habit in the open is to grow into an oval bush. Specimens thus shaped expend their energies harmlessly or comparatively so, although each season the base of the bush increases. Should, however, one of these tall cones be burnt, spread is accelerated laterally; huge horizontal shoots are sent forth, tentacles by which the victim is seized. A sheep but newly caught and still but loosely gripped exhibits an instance of inert brainlessness almost unimaginable; although one determined pull would free the animal, he yet suffers himself to remain anchored by a single strand. Tethered thus, further entanglement is but a matter of time; wool and bramble shoots become woven and twisted into a rope, until finally the sheep dies and its carcase goes to feed the triumphant plant. Perhaps unlimited time only is required to develop out of Rubus fruticosus a sheep-catching plant with more enormous shoots and yet stronger thorns.

No good word can be urged for the unhappy plant; not even its fruit, borne in vast quantities but lacking flavour, can excuse or even condone its iniquities. How and when the blackberry reached New Zealand I know not. Its importation is often, I believe, erroneously ascribed to the much-abused missionary; certain it is that the weed has not come into Hawke's Bay from the north. Its local origins are Petane and Tangoio, where long prior to my time stretches of blackberry hedge had been planted.

We can now follow inland the march of this terrible pedestrian. After leaving Petane the road for some distance ran parallel to one of these planted fences, a brazen example of a vested interest, for when at a later time blackberries were attacked with poison and spade, this hedge, grey in its hoary iniquity, was spared. There were several bushes scattered about the sandy hummocks of flood-silt in the Esk river-bed. Throughout the native cultivations, where there are now hundreds, I do not remember a specimen; two there were, however, on the shingle flats near the Coastal Hill. On it were established other two bushes: a single specimen grew at the base of Pane-Paoa, the County Boundary Hill; another huge plant grew where the road strikes sharply inland from the beach. Between that and the Tangoio homestead another

Pioneer planting blackberry on the Napier-Tutira-Maungaharuru trail.

hedge had been deliberately planted, seedlings dibbled in at regular intervals. Blackberry bushes were scattered here and there about the Tangoio homestead and along the bridle-track till it began to rise to the hills. Half-way between the Tangoio Flats and First Fence there was one bush. Between First Fence and Kaiwaka boundary gate there were none, and but a single stunted specimen on the high pumiceous tops. On the limestone edge overlooking the Waikoau valley flourished the furthermost inland centre of mischief, a colony of six or seven immense bushes. Another blackberry grew within half a mile of the crossing, another immediately on the Tutira side of the ford. There were none on the site of the disused Maori cultivation-grounds on the Racecourse Flat, pretty good evidence that the plant was a genuine pedestrian sticking to the road, that it had not been deliberately brought up as a fruit, and finally, that it could not have been in the province at an early date. There was a plant on the old native trail half-way to the Maheawha crossing, another at the ford itself. The westermost bush on Tutira proper was established just above the gorge separating Tutira and Putorino.

During the early 'eighties, in fact, except about the plague-spots

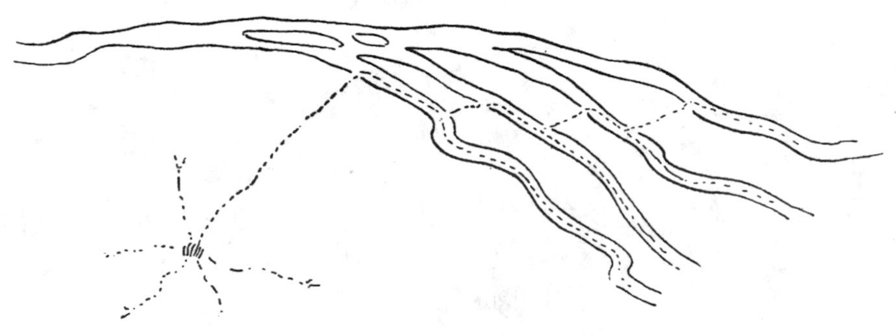

Blackberry roots tapping sheep-paths.

Petane and Tangoio, blackberries could almost be reckoned on a man's fingers.

There were, however, even at that date, dotted along the road, bridle-track, and pack-trail, a sufficiency of bushes to fix it definitely as a line of human traffic. The pioneers of the east coast had in fact marked their pilgrim path in blackberries, for it is man himself who first carried up-country the fatal seed. Each offering deposited at each improvised temple of Cloacina on the road has erected itself a living monument to the goddess; whilst intermediate bushes could still be individualised, they were to be found more thickly in proximity to the parent plantations, more sparsely at longer, or as I may say, more costive distances. Owing to its ensnarement of sheep, the blackberry is the most dangerous, perhaps the one truly dangerous, alien in New Zealand. On hill land impossible to plough on account of gorges and land-slips, the only method of eradication is by spade and mattock; even then these diggings have to be gone over again and again: the smallest rootlet grows, even half-buried leaves will root strongly in damp spots. The plant, moreover, possesses an intelligence and energy worthy of a better cause. Again and again I have dug out bushes, especially on light lands, sending forth roots which a few inches beneath the surface have followed exactly the lines of sheep-tracks within range—tracks enriched by manure carried from contiguous camps, —removal of the soil has revealed a subterranean root-system corresponding to their sinuosities. It only remains to add that after sheep had acquired a taste for the fruit,—I have seen their paunches black with the berries,—and more especially after the arrival and increase of imported birds, who carried the seed everywhere, the bramble increased in a most alarming way. Although a fortune awaits the inventor, no weed-destroyer has yet proved efficacious. It is impossible not to look with grave concern at the future of many hundred thousand acres in northern Hawke's Bay.

Bermuda grass (*Cynodon dactylon*) was in the 'eighties plentiful about the Port Ahuriri roadsides. It grew also in many spots along the clay blinding of the beach roads. There was a great patch on the sandy land opposite the old Tangoio homestead, between there and First Fence another carpet, then no further signs of the plant until Tutira. There this old friend still resides on its original site, where the trail strikes—or rather used to strike—the lake, close to the modern sheep-dip. It has survived changes necessitated by the widening of the pack-trail into a bridle-track, and that again into a dray-road. This grass was plentiful, too, about the sandy estuary of the Waikoau, from the river-mouth to the Arapawanui homestead. It could therefore have come up *viâ* the Maungahinahina by the native track. As traffic, however, was a hundredfold greater by the Tangoio route, that route has been given the benefit of the doubt.

Centaury (*Erythræa centaurium*) is one of several plants first seen by me on that great stock-camp the Coastal Hill. In due course it reached Tangoio and afterwards Tutira, arriving at the latter place in one great stride and immediately taking possession of the track between the wool-shed and homestead. Thence, probably carried in sheeps' feet, it rapidly skirmished along the main stock-routes of eastern Tutira. It is a weed that has never taken kindly to a diet of pumice, preferring the roadside soils of the limestone area. Like many another pedestrian plant, it does not freely reproduce itself, lacking such stimulants as the trampling and treading of stock, the stir of soil by their feet.

Centaury.

Only these three plants managed to negotiate the old pack-trail before it was transformed into a road. Their small number testifies how well-guarded were the passes into Tutira—estuaries, salt beaches, barren shingle strips, unbridged rivers, close-cropped hill-tops, high, cold, lean summits of pumiceous ground. On each the highway of the 'eighties cleansed itself of seeds as an animal rids itself of parasites.

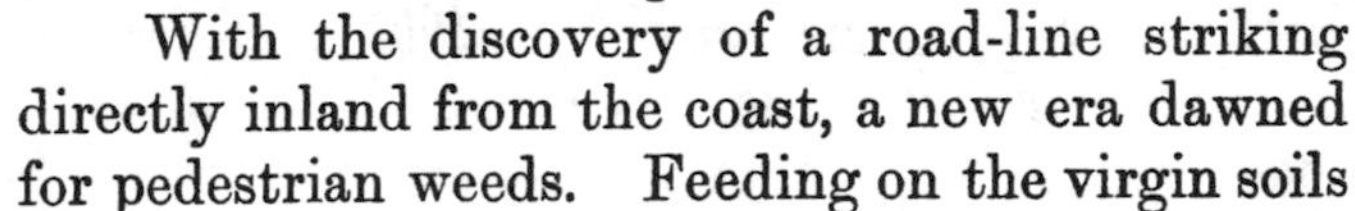

With the discovery of a road-line striking directly inland from the coast, a new era dawned for pedestrian weeds. Feeding on the virgin soils displaced by pick and shovel, basking on the dry banks of loose soil, wading along the water-tables, battening on the sheep-camps, they moved inland in numbers. One of the earliest to take advantage of the easier conditions afforded was ox-tongue (*Picris echioides*). Its local origin was the Coastal Hill sheep-camp, where for two seasons the plant grew in dense masses like a sown crop. Later it appeared on the Maori cultivation-grounds about Tangoio. There it had stayed its course, one of the many sybarites which had not dared to face the wilds or which had failed through want of stamina, but which was now again tempted to advance by the presence of stirred soils and the warmth of friable slopes. Keeping pace with the road-making operations, for two or three miles inland it grew plentifully, then as the distance from the coast increased it was less often to be seen. It failed completely on the pumiceous heights, but reappeared, though scantily, on the warm northern slopes of Dolbel's Big Face. First specimens noted on Tutira flowered in one of the old gardens close to the roadside. It is

Ox-tongue.

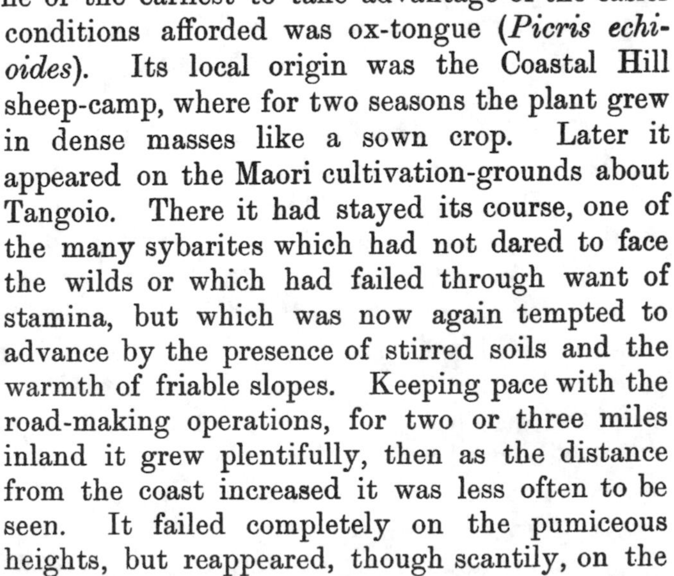

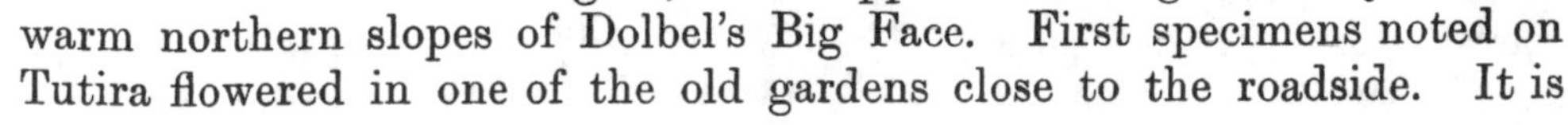

now comparatively a rare weed on the station, oftenest appearing, but always scantily, after an unusually warm, dry summer. Like many another alien, it has quite gone from its original site, the Coastal Hill.

Hare's-foot clover (*Trifolium arvense*) was first known to me in '82 as one of the most common weeds of shingle beds in lower Hawke's Bay. I noticed a plant or two the same year between Napier and Petane. It then took possession of certain dry shallow soils between Petane and the Coastal Hill; thereabouts, for several years, its progress seemed to cease. Later I noticed it almost simultaneously on the north-facing baked bridle-track leading to Arapawanui and on the Tangoio-Tutira road growing rather wretchedly on two spots along its higher pumice-sprinkled levels. A year later plants appeared on an arid road-cutting close to the station wool-shed. It then followed the main stock-routes, and lastly, with a very pronounced pedestrianism, the sheep-paths. Now it is prevalent everywhere in the trough of the run, and in light lands affords a winter bite of no inconsiderable value.

Narrow-leaved Cress.

Narrow-leaved cress (*Lepidium ruderale*) is a pedestrian I have followed up from the very streets of Napier. It reached Tangoio early in the 'eighties, and there for many years rested about the homestead drafting-yards. Only as the road progressed did it move inland, not infrequently in its travels choosing the angle of a hard road-bend as temporary camping-ground. It is a somewhat lonely plant appreciating trampled ground, liking to bake itself on almost naked clays, enjoying the dust of sheep-yards and the vicinity of road-edgings. In the home paddock and along tracks much used by the station collies, this sturdy, stubbly cress is largely used as an object upon which are recorded their more solid observations. It must be a very tickly plant.

Mallow (*Malva verticillata*), a southern Europe and central Asian species, first seen at Tangoio sheep-yards, advanced up the road without stop but without haste, never covering great distances at a stride, a single plant here and another there. Whilst on the march it grew nearly without exception on the highway's very edge, half of it clean and green on the wayside grass, the other half wheel-bruised and hoof-trodden in the mud. About sheep-yards, too, provided there is a sufficiency of manure, it is content to lie squeezed along the lowest rails of the dusty pens.

Another mallow, small-leaved mallow (*Malva parviflora*), has closely resembled its relative in mode of travel and preferences.

In the late 'eighties buckshorn plantain (*Plantago coronopus*) appeared about the railway crossing at Port Ahuriri. A year later it was thickly spread on all suitable sites along the beach road as far as the dry flats beyond the Petane river, where it grew in vast profusion. After a year's absence in England I found the plant densely carpeting the Coastal Hill camp. There was in the same year so thick a sward of it also on the Tutira home paddock that probably the weed had been missed by me on its first arrival on the run. An alien, as I have before said, suited by soil and climate, appears in the ratio of unit, hundred and tens of thousands. I had overlooked the unit stage. The plant was probably flourishing in hundreds during my year's visit to the old country. By the date of my return it had again increased enormously. It has now, after the manner of so many new-comers, died back to normal growth, and is chiefly noticeable during dry seasons when the turf is brown and withered.

Buckshorn Plantain.

Vervain (*Verbena officinalis*), a species few would suspect of harm, but which by overrunning grass has proved a nuisance in a small way, is an alien of note. It has invaded us from the north, a quarter from which comparatively few road-plants have come. Its proper name has been temporarily lost, it has been renamed, and, what must be almost unique in popular botanical nomenclature, that name again lost and the plant locally rechristened; finally, it has reassumed its correct designation. This alien from southern England appeared near Frazertown in the late 'seventies. It was first noticed on a station belonging to Mr Nairn; there it began to spread, and there it became known as Nairn's weed; the station then became the property of the late Mr Griffin. Now Mr Griffin, as chairman of the Wairoa County Council, touched the business and bosom of every up-country settler in the district. People knew him who did not know and had never known his predecessor; his office advertised the plant; presently, even in the immediate vicinity of its origin, the earlier designation was dropped, the later one assumed. It became known up and down the east coast as "Griffin's Weed." From Frazertown it radiated in all directions—northwards to Gisborne, westwards to Waikaremoana, eastwards to Wairoa. From the last-named centre it advanced slowly—the plant was known to me fifteen years before it reached the station—*viâ* Waihua, Mohaka, Waikari, and Putorino towards Tutira. Like many another road-plant, it thrived best on trodden, trampled soils, flourishing most luxuriantly where travelling stock had thoroughly stirred and scuffled the ground. Native village cultivation-grounds, too, such as those of Whatatutu and Mohaka were completely overrun; on the other hand, for some plants are most fastidious and precise in their requirements, it eschewed long-established sheep-camps as too rich, ploughed lands as too loose, and marl as too stiff. As the best portion of the station has proved too good and the worst too bad, it has given no trouble on Tutira, the few plants establishing themselves about the homestead having been from time to time dug out and burnt. Elsewhere also the plant seems to have shot its bolt,—less is heard of it each year. Curiously enough, too, with an extension of range the weed has in great degree managed to shake off the provincialism of its early designation: it is becoming known by its proper name, vervain.

Vervain.

In the early 'nineties, passing the little roadside accommodation-house of Marumaru, my attention was attracted by the gaudy, vulgar yellow blossom of ragwort (*Senecio Jacobœa*), fifteen or twenty specimens of which were in full bloom. As I came across but one other plant, and that within a few score yards of where the others grew, and as I ascertained by inquiries that the plant was still elsewhere unknown, there is good reason to believe that Senecio Jacobœa sprang into local life at Marumaru.[1] The plant, from its conspicuous blossoms, was impossible to miss, so that every step in its progress towards

[1] I then represented the Mohaka riding on the Wairoa County Council,—the time was well spent. What I learnt of wayside weeds and their habits could hardly have been acquired otherwise.

Tutira was distinctly marked. Nowhere plentiful and nowhere wandering from the road, though with ample opportunity of germination on lands unstocked by sheep, it slowly travelled *viâ* Opoiti, Frazertown, Wairoa, Waihua, Mohaka, Waikari, and Putorino towards the station.

In spite of winged seed, ragwort in its wanderings has seemingly been but little assisted by wind. As I have said, during its migration it never left the roadside, neither have seedling plants in my experience sprung up thickly in the lee or in the immediate vicinity of the old seed-stalks. The agent of dissemination has been the horse, patches which have now and again appeared in the homestead paddocks bearing indisputable marks of passage through the equine stomach, scores of seedlings germinating directly in the droppings. Horses, however, do not willingly touch the yellow flowers or mature heads; in ordinary circumstances the plant is left severely alone. Ragwort has either been spread directly by drovers' hacks starved into abnormal tastes, or, may be, the shed seed has been swallowed amongst herbage cropped round about the plants by horses.

Hyssop loosestrife (*Lythrum hyssopifolium*) I first met many miles south of Napier. It was the earliest of several species of waders which have taken advantage of the roadside water-tables to reach the run. The limestone hills from Tangoio to Tutira are rich in rills which never cease to flow; along them the plant has paddled its way. Advancing from one to another such site it took several years to reach its goal, on down grades the weed advancing perceptibly faster than on uphill stretches.

Mayweed (*Anthemis cotula*) was one of those kenspeckle strangers which could not but be observed during its inland journey. During its up-country tramp single specimens were never seen. Little companies travelled together, halting to breathe and breed on dusty trampled ground, on stock-camps in the making, rather than on those well established and densely turfed. It never grew, for example, on the very highly manured Coastal Hill camp. As in the case of many of these wayfarers, considerable stretches intervened between settlement and settlement, for species are either often more exacting as to environment than would seem likely *a priori*, or else the many agents by whose assistance they advance take up, retain closely, and as suddenly drop the seeds. Mayweed was first seen by me near the Hastings railway station, then on reclaimed land in Napier, then near Petane, then about

Tangoio *kainga;* then in a single season three separate patches appeared on the remaining length of road, while a fourth established itself on the run: it had moved from Hastings to Tutira in four years. Established as a huge patch near the wool-shed, it made during several seasons a half-hearted attempt to colonise areas of pumiceous land then under the plough; for a time scattered plants appeared here and there, but without the instant multiplication of a species thoroughly suited to its environment.

Strawberry clover (*Trifolium fragiferum*) has reached us also from the south—I believe from the Taradale district. A couple of thick mats, one close to Tangoio and the other a mile or so nearer, established themselves on sandy spots alongside the road. It then skipped the marls and pumiceous portions of the road till the sandy silt it loves was again available near the Twenty Acre Paddock and on the lake margin near the homestead. Many years later it was purposely sown, but has done nothing to justify itself as a fodder-plant.

Fennel.

Although fennel (*Fœniculum vulgare*) was indubitably introduced at a very early date on the Heru-o-Tureia block, it nevertheless merits mention as a road weed. In the 'eighties, and probably much earlier, it grew thickly in southern Hawke's Bay on the alluvial banks of certain rivers, on the waste lands of Maori villages, along railway embankments and other such places. There existed also in these days scattered plants between Napier and Petane and between Petane and Tangoio. They grew wretchedly on the clay blinding of the roads traversing sand and shingle; the plant dwindled in stature, too, on the poor hills along the coastal bridle-track. The nearest specimen to Tutira had early managed to establish itself thirty yards below the abrupt drop of one in three, down which the station pack-team in wet weather used to skid. There for years has that solitary plant, marooned in a green sea of grass, watched other passengers press forward to the goal of their high calling; there, indeed, it remains to this day. The actual establishment of fennel on Tutira proper only occurred in 1906; the plant, notwithstanding its early efforts to win an entrance from the south, actually did reach Tutira, not from that quarter but from the north. A mob of horses driven through the run from Gisborne sowed it thickly in their droppings over the paddock lent to their drovers for the night. Fennel is indeed largely spread by the horse, for just as sheep, if run on bramble lands, will devour the fruit, so horses in fennel paddocks will eat the heads and amplexicaul leaves.

Fiorin (*Agrostis alba*), though to my knowledge sown on the neighbouring run of Arapawanui in '85, for long and in a very extraordinary manner kept off the highway. It was not until twenty years later, and then almost certainly not from Arapawanui, that it began to move on the roads; established on the waste lands of the Tangoio *pa*, it then travelled inland at a rapid rate. The slow spread of so comparatively innutricious and unpalatable a grass can only be accounted for by the desire of sheep for change of diet—any change of diet. Surfeited with a superabundance of ryegrass, cock's-foot, and clover, fiorin stools, instead of being severely let alone and consequently free to mature their seed-stems unchecked, were cropped bare. I am the more sure of this as on fertile bush-land, sown only with grasses of the best quality, stowaway fog (*Holcus lanatus*) has in several instances to my knowledge been likewise cropped bare. Yarrow again (*Achillea millefolium*) has been established since the early 'eighties close to the Tutira-Arapawanui fence. Though a free flowerer, and possessing seeds so minute that several million go to the pound weight, the plant never increased. I have never known it allowed to blossom. For the same reason as fiorin failed to spread, yarrow has failed to spread. Until purposely sown the only specimens were the half-dozen near my eastern boundary.

The pearl-worts (*Sagina apetala* and *Sagina procumbens*) are so much alike that for present purposes they can be classed as one. Like Lythrum hyssopifolium they also, during their inland march, have largely utilised the water-tables of the road. About the margin of runnels and well-heads the plant often becomes a handsome cushion of green; one I recollect over whose verdure the bright drops of a little spring used to course like living pearls; its surface was so compact that each drop moved with as little loss of bulk as quicksilver on polished wood; the spring water was so clean that for the best part of a dry summer the surface of the clump was unsaturated and unsoiled. Each of these weeds has reached Tutira from the south.

Barley grass (*Hordeum murinum*) has flourished ever since I can remember against the northern edge of the Bluff in Napier, along the clay blinding of the beach road between Napier and Petane, and again at intervals and in similar situations between Petane and Tangoio. For years at the latter place it halted, unable to subsist on the colder clays, grassy tops, and higher altitudes of the old pack-track. With the new road, however, it crept up, selecting spots, the very poles apart from those chosen by pearl-wort and other waders. Sun, dust, and drought exactly suit this wild barley; cliff bases and sun-baked cuttings harbour it; now that the plant has reached the run it delights to live facing the north, squeezed up for choice against the lowest rails of arid sheep-yards, for though it can subsist in almost any desert, it is no scorner of good food.

Beard grass (*Polypogon monspeliensis*) I have watched travelling up from Taradale, five miles beyond Napier. When first seen in '82 this grass was thriving on wretched-looking land scarcely above the reach of the tides; then years later it appeared on the edge of the saline lands near Petane, then again after a long interval about a small surface pond at the wet base of the Coastal Hill, where so many travelling aliens have found a temporary resting-place and recruiting-ground. Another halting-place was the water-sodden land about Tangoio. From there onward it became a wader, paddling ankle-deep along several of the more suitable damp spots on the dray-road, three or four comrades together, never in large companies. Its first grip of the run was on the marl water-tables a few hundred yards from the Waikoau Bridge. Its chief hold on Tutira now is about the broad, wet, shallow crossings of the Kai-tera-tahi swamp.

Beard Grass.

Another beard grass (*Polypogon fugax*), which reached Tutira several years later, was first noticed on the salt marshes of Petane; it also, like its relative, whilst moving inland took advantage of wet ditches and water-tables.

Reversed clover (*Trifolium resupinum*) was earliest discovered where, rising from Tangoio, the road emerges on to pumiceous hills from the White Pine Bush. There for some years a fence crossed the dray-road, the closed gate of which, until tossed from its hinges by an irate drover, temporarily delayed the progress of travelling stock. This brief stoppage it may have been, with its consequent turf trampling and manuring, that was responsible for the arrival of this delicate refined little plant. Some years later I found a specimen on Tutira near the south end of the lake; now, though always a rare plant, it has passed the homestead still travelling north.

Pennyroyal.

Few of my pedestrians have had a longer tramp than Pennyroyal (*Mentha pulegium*). About '94 I first met the plant on the alluvial flats of Poverty Bay; then it appeared at Tinoroto, and lastly Wairoa. From Wairoa it came down to Mohaka, then, following the inland road behind Waikari, reached Putorino, and later again Tutira, the first-seen plant on the run appearing close to the bridge over the Tutira stream. Pennyroyal, doubtless, is of missionary origin, and might have almost equally well been included in "Children of the Church" group.

Bathurst burr (*Xanthium spinosum*) I suspect to have had its local origin in the north. I have found it, at any rate, growing more freely in Poverty Bay than elsewhere. Probably, therefore, the crop which at one time densely covered the Coastal Hill had been carried southwards from Poverty Bay by travelling stock, the barbed seed entangled in the tails of cattle and in the manes of horses,—I have seen a feeding horse touching a burr plant in an instant get his forelock covered with seed,—and in sheep's wool. On this great camping-ground for a season or two Bathurst burr grew with enormous luxuriance. As, however, it travelled inland, and towards greater cold and wet, the plant seemed to lose its vitality. It grew sparsely on one or two disused gardens in the Tangoio *kainga*. A single specimen appeared on the roadside between Tangoio and Tutira; only on two occasions has it germinated on the station, each time appearing on dug soils facing the sun.

Burdock (*Arctium lappa*) I first met in the middle 'nineties eighty or ninety miles from Tutira. It is one of several species which I think have of late years come from great distances, whose sudden appearance is attributable to motor-car traffic. Like the passengers themselves, seeds are carried by these swift machines greater distances in shorter periods of time. The foremost pioneer on the road appeared midway betwixt Tangoio and Tutira. In spite of my desire, however, to be able to register the plant as a station alien, the sheep-farmer triumphed over the weed-observer. I destroyed, though not without a pang, this solitary traveller while still at some distance from the run. Next year, however, burdock seedlings again had moved along the road nearer to the station. Now in 1920 they are close to my boundary. Having thus confessed that the plant is still an uitlander, the reader will condone its inclusion in my list of Tutira plants.

St John's wort (*Hypericum linarifolium*) has sauntered southwards in a very leisurely fashion, colonising one after another of the clay road-cuttings between Mohaka and Upper Waikari. The earliest specimen actually gathered on the station was taken near the bridge over the Tutira creek in 1913. Since that date the plant has again moved coastwards several miles.

Another St John's wort (*Hypericum perforatum*) appeared at Tutira in December of 1913 on the edge of the main road. Upon my return after the war the original clump had vanished, but the species had on two separate spots on the roadside re-established itself. For years it has grown thickly on the railway track far south of Napier.

During the war three new roadside plants have reached Tutira—one, a flax (*Linum* sp.), I had already known of near Napier and near Petane; another, Mugwort (*Artemisia vulgaris*), first noticed close to the station stables, was a complete stranger. Reaching us from the south, it seems to have been carried on certain tools, for small groups of the plant have appeared only where culverts have been recently deepened and cleaned; no intermediate specimens or groups have sprung up,—the species has confined itself strictly to soils shovelled from choked culverts. Lastly, there has appeared a very worthless grass called Reflexed Poa (*Poa distans*). I had first seen it fifteen seasons ago on the Tangoio coast road.

The reader has now seen how on Tutira each accretion in its growth, each phase in its development, has been marked by the establishment of one or more aliens seemingly sympathetic with the particular change. Pioneer work, cessation of tribal war, stocking, seed-sowing, settlement, introduction of machinery—each is stamped on the surface of the station in the shape of a corresponding weed; the very commerce of the Dominion indeed may be inferred from the popular names. Of these station aliens, England has supplied *par excellence* "English grass," that is ryegrass, white clover, and cock's-foot; Scotland the "Scotchman" (*Cnicus lanceolatus*). "Chili grass" (*Sporobolus indicus*) recalls the period of considerable trade with South America, "Californian stinkweed" (*Gillia squarrosa*), commerce with the United States, "Indian daub" or Bermuda grass (*Cynodon dactylon*), transference perhaps of troops from that great dependency, "Canadian thistle" (*Cnicus arvensis*) and "Canadian groundsel" (*Erigeron Canadensis*), exchange of goods with Canada; lastly, record of early connection with South Africa, with the Cape of Good Hope, is preserved by such plant-names as "Capeweed" (*Crepis taraxacifolia*), "Cape gooseberry" (*Physalis peruviana*), and "Cape barley," not improbably from its hardihood, the old Scottish bere (*Hordeum* sp.).

Had the vast change sketched in preceding chapters been fulfilled according to the inclination of man, only grasses and fodder-plants for his domesticated beasts, shrubs, flowers, fruit for his taste, and forest-trees for the pride of his heart, would have been acclimatised—Tutira would have been as the Garden of Eden, nourishing nothing but what was good for food and pleasant to the eye. Such an ideal condition is impossible to maintain; the pioneers of every colony set in motion machinery beyond their ultimate control; no legislation can regulate the dissemination of seeds. As the sun shines and the rain falls alike on the just and the unjust, so fleets, railroads, and highways convey seeds good and bad to a like common destination.

CHAPTER XXXI.

THE STOCKING OF TUTIRA BY ALIEN ANIMALS.

PRECEDING chapters have shown how in the vegetable kingdom useful and ornamental plants have been outnumbered by less worthy species. In like manner animal aliens, parasitical and predaceous, have come to exceed purposely-imported breeds.

It will be convenient firstly to consider the history of four of these self-invited strangers, prior to the days of ordered government, prior to the establishment of acclimatisation societies.

Of this quartette, three have been established in the land of their adoption for over a century, and one for scarcely less. Their dates of arrival, their journeyings, can now only be inferred; all that can be positively stated is the fact of their presence. All of them are members of the rat family.

The rat, in fact, seems to be almost parasitic to mankind, travelling in his shipping, feeding on his crops and stored goods; equally with his overlord and host, stocking the four quarters of the globe.

The four prominent steps or stages in the annals of early New Zealand were, firstly, the arrival of the Maoris; secondly, the discovery of the country by the navigators Vancouver, Malaspina, and Cook; thirdly, the exploitation of its seas by the sealer and whaler; fourthly, the initiation of commerce by the mercantile marine. Each of these periods has been responsible for its own particular mammal.

The first of these, Mus maorium (*Kiore maori*), in size rather resembling mouse than rat, was at one time an important article of diet amongst the natives. Except, however, that it was in disposition "tame" and "stupid," and that it subsisted according to native statement entirely on roots, berries, and woodland fruits, little seems to have been registered concerning its habits. By Colenso, who vainly tried to procure specimens, it was pronounced to be extinct in the 'thirties, but the forlorn honour of annihilation has so frequently been conferred on New Zealand species that it must always be received with caution. Very rare though this ancient breed may be, it has yet representatives in the land; it is neither extinct in New Zealand nor yet altogether absent, I believe, even from Tutira. Although my knowledge is not personal, and although again and again there has been confusion betwixt this small creature and dead specimens of the immature black rat (*Mus rattus*)—dead specimens be it noticed—I give the facts for what they are worth.

In '79 Harry Young, then engaged in a bush-falling contract on the sea-cliffs of Arapawanui, caught, or rather secured, for the seizure met with no resistance or attempted escape, what he has described to me as a minute "rat." When first seen it was noticed to be feeding on the yellow oval drupes fallen from a grove of karaka trees (*Corynocarpus lævigatus*). The little animal, entirely unperturbed, was passed as a curiosity from one to another of the half-dozen workers and then liberated. Many years after this, during a drafting of sheep at the Conical Hill yards in central Tutira, Jack Young, the brother of my first informant, seated on the fence rails eating his lunch, noticed a small dark-coloured "rat" with back "arched like a rainbow," as he described it, feeding on the seeding docks. Crumbs which he threw towards the little rodent were taken; as in the former case, it was secured without the faintest attempt to escape, and placed temporarily in one of the long narrow coffee-tin "billies" carried by shepherds. This receptacle unfortunately was overturned, and the rat, which was to have been brought in for my inspection, escaped.

It is impossible to believe that such a mercurial lively animal as Mus rattus should have at any period of its growth remained in the vicinity of a drafting-yard where work was in progress, much less allow itself to be caught with facility and handled without alarm. The rush of driven sheep, the shouting of shepherds, the barking of dogs, would scare the seven sleepers from somnolence; emphatically the specimen caught and then lost by Jack Young was not a representative of Mus rattus.

The third instance of the presence of a *kiore maori* occurred in 1906. One of my shearers, then resident on the wild rough country immediately behind Maungaharuru, saw and secured another of these small rodents.

KIORE MAORI (*Mus maorium*).

Photographed from specimen obtained by Captain Donne near Waikaremoana.

Amongst the natives—whatever may be the value of their opinion—it was accepted as one of the long-lost race. It was taken down to the coast, and there, living inside a buggy lamp, was for some time exhibited as a curiosity.

In each of these three cases of capture there was the same insistence on the particular trait of absolute fearlessness and tameness,—terms which may well be synonymous with stupidity, the word used by early writers quoting memories of Maoris who had themselves eaten the dainty, uneviscerated, stuffed with its natural diet of native berries.[1]

Prior to 1730—thirty-nine years, that is, before Captain Cook dropped anchor in New Zealand waters—the old English black rat (*Mus rattus*) held undisputed possession of Britain; it was only after the date mentioned that the brown or Hanoverian rat (*Mus norvegicus*) appeared. These species, however, as events were speedily to prove, will not live together, the one dominates and destroys the other; it is therefore in the last degree unlikely that Mus rattus and Mus norvegicus reached New Zealand cribbed, cabined, and confined in the same vessel. The limited space in such cockle-shells as the *Endeavour* and *Resolution*, and the duration of the voyage—over 300 days—alike negative the idea; the brown would have devoured his milder-mannered congener. It is, in fact, impossible that the black rat could have reached the colony otherwise than in the company of Cook, for by the date sealers had established themselves in New Zealand waters Mus norvegicus had overrun Britain and had destroyed his rival. When, therefore, the circumnavigator passed through Queen Charlotte Sound and careened his little vessel in Ship Cove, and later, when he moored the *Resolution* in a small creek "so near the shore as to reach it with her prow," the rats which vacated the vessels were doubtless representatives of the black species. Rats had reached New Zealand; they began at once their evil work. Cook himself has left it on record that on revisiting his clearing in the forest he found that "although the radishes and turnips had seeded, the peas and beans had been eaten by rats." We have still, however, to trace Mus rattus to Tutira.

[1] The likelihood that we still have representatives of this exceedingly rare little animal on Tutira is increased by the capture in the district of a fourth specimen by Captain Donne. Captain Donne's description of the conduct of his "rat" after capture bears out the testimony already given of the absolute fearlessness of the breed. It was taken in the forest path between Waikaremoana and Waikareiti not many years ago.

There are many alternative routes by which the species may have attained the North Island. The least improbable is that descendants of the original *Endeavour* and *Resolution* rats may have again taken shipping, may have at very early dates boarded native craft laden with food plying across Cook's Straits.

Although there is but little to tell, we can now proceed to Mus norvegicus.

A vast change had occurred in the fortunes of this breed between the sailings of the *Endeavour* and *Resolution* late in the eighteenth century and the rise of the sealing and whaling industries early in the nineteenth. It had in England completely ousted the old English black breed from pride of place. Ports and shipping centres where once the black rat had swarmed were now overrun with the brown. As had formerly happened in the case of the black rat, the brown breed also took shipping and was carried to New Zealand. Except that this did occur, nothing more can now be certainly known; probably the brown rat arrived by many routes, by different ships, to different ports.

In New Zealand, however, as Lot and Abraham parting from one another established themselves, the one in the plain of Jordan, the other in the land of Canaan, the two breeds separated themselves and divided the territory lying unstocked before them. The country chosen by the brown comprised the coastal belts, the settled districts, homesteads and cities. The black, as became his more adventurous spirit, possessed the native woods everywhere, and especially the high wet forest ranges.

The line of demarcation is, however, nowhere exactly drawn.

The brown rat, though sparsely scattered, is to be found in high country, whilst the black will here and there thinly colonise portions of a settled district, tempted thereto by specially favourable conditions; it will even on occasion breed in men's houses. It is true, nevertheless, that to Mus norvegicus belong cultivated land, crop and barn, whilst to Mus rattus appertain the wooded wilds, the rainy forests of the interior, wild fruits and drupes, forest birds and birds' eggs. On Tutira the homestead and the warm fertile coastal hills are the headquarters of the one; the other possesses the trough of the run, the hinterland, the forests of Opouahi, Maungaharuru, and Heru-o-Tureia.

The black rat's domicile in outward form resembles closely the untidy nest of the British house-sparrow; unconcealed and obvious, it may be found in masses of "lawyer" or native bramble (*Rubus australis*) wrapping some tall shrub, in thickets of supple-jack (*Rhipogonum scandens*), in dense shrubberies of tutu (*Coriaria ruscifolia*). Within this rough, rude, careless structure extends an elongated dome, tidy and warm, usually built and lined with a single material. Oftenest at Tutira the leaves of the tutu, or the leaves of the native bramble, are worked in as scales and shingles, and so made to curve and overlap one another as to produce a rainproof roof. The black rat is comparatively harmless to man and his property. In camp, where the brown will in a night rip and tear to pieces a flour-bag, the black breed will nibble rather than rend and waste. It is Mus rattus that is probably chiefly responsible for the disappearance of the Polynesian species, whose ancient feeding-grounds lay in the woods and forests. Although, I am told, practically extinct in Britain, the black rat is common throughout the uplands and wilds of New Zealand.

About the habits of the brown rat there is nothing that calls for special comment. It seems to live on Tutira as its forebears have lived in the old world: going forth during summer to the open lands, and during winter-time in some degree returning to the shelter of buildings. The brown rat is as deadly to native birds in the lowlands and swamps as is his black relative to species inhabiting the highland forests.

The third member of the Mus family, the mouse (*Mus musculus*), seems to have arrived at a considerably later date. The late Archdeacon Samuel Williams has informed me that he has no recollection of mice in New Zealand until the 'thirties. About that date they were noticed at the Bay of Islands Mission Station. Vessels were then beginning to reach the colony, laden or partly laden with cargoes of a kind that for the first time offered shelter, harbourage, and breeding accommodation for the small creature.

It was in association with toys for white children, printing paper, printing-press machinery, ironmongery, clothes for English ladies, seeds for the Mission garden, cereals for the Mission fields, linen and cotton goods, books, bells, glass, and crockery, that I imagine the mouse to have reached New Zealand. At any rate, only in the intracacies of a miscellaneous cargo could the mouse have obtained during the long voyage shelter and safety from his voracious kinsman the rat.

Although twice during my occupancy of Tutira irruptions of mice have overrun the station, it is unlikely that the place has been directly stocked from the north. The distance is too great, the obstacles too serious. Mice are not great adventurers; their pilgrimages are comparatively short and unsustained. A violent storm would in a few hours destroy the movement, for after a single night of three or four inches of cold gale-driven rain I have found them dead in scores. They can, in fact, no more stand heavy weather without shelter than can the sparrow. We may take it, therefore, that the local origin of Tutira mice was almost certainly the port of Napier. Thence there may have been a migration sufficiently sustained to reach the station. On the other hand, mice may have been directly packed up on horseback amongst station stores. Certainly from the homestead they have been carried over every part of the run in grass-seed bags; I have myself sown the half-smothered little creatures from sacks of grass and clover seed.

To recapitulate: there are or have been four species of rat on Tutira in my time; firstly, a species—four instances of which I have given at second hand—which may or may not be the *kiore maori* (*Mus maorium*). These four captures have taken place at intervals of years. Certainly it is not impossible that the little native rat should still exist on a run whose miles of cliff and crag offer such extraordinary chances of harbourage to a persecuted race.

Secondly, there is the breed often known as the bush rat. Of it I showed a dozen skins to the authorities at the South Kensington Natural History Museum. They were skins, I was told, of Mus rattus, the old English black rat.

There is the brown rat (*Mus norvegicus*); lastly, there is the mouse (*Mus musculus*).

The first chapter in the modern history of New Zealand was the arrival in the fourteenth century of the fleet of native canoes known as the *heke* or great migration. It was marked, according to tradition, by the advent of Mus maorium.

The second chapter was the landing of Europeans. In the annals of natural history it was marked by the appearance of the black or bush rat (*Mus rattus*).

The third chapter in the development of the colony was the rise of the sealing and whaling industries. They were responsible for the appearance of the brown or Hanoverian rat (*Mus norvegicus*).

The fourth chapter was the initiation of the mercantile marine. It was marked by the appearance of the mouse (*Mus musculus*).

This volume is the history of one sheep-run only; we can, therefore, localise events and say that Tutira owes its *kiore māori* to the Polynesian fleet; its black rat to the Royal Navy; its brown rat to the sealer; its mouse to the mercantile marine.

CHAPTER XXXII.

OTHER ALIENS ON TUTIRA PRIOR TO 1882.

According to the Wellington Acclimatisation Society's account,—an exact and democratic document, the courtesy prefix and initials of the bullock-puncher responsible for their carriage inland being given,—red-deer were introduced in '62. At the suggestion of John Morrison, then New Zealand Government agent in London, Prince Albert had presented six deer to the colony—three for Wellington and three for Canterbury. Two stags and four hinds had been captured in Windsor Park and there housed for a short period in preparation for their long sea voyage. One stag and two hinds were shipped by the *Triton* for Wellington, where on 5th June one stag and one hind arrived, the other dying during the voyage of 127 days. About the same date the other three deer were despatched for Canterbury, one hind only reaching Lyttelton alive. This hind was reshipped to Wellington. For some months the survivors were kept in a stable near Lambton Quay, where, according to the Society's report, they appeared to have been regarded somewhat in the light of white elephants. There was considerable grumbling by the public and by the members of the Provincial Government at the expense of their upkeep. Eventually J. R. Carter, then M.H.R. for the Wairarapa, offered to defray the cost of their conveyance to that district. To this the superintendent of the province agreed; the deer were replaced in the boxes in which they had travelled from England and carted—he shall have his prefix to the last—by Mr W. R. Herstwell over the Rimutaka Range to Carter's station on the Taranaki Plains. There they were given into the charge of James Robison. After several weeks further detention they were liberated early in the year '63; crossing the Ruamahanga, the little herd of three took up its abode on the Maungaraki Range.

Four years later according to one observer, five years later according to another, a red-deer stag had reached Tutira. George Bee, then working his father's station on the Heru-o-Tureia, believes he saw it first in '68; MacMahon, at that date managing Sir Thomas Tancred's Maungaharuru property, thinks it arrived in '67. Aparahama, Anaru Kune, and other natives give no specific date, but mention contemporaneous events which, however, I can only fix as having taken place also "about" the late 'sixties.

The attraction of the stag to the spot chosen was doubtless the small herd of wild horses strayed from native villages deserted and never afterwards repeopled. With them the lonely deer formed one of those curious animal friendships that strayed creatures make, a companionship similar to that of another stag which, at a much later date, consorted with the Black Head stud bulls,[1] or to that of the first rabbit seen north of Petane, which for several seasons accompanied a flock of "wild turkeys" on the Tangoio run. The locality otherwise was in no way suitable. There was not an acre of grass-land in the neighbourhood; it was covered with tutu thickets and tangled bracken.

There, however, the deer remained for several years, a source of speculation to settlers and shepherds and of wonder to parties of natives pig-hunting or pigeon-shooting. Although the late Mr J. N. Williams and other friends and correspondents in Wairarapa and Hawke's Bay knew of the whereabouts of this wandered stag, the stages of its journey, as might have been expected, had been unmarked. That it followed for a great portion of the distance the mountain-top route I have little doubt. By this line the forest lands and densely wooded gorges running athwart its route would be avoided. Had the stag travelled by the coastal route, consisting then of strips of sea-shore and narrow native paths connecting station with station, sooner or later he would have fallen in with cattle or horses and stayed his career on one or another of the coastal runs. That he did not do so proves, I think, that

[1] Mr Leslie M'Hardy of Black Head writes as follows: "The stag you inquire after used to come here about the month of April every year and stay for about four months. I remember him for three years, and the last year was the year of the Tarawera eruption—'86. He always stayed in the bull paddock at the cattle station, and was always to be found in the company of one old white bull. He seemed quite tame, as we could ride quite close to him, and on one occasion he followed the bulls into the stockyard. There was no doubt about him being a Wairarapa deer, as he was a typical Windsor Park specimen. He used to be very cruel to our bulls, but they got so used to him that they would not fight with him, and would lie down when he came near them."

the wanderer must have passed through regions uninhabited by the larger domesticated animals. It is, in fact, as certain as any matter of this sort can be, that the first red-deer on Tutira reached the station by the ranges connecting the Wairarapa with Hawke's Bay.

The distance of the journey, about 150 miles, gauged even in mileage, is respectable; in regard to difficulties surmounted it is little less than marvellous. Loose rock, snow, pitfalls camouflaged by herbage uncropped and unburnt, want of water, entanglement and poisonous plants such as tutu and rangiora (*Brachyglottis rangiora*), must have at different times imperilled every mile of the distance. Only those who know by experience the enormous percentage of loss among heavy beasts running on country previously unstocked, can fully appreciate the risks of the long journey, the small chance of ultimate survival. Though nothing further can now be known, the identity and age of the stag, the exact course of his trek, the time taken in the journey, are interesting subjects upon which to speculate. It is difficult, for instance, to believe that the progenitor of the North Island herds should have deserted his hinds. It is more likely that the Tutira stag may have been the first-born male of the Windsor Park importations—the first red-deer calf born in New Zealand—either expelled by his sire, the master stag, or a voluntary wanderer in search of a harem of his own. Excepting likelihood, however, there is nothing now upon which to base theories. Those who saw the stranger were fully occupied with their own affairs; they had other matters in mind than the noting of his horn growth; he was merely the stag that had wandered from the Wairarapa. He was, moreover, apparently undetected during his journey; had he been anywhere seen by bushmen or shepherds, so curious an occurrence would have come to the ears of the run-holder near whose country he had been spotted, the fact would have obtained circulation at shows, race-meetings, county clubs, acclimatisation society meetings—in short, wherever country folk do most consort.

Although, however, the details of the journey can never be precisely known, there is one fact in connection with it worth recording,—it is this, that when forty years later rabbits in alarming numbers reached Tutira, they appeared in greatest profusion on the very same locality as that tenanted by the stag: it is probable that they had followed the same general line taken by the deer, threaded the self-same passes and moved along the summits trodden by him so many seasons before. I have always felt an interest in this exiled deer, in his wonderful trek, his solitary existence, his pathetic affection for his unresponsive neighbours the horses, his banishment from kith and kin, and lastly, his tragic fate—tangled by his antlers among supple-jack and slowly starved to death.

Until the 'nineties no other deer reached Tutira. A hind was then seen on the station, but by that date the local herd of red-deer liberated in the highlands of Hawke's Bay had been established; the wanderer had strayed but fifty or sixty miles over open, well-roaded country. Twice since that time a single deer has been on Tutira for brief periods.

The line of the stag's journey, it is hardly necessary to state, is merely conjectural. We can only surmise that he did not—for reasons already given—travel by the coast; that he could not have travelled through breadths of forest land, dark dense jungle, seamed, moreover, with ravines running at right angles to the line of march. The route suggested is that of general likelihood, compiled from bushmen, shepherds, and surveyors, each wise in the lore of his particular beat.

There have, in all likelihood, been many liberations in New Zealand of black swan (*Cygnus atratus*). The earliest I can hear of were those freed in the 'fifties at Kawau by Sir George Grey. They were imported from Australia, the story goes, to destroy the water-cress, then considered as a menace to some of the New Zealand rivers. It is quite as likely that Sir George acclimatised the breed on æsthetic grounds. Be that as it may, swan cannot have spread fast, for Archdeacon Samuel Williams has told me that they reached Hawke's Bay—at any rate, that they became conspicuous—only in the 'seventies; in the early 'eighties I remember them very plentiful in the lagoon whereon Napier South is now built. Swan have always been scarce on Tutira, where there are no suitable feeding-grounds; a few pair remain for a few days each season.

Pea-fowl reached Maungaharuru in the 'seventies. Mr MacMahon, for long resident in the district, first as manager of that station and later of Waihua, has told me of a hen which for years lived a solitary life on the former station. Another hen had been resident not far from the Tutira homestead five years before my arrival, and continued to live for another four years in a patch of open bush west of the Natural Paddock hill. A third hen appeared on the run in 1900, and survived for three seasons. Each of these female birds had a beat from which she never wandered, for pea-fowl are too large not to be dangerously conspicuous. The plumage of the male, indeed, may account for the fact that hens only have reached Tutira; male birds migrating may have been at once destroyed in consequence of their brilliant trains. Persecution of strangers, and especially of conspicuous strangers, is one of the minor factors by which nature emphasises limitation of range. The handsome native Paradise duck (*Casarca variegata*), for instance, which until 1917 but rarely visited the run, used to be unceasingly persecuted by hawks (*Circus Gouldi*) working in parties of three and four, whilst it is scarcely an exaggeration to state that the first arriving rabbits and hares were plucked alive.[1] Pea-fowl have never been domesticated on the run; the birds reaching it from time to time were stragglers from a flock belonging to a settler twenty miles distant.

Pheasants have been imported into Hawke's Bay by private enterprise, by the Hawke's Bay Provincial Council, and, I believe, by the local Acclimatisation Society. Though never plentiful even in the 'eighties, it was possible to obtain three or four brace of cock birds in a day. They are now so scarce that, but for their crowing during earth-tremors, we should scarcely know of their presence on the station. Pheasants are either peculiarly sensitive to vibration or particularly noisy in their comment on it, for earth-tremors of even the faintest kind are invariably registered by the cock birds, tremors so slight as to be barely sensible even to persons perfectly quiescent; if by comparison of watches it is discovered that pheasants have in several parts of the run simultaneously chirruked, it is safe to infer that a slight shake has taken place.

A brace of partridges were sprung by myself and partner in September of '82, during inspection of our eastern boundary. They had been liberated on Moeangiangi by its then owner, Mr John Taylor.

Of alien insects on Tutira before my day, the mason fly (*Pison prumosus*), the black cricket (*Gryllus servillei*), and the honey bee were the most remarkable. The first-named, believed to have reached New Zealand in chinks and knots of Australian lumber, was noticed by the late Mr J. N. Williams in the late 'sixties. Unlike the black

[1] After the great flood of 1917 a large flock of Paradise duck remained for months about the mud-submerged flats in the vicinity of the lake. During that time the harrier hawks became used to them. As a strange species they no longer attracted particular attention. Now in 1920 the few pair that remain to breed are left comparatively unmolested.

cricket, it seems never to have received a Maori name—a fact in itself pointing to a comparatively late naturalisation, to a period when the mind of the native had become surfeited with novelties, his intelligence sapped by ill-digested alien knowledge, his old-time interest lost in forest life and lore.

In the open the mason fly plasters its cells on to the pitted surfaces of limestone crag; within doors its vermiculated clay chambers are fitted into every available crack and chink, into key-holes, beneath projecting laps of weather-boarding, in folds of suspended garments. A situation particularly favoured is an oilskin coat suspended on a verandah—such an article, if shaken after prolonged disuse, always precipitating a rain of broken clay chips and flaccid spiders. Every chamber contains cells of different sizes, in each of which an egg is deposited, and the compartment then filled in with spiders, which for long retain their freshness, and which appear to be torpid rather than dead. In due course the eggs hatch, and the grubs feeding on the stores provided, become white maggots. Later again—unless, as not infrequently happens, destroyed by parasites—the mature insect, dark, slender, and elegant, emerges and completes the circle of life.

The black cricket, *puharanga*—"bush-ranger"—of the natives, whose faint musical trill tells us that autumn has come once more, is reputed to have reached New Zealand either in matting from the islands, or in the bedding of troops from India. It has never been plentiful on Tutira; the rainfall is too great for a semi-tropical insect, the soils of the run too porous. Only in localities where alluvial clays fissure and crack in summer can the insect become a plague, but on such lands I have known its numbers multiply into millions.[1]

[1] Whakawhitira, a Poverty Bay farm, owned by my brother, Harry W. G.-S., in the 'nineties, was on one occasion so stripped of grass that not a green blade remained over several hundred acres. Each stool of ryegrass was nibbled as close as the night's stubble on a man's unshaven chin. The season had been unusually dry, and the soil—an exceedingly stiff "papa" alluvium—had fissured in innumerable deep cracks which afforded cover to the crickets, and where they bred in enormous quantities. Their numbers, vast in themselves, were reinforced by a general move coastwards from the interior, a movement increasingly noticeable during autumn. It was indeed only on the approach of winter that the crickets loosened their grip on the ravished farm; finally, probably in search of warmth, they perished in the sea, at any rate on two occasions whilst on the bay I noticed them thick aboard the steamers. Besides ruining the ryegrass fields, the boles of the lemon and orchard groves were barked as rabbits in snow bark ash and sycamore. As rabbits, too, cleared certain districts of cabbage-trees (*Cordyline australis*), first falling and then eating them, so the black crickets felled my brother's nine-feet maize crop, nibbling each stem through at the base, and then on the ground consuming stalk, leaf, and milky pod. Leather bands of machinery, kid boots, wall-papers and men's coats, were attacked. Ducks, fowls, and turkeys, gorged with insects, laid as if spring

The honey bee was liberated about the same date at the Bay of Islands and at New Plymouth; probably from the former, Hawke's Bay has been stocked directly or indirectly. The newly-imported insect had no enemies to contend with; there were no diseases and no competitors. The winters of the North Island are brief, or non-existent; there is no single month of the year when some native shrub or another is not to be found in blossom. Local conditions were extraordinarily favourable too; portions of eastern Tutira, viewed even from considerable distances, were during spring-time actually grey with the profusion of white clover-heads. Everywhere then also the purple-headed prickly thistle possessed the land. There was not a hollow tree or crannied limestone rock which in the 'eighties did not contain a hive; colonies were established even in the open, though from these unsheltered swarms no great store of honey was obtainable, dews and rains diluting the nectar gathered, and washing it from the uncapped cells.

The exuberant prosperity of the bee has passed away with the disappearance of the white clover and the thistle. Few indeed of the hollow crags now harbour colonies. One rock only—a vast square projecting from the highest tier of ocean floor on the Racecourse Paddock—has never to my knowledge during forty seasons been untenanted. Bees are now again on the increase, owing to ploughing, the use of artificial manures, and consequent revival of white clover.

Only these few insects, birds, and mammals had reached Tutira before my own occupation of the station; it was the good fortune of the writer to witness personally a later and greater trek of living things.

were again come. Poison was of no avail: grain phosphorised, or soaked in strychnine, or soaked in arsenic, was apparently innocuous to these terrible insects. The damage done—the utter destruction of grass and crop—was a revelation to me; having witnessed the ruin wrought in a few weeks, it has been easy to sympathise with the demands of early agriculturists for the importation of birds.

CHAPTER XXXIII.

ACCLIMATISATION CENTRES AND MIGRATION ROUTES.

MANY factors — sport, sentiment, and business — entered into the jubilation with which the project of acclimatisation was acclaimed in New Zealand. It was the age of enthusiasm: the possibilities of small settlement were then undreamed; the land was still parcelled out in great estates, whose owners, I daresay, thought of founding families, of game preservation as at home, of fox-hunting,—it is a marvel that New Zealand has escaped the importation of the fox, — of all the jolly old-time country life that for good and evil is passing away from the world. The protests of those in England whose comparative knowledge and experience could properly appreciate the dangers ahead were unheeded. If heard at all, they were passed over as the remonstrances of persons without practical knowledge of conditions obtaining in the Antipodes.

Yet, if not a failure, acclimatisation in New Zealand has not at any rate been an outstanding success.

In justice to its founders and supporters, however, it must be conceded that failure has in some degree been consequent on an alteration in the social fabric which could hardly have been foreseen. Game preservation is an abomination to a democracy each of whom is a freeholder, and each of whom, not unnaturally, desires to shoot his own game—equally naturally to shoot it without delay lest it cross into his neighbour's territory. Each bird, in fact, is treated as were those coveys of grouse found in pre-driving days too near the marches of a moor—slain to save them.

Then, too, another outcome of close settlement—a multiplicity of cats and dogs—spells death to game. It is probable that deer-stalking is in the not far-distant future also doomed. The finest heads—park type—have hitherto been obtained from lands that cannot be, or cannot much longer be, held in large areas. Elsewhere, into tracts of poor high country where indeed deer might subsist without detriment to the State, the rabbit has of late penetrated, and with him the rabbiter. These men know the countryside as they know their huts. They possess in full the sporting tastes of their fellow-mortals. I am given to understand that the best heads disappear before the season opens, or instantly afterwards. Rangers are few and far between, and fines for poaching inadequate even were evidence forthcoming. It is true there are areas still free of rabbits; but it is to be feared that, sooner or later, these regions too will be overrun, that once more in the wake of the rabbit the rabbiter will follow, that in any case the deer will be eaten out, that the best trophies and the glorious loneliness of stalking will be gone.

There can be no two opinions as to the importation of vermin such as stoats, weasels, polecats, and ferrets. Only the value of the avifauna brought to New Zealand for sentimental and utilitarian reasons needs to be considered.

On this topic much can be said for and against acclimatisation. It is true that some of the birds are already troublesome, and it is likely that others may become so. It must nevertheless not be forgotten that good has been done as well as evil; that if, for example, the sparrow takes a proportion of the farmer's ripened grain, it is but a fraction of what was robbed from the pioneer by plagues of caterpillar, grasshopper, and black crickets. Only those who are aware of the enormous depredations of insects in the early days of New Zealand agriculture can properly adjust the balance.

The importation of the Salmo tribe seems, although little or nothing is known of their habits in the sea, to have been a genuine success.[1]

[1] Compared with results obtainable elsewhere in New Zealand, rod-fishing in Hawke's Bay is not first-rate, the great floods that now and again pass over the province destroying the ova and drowning the trout. This deprivation from an angler's point of view may, however, be perhaps remedied. There are indications that trout are coasting the shores and already in a small way running up the rivers. In the Waikoau, for instance, fish which are practically sea trout in their silvery appearance and red flesh have been taken five miles inland. These fish are quite dissimilar to the river fish, good as is their condition also. The little Moeangiangi stream reaching the sea two or three miles north of the Waikoau may also be cited. It has never been stocked, yet up it trout from the Pacific are running, trout of three and four pounds' weight. Possibly in the Waikoau it might have been thought that river trout washed out in floods had taken to the salt water and later returned to spawn. In

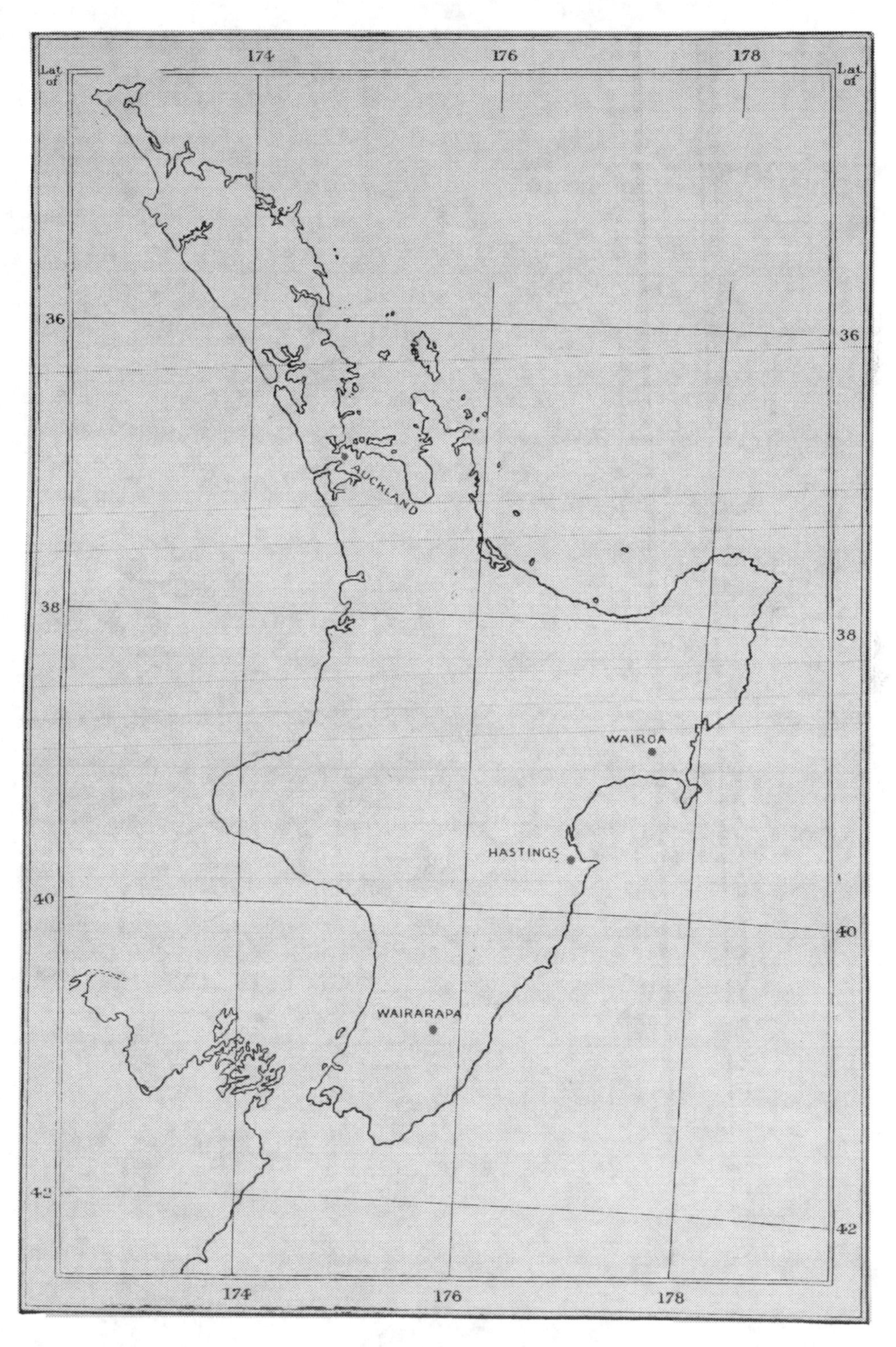

ACCLIMATISATION CENTRES NORTH AND SOUTH OF TUTIRA.

Proceeding from this disquisition on New Zealand's acclimatisation in general to our immediate subject, the accompanying map shows the four centres from which animal aliens have reached the run. They are Wairarapa and Hastings to the south, Auckland and Wairoa to the north.

For observation of aliens passing north and south, Tutira has occupied an exceptionally fortunate geological situation. Between Wairarapa and Hawke's Bay stretches a broad belt of fertile land running roughly parallel with the coast. This band of limestone and marl country begins, however, to narrow immediately south of the station into a sort of tongue or peninsula. Tutira, in fact, may be considered the terminal portion of a fertile belt that stretches the whole distance from Wairarapa.

To this belt migrants have clung, repelled by the poorer soils and grassless lands impinging on it from the west. They have, moreover, closed their ranks as the band of good land shrunk in width, and have therefore not only passed through Tutira, but have passed through it in relatively large numbers. Aliens, therefore, moving northwards had perforce to pass through the run. Animals, again, moving southwards, and coming during the last portion of their trek by way of the fertile Waiapu and Poverty Bay districts, have followed likewise the coastal route. For precisely similar reasons to those which, immediately south of Tutira, compelled the concentration of northward-moving migrants, about Wairoa has occurred another contraction. Westwards of that district extend large areas of dry hungry lands over which no creature accustomed to such fertility as that of Poverty Bay would be likely to straggle. All southward-moving migrants, too, have passed through Tutira, and, because of their concentration, have passed through it in relatively large numbers. The run, therefore, may be said to have stood in the centre of a double current of aliens—some moving south, some moving north; it has been the waist of the sand-glass, through which each grain was bound to flow.

Before proceeding to relate the history of the different living creatures that have managed to reach Tutira in my day, it will be

the Moeangiangi that is not possible; there beyond controversy trout have entered virgin water direct from the ocean. The earliest trout liberated in Hawke's Bay (*S. fontinalis*) were brought from America in great baths on board one of the paddle steamers that used to ply between the two countries.

well to give a general account of the routes by which these uninvited guests have invaded the run.

The movements of masses of living things may or may not be blind and involuntary. In any case, it is certain that some roads of expansion are favoured over others, and that, in the selection of these, birds and animals follow, like man, the law of the line of least resistance. That line may, in the North Island of New Zealand, be generalised as the line of light.

There are three natural highways by which imported animals have chiefly travelled from centres of acclimatisation north and south of the run. Each of them is a line of light fringing or piercing the dense vegetation of a fertile, warm, well-watered land. They are the coastal route, the hill-top route, the river-bed route.

During their journeyings to the run from different centres of

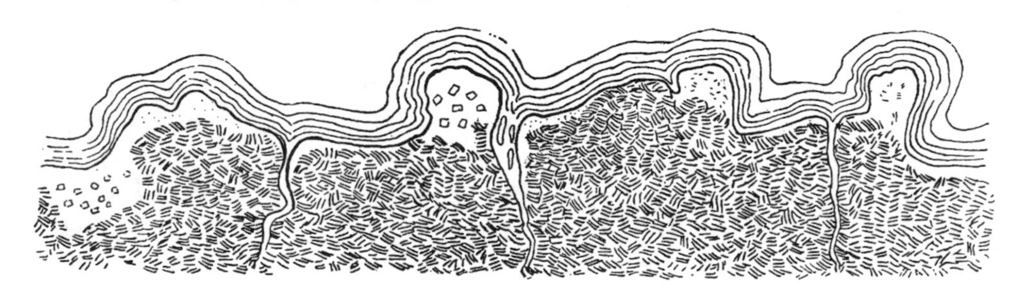

Line of light—coast—showing native clearings.

liberation, many of the migrants—especially the bird migrants—have at different periods, passing through different districts, used all of these ways. The coastal route on the whole has been most helpful; upon it certainly the final laps of many species have been accomplished. It has, in the first place, offered superior attractions to any other line of ingress in the matter of warmth. By following the coast line, which in New Zealand happens, generally speaking, to be also the line of human settlement, Maori or European, migrants have not only procured food more easily, but have, through man's tillage of the ground, found food of a suitable sort. It must be always remembered that the alien vegetation of New Zealand was well established on pioneer plots before the majority of the alien animals and birds had arrived; that there were procurable the seeds and tender leaves of imported garden plants, grasses, and weeds.

Furthermore, it must be taken into account that there existed along the route insects, grubs, and blights also of European origin.

It is possible there may have been another inducement to follow the coastal route; there may have been an instinctive attraction to the sea, that great plain over which the old-world ancestors of imported species for generations have ventured.

The second of these animal highways was the hill-top route. Wide belts of impenetrable forest covered the slopes—forests still standing as late as the 'eighties. They extended from ocean almost to mountain-top, heaviest timber growing along the coast zone, trees lower in height roped in tangles of "lawyer," vine, and supple-jack on the foothills and lower slopes. There was another reason why migrants were forced on to the tops. It was the only route by which the river gorges that furrow the flanks of the main ranges could be avoided, gorges always difficult

Line of light—mountain-top route through forest.

of access and often impassable. The line of summit was the line of comparative light and comparatively open ground. Although there existed nothing approaching a continuity of bare ground, although the line of light was broken and checkered, there were here and there at least reaches of uncovered summit, rockfalls of jumbled stone, windblows where gales had swept off the top soil, lucid intervals of turf. These spots and stretches upon which, at any rate, the sun could shine and where migrants could touch ground unshaded by foliage, were as stepping-stones encouraging advance across a river ford. Compared with the gloom and tanglement of the damp forests they must have been oases pleasant to reach. They offered freedom and light as against unknown possibilities of danger and darkness.

The third line of migratory movement was the river-bed route. Particularly alluring has this route proved to migrants where a confused jumble of valley, slope, and summit devoid of any open connection with one another, and covered alike with dense tangled greenery, offered no place of alightment, no city of refuge, no perch for the sole of the foot, no bare ground whatsoever. This route was to birds the primrose path, the line of sun and warmth, of water in hot summer-time, of open ground, of minute alluvial plains, of grass and grit and dry clean sand. It is a line which, although less used than the hill-top or coastal routes, has nevertheless been of considerable aid to several migrants during stages of their wanderings. Some of the species which have reached Tutira from the north have thus passed through forest and fern lands otherwise impenetrable; they have been guided through the wilderness by sunlight on bare ground. Whereas, moreover, the river system to the south was a bar, northwards it was in some sort a key to Tutira. Southwards the river-beds ran athwart the line of migration; northwards

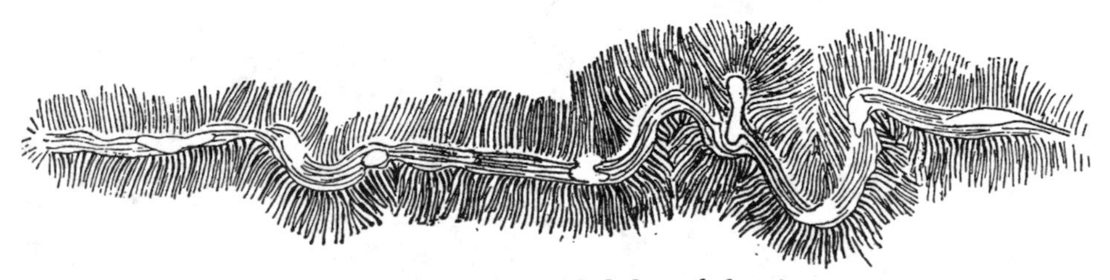

Line of light—river-bed through forest.

several of them served under certain conditions as short cuts to the station.

There is one other migration route which, albeit artificial, has also been largely utilised. It is the highway which man, proud man, believes built by himself for himself only. No belief could be more erroneous; slightly paraphrasing the proverb, it can truly be said that man builds roads and wise animals use them.

The highway of man is after all but a track better graded and more evenly trodden than that of the sheep, the penguin, the kiwi, the petrel, or the pig. Roads are, to animals as to men, lines of easiest access. In the Old World all lines of migratory movement had been established long prior to road construction; in New Zealand the peculiar conditions of importation and subsequent spread have lent to them a novel function. There is a mass of evidence in support of the view that they have been largely used by travelling animals and birds. The obvious demurrer to the connection of observed animals and roadways—that it is because the observer is more on the road than off it—will not hold good at Tutira. Life on a rough run is spent not on but off the road ; for every ten miles ridden by the author on roads, a hundred has been ridden in the wilds. If, therefore, migrants have been relatively more often and at earlier dates seen about highways than elsewhere, it is because they have been followed at first as lines of least resistance and clung to afterwards as guides to further goals.

Their importance, indeed, as a factor in the spread of aliens can hardly be estimated, except by realisation of what is called the inert mentality of animals, but which might, perhaps, be better termed their prudent conservation of energy. It is a truism that sheep in a field will avoid obstacles as man himself does ; thistle-groves, clumps of fern, logs, soft ground, will all cause deviation ; in fact, as I have said before, a straight sheep-track is no more discoverable than a straight path between village and village. Advantage is taken of the smallest saving of toil, even such as that entailed by the stepping over a fern frond fallen out from a clump, or the avoidance of a prone spar a few inches high. One particular instance out of thousands occurs to me : On certain low hills on Tutira a foot-wide track had been roughly slashed through low fern. Such a track it is comprehensible sheep in "working" a paddock might discover and prefer, but that a year later, when fire had run over the whole area, they should rediscover a line by that time indecipherable to human vision, shows how the smallest saving of exertion is appreciated.

If advantages in ease of progression of so infinitesimal a kind are thus taken into account, it is not difficult to understand the allurements of a road,—an entrancing surface free of logs, free of tangled grass, free of hummocks,—an open way allowing full vent to that mental sloth which is the lower creatures' bliss. From four centres of liberation and by four lines of ingress—the coastal route, the hill-top route, the river-bed route, and the highway of man—has the modern alien invasion of Tutira been accomplished.

CHAPTER XXXIV.

THE INVASION FROM THE SOUTH.

We can now pass to the history of aliens which I have myself seen establish themselves on the run. We can deal first with strangers reaching us from the south, from the Wairarapa and Hastings.

From the former—better that a millstone had been hanged about that district's neck, and that it had been cast into the depths of the sea—have come rabbits and weasels.

From the Hawke's Bay Acclimatisation Society, on the other hand, nothing harmful has emanated; on the contrary, that society highly disapproved of the liberation of vermin; it could do no more, however, than protest and offer rewards for every weasel, stoat, or polecat taken in its domains; but, alas! the harm had been done—not all the poppy and mandragora, not all the guineas in the world, could stay the plague.

Taking, first of all, species liberated in Hawke's Bay, evidence is unanimous that the goldfinch (*Carduelis elegans*) proved able at once to adapt himself to his new surroundings. It would, in truth, have been surprising had the bird not flourished; fern-crushing had passed the experimental stage; the far-famed fertile plains surrounding Hastings had been drained; bush-felling, though still in its infancy, had been proved a success. Everywhere fern, flax, and forest were giving place to English grass and those self-imported weeds that flourish with peculiar luxuriance in virgin soils. Amongst them were two plants particularly affected by the goldfinch—the one the sow-thistle, the other the prickly thistle. On every run in Hawke's Bay during the 'seventies and 'eighties the latter flourished over hundreds of acres.[1]

[1] I myself have seen, early in the twentieth century, on the wooded ranges of Poverty Bay, an equally luxuriant growth. These hills were composed of fertile marl, enriched further-

From descriptions given me by the late Mr J. N. Williams and others, it was into a countryside where each season new blocks of land were being handled that the goldfinch was freed; it is no matter of surprise that the species flourished exceedingly.

It was in the summer of '83 that I first saw a goldfinch on Tutira. Shearing was in progress; my duty as wool-classer obliged me to be in the shed before five; at that early hour, in scrub close to the wool-shed, the bird was noted, and there for six days he remained about the same spot. I say "he," for though the plumage of the sexes in this breed is almost indistinguishable, yet I have always believed the bird to be a male,—his feathers were so magnificently resplendent in their sheen and depth of colour, the reds and yellows so deep and pure.

Weather permitting, shearing on a sheep-run starts each season at the same date. Just one year later I was again wool-classing, again walking shedwards, a few minutes before five in the morning. No scrub had been felled in the vicinity of the shed—the clump of manuka and tutu remained as it had been; now, however, where one goldfinch had been observed, a pair had taken possession of the locality. They were not so constant to the little thicket as the pioneer bird had been, but remained within, say, a hundred yards of the original spot for a few days. After that I lost them till autumn, when again a party was seen in the neighbourhood of the shed. We supposed then that the pair must have been male and female, that they must have reared a brood, or couple of broods, and that the little congregation observed must have been parents and offspring in flock. Now, for reasons to be given later, I am not so sure of this; the small party seen may have been with equal likelihood composed of birds following in the wake of the first-seen specimens. At any rate, for several seasons, a thin wedge, or narrow

more by centuries of leaf-mould; lastly, there lay on the ground a heavy top-dressing of potash, from timber consumed in the clearing fires. The result was a marvellous thistle growth; over thousands of acres it was impossible to walk without body armour of sacking to protect the legs and chest, without gloves to defend the hands, without slasher to clear a track. For all practical purposes the country was locked up and the stock lost—if, indeed, such land was stocked at all—until late autumn. Thistles reached the height of a tall man's neck, and grew, not a plant here and a plant there, but in vast impenetrable thickets. In spring each prickly star raised itself into a tall plant; in summer, when the morning dews were dried, the ripe heads burst and liberated their packed seeds in millions of millions; hour after hour, like snow, this summer storm sailed airily amongst the charred black boles; in autumn the feathery pappus, shining and bright, lay in drifts feet deep; finally, what had been a green grove stood wrecked and grey—a sere forest of masts leaning against one another in swathes or flattened like rain-laid corn. During the following season thistles were almost absent, during the next there was a recrudescence of the plant, and after that normal growth.

spear-head, thrust itself more deeply into the run. Then at once the goldfinch became extraordinarily plentiful, enormous flocks tenanting every part of the eastern run. Lastly, normal conditions prevailed; as the soil became thistle-sick, the superabundant food-supply of the species began to fail. In fact, just as the bee diminished on Tutira with the disappearance of white clover, so the numbers of the goldfinch declined with the vanishing thistle.

A small number of minahs (*Acridotheris tristis*) had been liberated in '77 by the Hawke's Bay Acclimatisation Society. In '82 these birds were still regarded as novelties; the habits of mated pairs were still carefully noted; I well remember the interest attached to a pair breeding in '84 beneath the eaves of the late Mr W. Birch's house near Hastings. As the goldfinch on Tutira will always be associated in my mind with early efforts at wool-classing, so is the minah associated with another great event in station life—the docking and ear-marking of lambs.

Cuningham and I had by '84 built a cottage, and provided ourselves with a married couple to look after it and us—"to make themselves generally useful," as the phrase goes. A garden had been dug and fruit-trees planted; our semi-wild game-fowls had been brought over from the site of the original homestead. Instead of stravaiging hundreds of yards away, and flying with frantic eagerness from the hillsides when the milk-dish was scraped of its dough at the *whare* door, they lived well-regulated prosaic lives within a run of wire-netting visible from the dining-room. One day in November—docking was later in those times—we were breakfasting about nine after a long morning's work, when Cuningham, who sat opposite the window, drew my attention to a solitary bird crouched up against the netting of the hen-yard, as if endeavouring to chum up with the fowls. It was a minah, a solitary wanderer, attempting, as lost animals do, to associate with any other living creatures, however remotely connected. Next morning it was gone.

A year later, when we were again docking our lambs, a — or, as I have always believed, the — minah reappeared. As before, when noticed, it was crouched on the ground close to the wire-netting of the hen-run, exactly on the spot where Cuningham had seen it twelve months previously. Its whole appearance was that of a creature desiring association with some living thing, however distant in degree. Its forlorn air and close proximity to the wire suggested the idea of a wish to plunge itself amongst other fowl, even though only game-fowl, to bathe itself in their society, to wash away the feeling of appalling loneliness. The homestead, it must be remembered, in those days was a bare square of wood in a bare paddock. There was no shelter, there were no trees, there were no alien birds.

For the third time minahs appeared in '86, two pair attempting that year to build about the house itself; in their thorough investigation of the possibilities of the chimney, more than once the birds fluttered into the rooms beneath; whole mornings were spent on the shingle roof, inspecting the eaves and carrying little sticks into impossible places. Once or twice the incomplete nests were flooded out from the gutters; at last, after many weeks, the birds left without having succeeded in their efforts to found a family.

During the two following years there was no further attempt to colonise Tutira. In '89 I was at home; upon my return to New Zealand I found that the construction of the Napier-Tutira-Wairoa road had begun. On the route of the surveyed line navvies were camped under canvas in several different parties, each of which was attended by its special flock of minahs. The birds lived on the leavings of the meals thrown out; they were there for what they could get; for the very same reasons as induced the Jews to follow the Normans into England, the minahs followed the navvies into Tutira. A pair once more reached the Tutira homestead. Again, however, they were balked for want of building sites; again the only suitable-looking spot—the chimney—was explored, the birds as before, during the progress of research, fluttering into the rooms beneath. By this date, however, our plantation of Pinus insignis had reached a height of eighteen or twenty feet; lashed together, several of them were stout enough to support a box. A box accordingly was fastened into the pine clump, not the one selected by me, but another quite unsuitable in shape, and which exposed the sitting bird to strong light. So keen, however, were the minahs to build, if not actually on the house then close to it, that even this poor substitute was eagerly utilised. It was evident that if they could not breed in immediate proximity to man they would breed nowhere else.

Then began a process of emancipation—the minah became less dependent year by year; he outgrew the tie binding him to mankind. At first nests were built in rocky road-cuttings and beneath the woodwork of culverts and bridges, where, at any rate, there was the solace of man's countenance and support if but for a passing glimpse. Later again, even this slender tie was dropped; the minah became a wild bird, building far from any homestead in the cracks and fissures of dead trees standing in bush reserves. Nowadays, upon the approach of autumn, minahs largely use the roads, closing in on homesteads for scraps of fowl-feed and leavings of the gallows and kennels. The species has also of late developed a vulture-like habit of congregating near any sheep dead on the hills; in the vicinity of the carcase, awaiting the process of skinning, the expectant birds gather for their ghoulish meal.

"Australian magpies" (*Gymnorhina leuconota*) were liberated in Hawke's Bay during the 'seventies. A brace appeared at Tutira in '85; they had moved inland from Tangoio, where prior to '82 a small colony had established itself. The migrant pair were accidently destroyed. The Tangoio magpies three seasons later were purposely shot; their attacks on the sheep-dogs had become so intolerable during the nesting season that the wretched collies dared not follow their owners, who in their turn were unable to muster the run without canine assistance. Since that date, from time to time solitary birds moving north have been noted on the station.

The evidence of early settlers is unanimous that the acclimatisation by the Hawke's Bay Provincial Council of the yellow-hammer (*Emberiza citronella*) was an immediate success. The breed was perfectly suited by the multiplication of insect life and the increase of alien fodder-plants and weeds everywhere taking place; like the goldfinch, it flourished from the date of liberation. In '87 several pair were noted on the run, one at least of them rearing a brood. The history of the breed is unusual in this, that its numbers on the station nowadays are as great as at any earlier period. The yellow-hammer has not, as usually happens, appeared, rapidly increased to a maximum and then with almost equal celerity suffered diminishment.

The greenfinch (*Ligurinus chloris*), also imported by the Hawke's Bay Provincial Council, is reported to have succeeded at once. I have, however, no recollection of any specimens on Tutira prior to '90. From the fact, however, that I got half a dozen nests during that year, it is likely that earlier pairs had been overlooked; it is improbable that so many nests would have been discovered during the first year of the appearance of a new breed.

The Australian quail (*Synœcus australis*) was a private importation of very early date, birds being liberated on Rissington in the 'sixties by Colonel, afterwards Sir George, Whitmore. It did not, however, reach Tutira for more than thirty years. So tardy a spread can be probably ascribed to the uninviting nature of the intervening country, which, until improvements began, was one vast sheet of bracken. Over this inhospitable wilderness, moreover, huge fires raged from time to time, providing the harriers—who well understood the significance of the event—with a burnt-offering of lizards and small birds.

Another game-bird, the Californian quail (*Callipepla californica*), was imported by the Hawke's Bay Provincial Council, and again at a later date by the Hawke's Bay Acclimatisation Society. Californian quail reached Tutira in the middle 'nineties, and although there was at first an increase in their numbers, it was a limited increase and soon ceased. Their advent as game-birds had come, in fact, too late to admit of any great success. The competition of innumerable goldfinches, yellow-hammers, larks, sparrows, and native species, several of which had also increased with the enlarged area of open country consequent on the destruction of bracken, had already affected the insect food-supply; the Californian quail is now disappearing from the run.

Hares were brought to Hawke's Bay from the South Island. In '82 they were fairly plentiful on the river-bed country in the neighbourhood of Hastings. Eleven years, however, elapsed before the thirty-five miles betwixt that district and Tutira were traversed. Possibly the tendency of scared hares to run in circles has prevented a rapid spread; on the other hand, this trait had been in some degree neutralised by another—the remarkable use by hares of roads and road-cuttings. On the station, shepherds riding at dawn would sometimes strike a hare on one of the narrow cuttings that preceded the completion of the road, and for amusement course it with their collies for hundreds of yards ere the animal would dash off to right or left. The consequence has been that the spread of the hare has been in a peculiar degree governed by chance.

A third factor in the naturalisation of the breed has been the ceaseless persecution of its vanguard by the native harrier hawk. Too stupid to take cover, they could be seen hirpling about the run pursued by their relentless enemies.[1]

Allusion has already been made to the disgust felt at the freeing, firstly, of rabbits, and secondly, of weasels, in the Wairarapa. It was an outrage that any individual or local body should have been allowed to attempt to correct a blunder by a crime. The feeling in Hawke's Bay was evinced in the resolution—as I recollect it, for the minutes of the Society are missing—proposed and passed with ferocious unanimity, that a guinea a head—and guineas were not then as now to be gathered from every manuka bush—should be given "for every head of vermin dead or alive." No paper resolutions, however, can stay a plague once introduced; harm of this sort done cannot be undone.

Weasels, stoats, and ferrets were bred and liberated at Carterton, the first-named thriving on a great scale and quickly overrunning the countryside. "At least before 1901" my then neighbour, John Moore of Rakamoana, noticed weasels on his property. By 1901 they had reached Tangoio, by 1902 a specimen had been seen on Tutira.

I was then one of the three members who represented the Mohaka Riding of the Wairoa County Council. The township of Wairoa, where, once a month, the Council met, lay thirty miles northward from Tutira; a brother then possessed a farm near Gisborne, seventy miles from Wairoa; there, during the early and middle 'nineties, I was a frequent guest; in search for bush-land for my younger brothers, I had occasion to visit districts thirty and forty miles inland and sixty and eighty miles north of Gisborne. Between Tutira and Wairoa, therefore, and between Wairoa and Gisborne, I may say I had an intimate knowledge of roads made and in the making; northwards of Gisborne I knew something of them personally and more through observers acting on my behalf. During ten or twelve seasons, in fact, I was cognisant of the progress of aliens—birds and animals—over a distance not far short of 150 miles north of the station. Excellent chances were thus afforded of following the progress of the weasel movement.

Prior to their arrival at Tutira, during their approach through the southern settled districts of Hawke's Bay, terrible tales of the murder of young lambs, of the biting of babies and grown folk, of rape of hen-

[1] In South Canterbury I have seen a hare, dazed with buffetings, continue to follow for quarter or half an hour an open road, when within a few yards a gorse hedge, running parallel, afforded perfect cover and protection.

roosts, were rife in the daily papers, perhaps — for squatters had imported the horrid vermin—most prominently in papers hostile to the sheep-farmer interest. At the time I took considerable trouble in the investigation of several of these stories of attacks on grown folk, and believe that some at least were true, or at any rate that much evidence of a circumstantial sort could be adduced in support of them.[1]

The earliest weasel was seen on Tutira in 1902. Between that date and 1904 they had overrun the country between Tutira and the southern edge of the Poverty Bay Flat. Everywhere I heard of them. On every road and new-cut bridle-track during these two seasons I met or overtook weasels hurrying northwards, travelling as if life and death were in the matter. Three or four times also I came on weasels dead on the tracks. These weasels, alive or dead, were or had been travelling singly. The only party I heard of was reported by Mr J. B. Kells, then managing Tangoio. In firing a small dried-up marsh he dislodged a large number; according to his statement, they "poured out" of the herbage. For a short period weasels overran like fire the east coast between Tutira and Poverty Bay, and then like fire died out. I traced them by personal observation to the very edge of the Poverty Bay Flats, then, like the Great Twin Brethren, "away they passed and no man saw them more." Nowadays on Tutira I do not hear from shepherds or fencers of the weasel once in six years. I have not seen one for twenty years. There is something ridiculous in the fact that the weasel should have arrived on the station before the rabbit, and that later, when rabbits had become numerous, weasels should have practically passed out of the district—that the cure, in fact, should have preceded the disease.[2]

[1] I have myself known seven or eight healthy young lambs killed in a night within a short distance of one another, each with a small puncture in the throat.

[2] Returning in March 1919, after five years absence owing to the war, I found that pukeko (*Porphyrio melanonotus*) and weka (*Ocydromus greyi*) were practically gone from Tutira; the former, which used to feed in hundreds about the swamps, had been reduced to three pairs on one spot and three pairs on another; the numbers of the weka had declined in an equal ratio. There had been no poisoning with grain and no shooting, for during these anxious years my brother was never away from the place. The damage, I found, was generally attributed to weasels; that they had been seen here and there was cited in corroboration of this belief. It may be so, but there are facts that do not dovetail into this theory. I say nothing of not having personally seen either weasels or signs of weasels during twelve months since my return, but why, if they have destroyed weka and pukeko, have the numbers of the small pied tit (*Petroica toitoi*) hugely, astonishingly increased during these five years? Why have Californian quail certainly also increased? Why do starlings, blackbirds, thrushes, minahs still as formerly swarm on the station? Possibly light may be thrown on the problem by remembering that twice during my residence great irruptions of weka have passed through Tutira. Is it possible that for a third time wekas may have, as formerly, followed a mouse trek and not returned? Is it possible that for some

Starlings (*Sturnus vulgaris*) were imported from Otago by the Hawke's Bay Acclimatisation Society, though two years later additional birds were brought from Auckland. According to the late Mr J. N. Williams, the local success of this introduction was never in doubt. It was nevertheless not until ten years later that starlings were reported in the minutes of the local Society to be spreading rapidly.

Starlings were liberated in Auckland in '67. Next year the Acclimatisation Society of that district reports that the starling has "perhaps multiplied more rapidly than any other species introduced." The minutes of the Society give the further information that starlings—doubtless lost individual birds, "their plumage worn and shabby"—had been noticed "twenty miles from the point of liberation within a few days." At a meeting in the following year the wide spread of the starling was again remarked. In spite, however, of the Auckland Acclimatisation Society's belief in the rapid increase of the species, its forward movement cannot compare in celerity with that of other migrants liberated about similar dates. Starlings, for instance, were not in Poverty Bay in '96; when black crickets during late autumn of that year ravaged my brother's farm, no starlings appeared to combat the plague.[1]

Such are the general claims of south and north. There can be little doubt that starlings reached Tutira from the former direction, for only ten years after liberation in Hawke's Bay they were building on the Bluff in Napier, the huge pitted limestone cliff of Scinde Island. It was not until 1895, however, that they reached Tutira, a flock remaining on the station during the autumn and winter of that year. Then for several successive seasons each year they extended their winter range, large

reason unknown, pukeko have also migrated in a body? Were their runs at last overstocked? Lastly, is there any inference to be deduced from the fact that the three pair left in two localities were practically domesticated, that being protected, close to men's homes, they had lost in some degree the full force of the racial feeling, that in a migration, for whatsoever reason undertaken, they alone had declined to participate? The possibility of migration, at any rate, might help to solve a difficulty in another part of New Zealand over which I have puzzled for years—the absence of certain species from very extensive areas perfectly fitted to their wants in climate, breeding accommodation, and food supplies.

In the hinterland of Poverty Bay, for instance, in the fertile, virgin marl lands of Mangatu, kiwi and weka have always been unknown. Weasels, doubtless, there may have been in the district, but not in larger numbers than at Tutira prior to 1914, when, as we know, pukeko swarmed and weka and kiwi were very plentiful.

[1] Had the birds been in the district they would have gathered in multitudes, as once happened in later years on Tutira. One of my plough contractors had sown a patch of oats which, when flowering, was attacked by an army of caterpillars, millions of whom had appeared as if by magic. At once I rode out to note their horrid depredations, but was surprised to find not caterpillars but starlings in the ascendant; the latter had in a few hours collected in such numbers as entirely to save the threatened crop.

flocks visiting the hill-tops, flying in ordered companies, moving in the busy alert manner of the breed, probing the turf for beetle grubs. Still, however, not a pair remained to breed, each year the flocks deserting us before the nesting season. They had not yet adapted their habits to novel conditions: they still returned to breeding quarters suitable to the gregarious or semi-gregarious habits of the breed.

At first their colonisation of the run was confined to visitations during autumn and winter—upon the approach of spring the flocks returning to breeding-areas more or less protected and countenanced by man, to cliffs such as those of Roy's Hill, to escarpments such as the Bluff in Napier. Later they began to breed on Tutira not far from the shed and house, choosing as nesting-sites pollarded willows and the dense decumbent thatch of cabbage-trees. Lastly, quite away from mankind, they established themselves amongst the dead timber of station forest reserves.

Rabbits were purposely freed near Carterton in the Wairarapa; they are known also to have been furtively brought into Hawke's Bay at a very early date, and there also liberated in an attempt to ruin the squatters; they were on the Apley Run, for instance, as early as the 'sixties. After fifty years Mr Bernard Chambers of Te Mata still retains an indignant recollection of those days.[1]

The earliest rabbit known near Tutira was the celibate who, in the early 'eighties, travelled about Tangoio in the company of a flock of "wild" turkeys. From then onwards during the 'eighties and 'nineties there were rumours of rabbits, and traces of rabbits, the scrapings and droppings of wanderers from the pairs purposely put down with intent to harm the sheep-farmers. Fortunately, however, where a few pair only of any imported creature are scattered over a wide expanse, increase is prevented by normal wear and tear of life. Native birds of prey, harriers, wekas, and moreporks (*Spiloglaux Novæ Zealandiæ*), are quite capable of holding in check limited numbers of aliens, although in the face of a real invasion such puny barriers are unavailing. The rabbit at first failed in Hawke's Bay, not, as is often believed, because the earliest known specimens were of some less prolific strain, but for the same reason as the blackbird and thrush failed—the numbers liberated were

[1] He writes—I give initials only—"It was a fellow named J—— P—— who put rabbits on the station—the damned scoundrel."

insufficient; for the same reason, indeed, as *homo sapiens* failed at first on Tutira, from want of experience of novel conditions.

That was the main cause; there was another in the contraction of feeding-grounds, already fully explained, which took place on every half-developed sheep station. Rabbits, like sheep, were forced on to the tops, and thus in the open exposed to the hawks.

The earliest specimen killed on Tutira was secured by the late Sir Norman Campbell, Bart.,[1] then rabbiter for the Mohaka district. It was taken near the centre of the run, within a few yards of the road. Signs also were noticeable in different parts of the run, but it was a couple of years still before other rabbits were actually taken.[2]

Not, in fact, until the beginning of the new century did the long-threatened invasion begin. Then in a few weeks rabbits were found to have established themselves on many parts of the run. A percentage of these colonies and congregations were alongside the main road, but the chief gate of entrance was by way of the "Wild Horse" country. There, where the red-deer had struck Tutira in the late 'sixties, the rabbit vanguard was thickest; it had, I believe, followed approximately the route selected years earlier by the wandering stag.

Rooks (*Corvus frugilegeus*), which were imported by the Auckland Acclimatisation Society, did not, according to the minutes of the Society, immediately prosper; it was thought that the young birds were languid and weak, owing to the heat of their new surroundings. Be that as it may, the species seems soon to have shaken off this early malady; in Hawke's Bay, at any rate, the few pair liberated near Hastings have after many years increased to a considerable rookery.

In 1907 the financial barometer of Tutira had dropped to "stormy,"—wool had fallen to something under fivepence, the station had, owing to wet seasons, become overgrown with scrub, half of the run was held only by annual agreement with the native owners, the new lease was

[1] Although they have never actually bred on the station, there have been three instances of baronets remaining on Tutira for considerable periods. They have reached the run moving northwards—indeed, as in the case of the Paradise duck (*Casarca variegata*), Tutira seems to be about the northern limit of their range. They have never thrived, appearing unable to accommodate themselves to light soils. As ryegrass has done, after a few seasons they have disappeared.

[2] In those dark days when one squatter drew aside another for private talk, it was easy to guess the subject of conversation. It seemed too as if the old habit of snuffing was about to be revived. There wasn't a sheep-farmer in the province who could not produce rabbit-droppings from his waistcoat pocket, who would not tremblingly request his friends to smell 'em and affirm they were hares' or lambs' or sheep—anything, in fact, but what they were.

hung up. My brother Harry G.-S., Harry Young, and myself were debating the pros and cons of a much-needed plantation; should we, under these melancholious circumstances, fence or not fence—that was the question. Whilst the matter was in debate and we were gazing for inspiration at the hill-face in question, three pairs of rooks flew into view and settled in the pine plantation above Harry Young's house. They must have been extremely hungry, for rooks at home, even in winter-time, do not care to venture too near a house or to alight in small enclosures. These wanderers, however, dropped at once into the garden, although fenced by tall clipped macrocarp hedges and directly in view of the verandah; since then rooks have been in the neighbourhood but once again, a few pair having been noticed between Tutira and Tangoio.

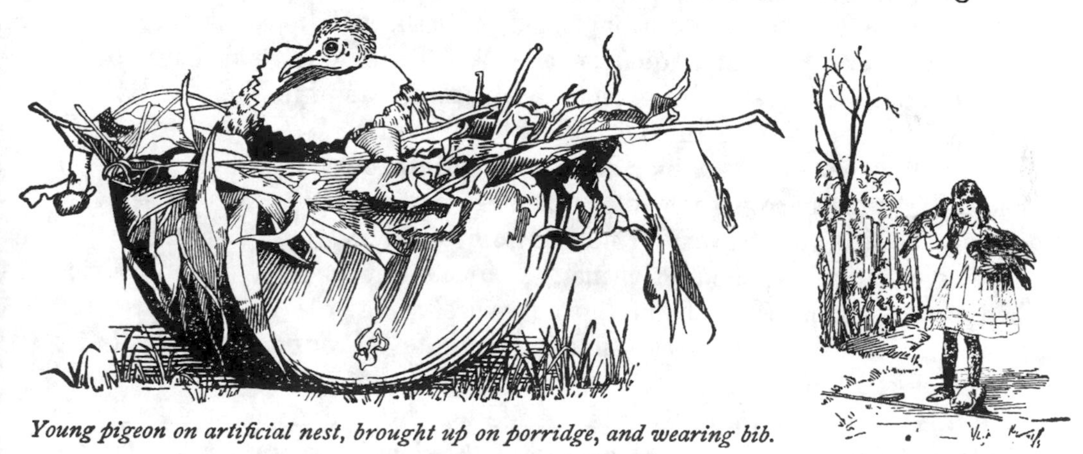

Young pigeon on artificial nest, brought up on porridge, and wearing bib.

"Wild" turkeys have on several occasions straggled on to the run. "Wild" geese, probably from the lagoons about Longlands, have twice visited the lake for a few hours. Carrier pigeons have at long intervals —the earliest in '83—rested on the roof of one or other of the station buildings. One, in 1912, took up his permanent abode until I had him caught, taken to Napier, and there liberated. I was then engaged in the domestication of native pigeons (*Carpophaga Novæ Zealandiæ*), and feared the stranger might some day or another lure them away.[1]

[1] The domestication of these magnificent birds was an entire success. For seven or eight years now they have bred in the homestead plantations; their confidence in their friends can be gathered from the accompanying sketch of a child feeding them in the open.

CHAPTER XXXV.

THE INVASION FROM THE NORTH.

THE policies of the Auckland Acclimatisation Society have been the chief centre of dispersion from which northern aliens have reached Tutira. From the same quarter the Wairoa district has contributed certain species liberated by private enterprise; a single alien, lastly, has reached us from the neighbourhood of the Thames.

In October of '82, a month, that is, after our arrival at Tutira, a small flight of sparrows rested for a brief space on the wood-heap, that inevitable adjunct of every primitive homestead in New Zealand. The species had reached the station neither by mountain-top, coast, or river-bed, but by road. They had followed—surely one of the most interesting treks in natural history—the highway of man through the very heart of the North Island.

Sparrows were imported and turned out by the Auckland Acclimatisation Society in '67. Two years later the Society reports: "Sparrows have increased largely, but seem reluctant to go far from home, though stragglers are occasionally met with." A few years later their migration must have begun, for in '76 they were suspected to be at Opepe on the Taupo road. In '77 we find the Hawke's Bay Acclimatisation Society requesting their Committee "to take any necessary steps for the destruction of sparrows said to be in the district," a request, by the way, about as futile as that of King Canute to the flowing tide. In the "late 'seventies" sparrows were seen by Mr J. N. Williams at Te Puna. In '80 specimens were shot near Hastings; in '81 they had reached Napier; by '82 they were present at Tutira. In '84—that is, only seven years after the Hawke's Bay Acclimatisation Society was dubious about the very presence of sparrows in the province—this same Society "viewed with considerable alarm the enormous spread of small birds, but took the opportunity of reminding the public that they were not responsible for the introduction of linnets, sparrows, and larks." In fifteen years, therefore, the sparrow had travelled nearly two hundred miles through an uninhabited waste, had invaded the settled portion of Hawke's Bay, and had even begun to follow up tracks leading away from that district.

A chief reason for the choice of man's highway as his route of migration may be found in the sparrow's relation to and reliance on man. Passer domesticus is his name, and passer domesticus is his nature. Of all wild creatures that utilise our roads in New Zealand, none take advantage of them in so great a measure as the sparrow. He knows, perhaps instinctively, certainly through the experience of the older birds, that it is by the work of man's hands his race principally thrives; that it is man who provides for him shelter plantations, building sites, and food. The man-built road by which he moves is indeed in itself a provision house. There are to be found on it horse-droppings containing undigested oats, foodstuff thrown down by travellers, wheat, barley, and grass seed fallen from sacks. On either side of its white sinuous line, so conspicuous from above, so markedly dissimilar to surrounding surfaces, extend tilled earth and land in crop. Like the bee-bird, which guides the hunter to the hive, the sparrow in striking and following up a road foreknows the benefits that will accrue to him. Maybe in the neighbourhood of townships, between village and village, the sight of travelling sparrows is too common to excite remark; they are merely specimens of the most common bird in the country. On the far inland up-country roads of New Zealand, however, where ten or fifteen miles may intervene between homestead and homestead, travelling parties cannot but excite attention. During autumn it is impossible on the roads of the interior not to observe and not to wonder over these roving bands moving in search of winter quarters. It is hardly too much to say that there is developing in the sparrow something in the nature of an annual change of residence—a summering in the country, a return during winter to a town, to a village, at least to a large farmsteading.

The accompanying section of map will serve better than any description to show the nature of the countryside traversed by the migrant sparrow and the marvellous results of half a century's human toil. The "natives" and "wandering natives" are now our exceeding good friends; the "forest" and "scrub" have been transformed into sheep and dairy farms.

Auckland, where the sparrow was liberated in '67, is built on a narrow strip of sandy land; east and west of it lies the ocean. Northwards protrudes a meagre egress leading in the 'sixties towards land poor in quality and covered with scrub. In the opposite direction ran the only road of the period, the Great South Road, as it was called. By this route *viâ* Mercer, along the Waikato river to Cambridge, by the armed constabulary posts to Hawke's Bay, through uninhabited belts of forest, tussock-grass, and bracken, sparrows holding to the road moved south. Finally, debouching from the ranges of the interior and striking the open lands of western Hawke's Bay, they followed coastwards one of the bullock-tracks of that period, unmetalled, uncrowned, in winter a quagmire, in spring nor'-westers rutted deep and dry, in summer thick in powdered dust, but always distinct, always dissimilar to other surfaces, and always full of promise to the sparrow tribe.

In the orchard of Waipuna, the original homestead of the Rissington run, sparrows were first seen in Hawke's Bay. The late Mr J. N. Williams has often told me of the circumstances of his find. He was taking delivery of cattle in early spring. Snow had fallen in calm weather; lying on the ground and resting on the trees, its whiteness brought into additional prominence a party of sparrows perched on a plum-tree in the station garden and feeding in its vicinity. Two years later, "about 1880," sparrows had reached Frimley, Mr Williams' beautiful residence near Hastings. He has described to me how wild and shy the birds seemed to have become after their journey through the wilderness, and how difficult it was to obtain a proper view of them. They had, in fact, in some degree become a tree-top species and kept resolutely to the upper branches of the tall eucalypts. After considerable delay specimens were, however, shot for proof, for until they were actually handled and viewed, Mr Williams' friends, like Thomas, would not believe—"it seemed impossible to them that the sparrow could have reached Hawke's Bay in so brief a period over such a stretch of wild country."

The history of the sparrow, as observed on Tutira, will serve to show how its multiplication, rapid enough in all conscience, was made, by certain habits acquired during its pilgrimage through the desert, to appear a marvel, a portent; it will make intelligible how in the late

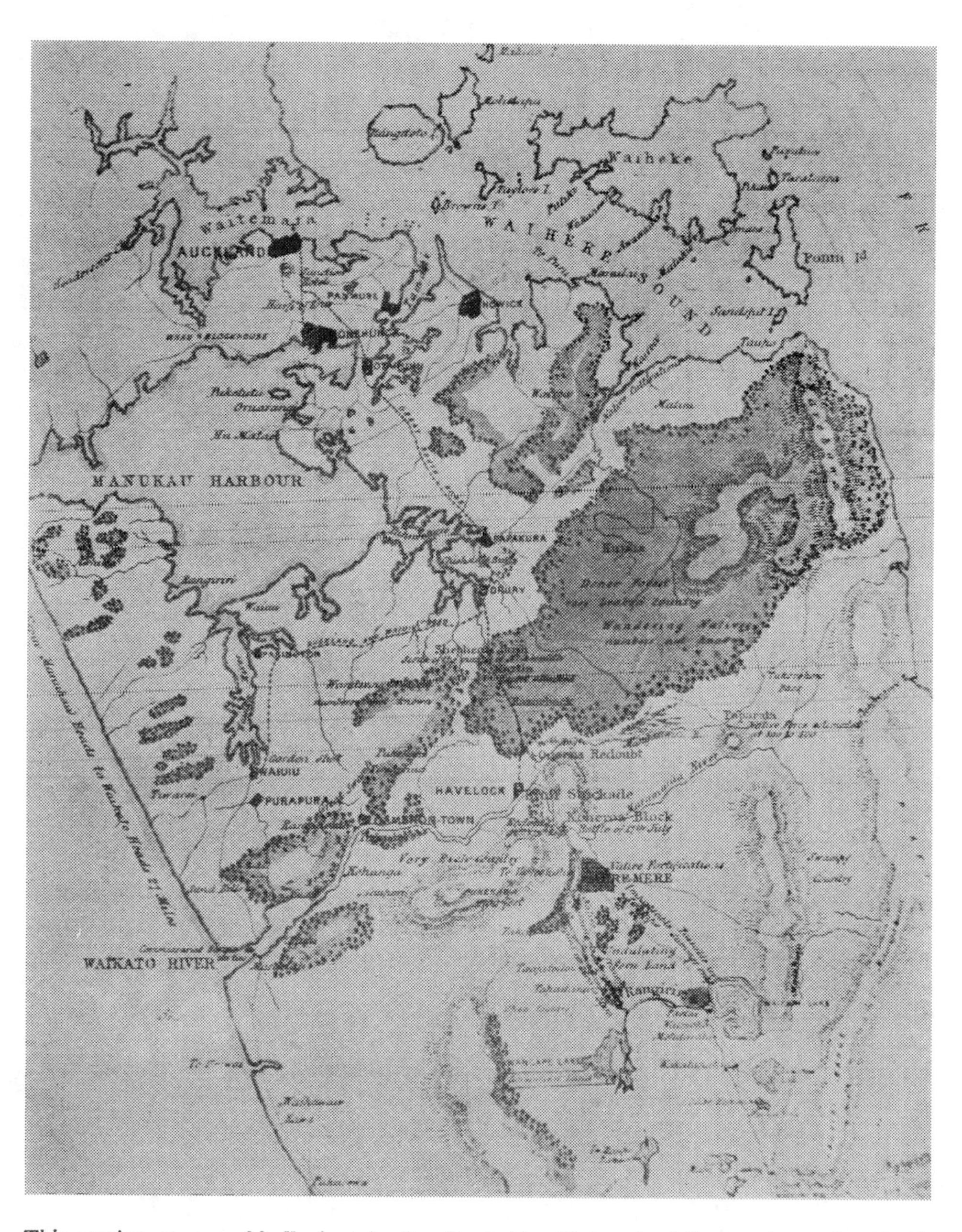

This section of map, kindly lent by the Cambridge University Library, shows the North Island of New Zealand as it was when the Sparrow began his migratory movement south. It would seem almost incredible that so short a time ago any Dominion map should be scored with such grisly particulars as "Battle of 17th July," "Scene of the murder of Mr Meredith," "Escort attacked," "Gordon shot," &c., &c.

'seventies the sparrow was still an object of interest and incredulity, whilst in the early 'eighties "its enormous increase was viewed with considerable alarm."

Sparrows, as stated, were first seen on Tutira in '82; they were not again noticed for many years. Tutira must have been in those days a most unattractive spot to such a species. These were the winter starvation times already described, when not a fat sheep or beast was to be had; when oats and chaff, milk and butter, were unknown; when the fowls went without grain; when, in fact, Tutira was no fit place for any decent self-respecting sparrow. Not only was there no food, but there was no covert, except three weeping willows, a species useless for purposes of nidification; there was not a tree about the homestead able to support a nest. No wonder the sparrow scorned the naked, treeless, poverty-stricken station.

By '92 conditions had somewhat altered; the sparrow then for the first time bred with us; two nests were built that year in an African box-thorn hedge which had been planted round the original garden. Later again, there was a large increase in the sparrow population; pines planted in the late 'seventies by the Stuarts and Kiernan had grown into trees big enough to provide ample nesting-quarters. In their vicinity a considerable patch of oats had been reaped; there, attracted by the cropping and by auspicious nesting-sites, forty or fifty pairs established themselves, their numbers certainly larger than any increase possible from the station-bred clutches of the previous season. The day for continuous cropping on Tutira had, however, not yet come; it ceased, and with its cessation only two or three pairs of sparrows remained at the homestead, building their nests as before in the box-thorn hedge.

Up to this date sparrows had bred within sight and hearing of man. In the late 'nineties a change came about in their habits and customs. They began to establish themselves in small congregations of five and ten pairs, miles from the homestead, though still always within a few score yards of the road.

A further step towards a summer feral state is the selection of breeding-quarters, not only away from the homestead, but—another stage in the emancipation of the race—away even from his much-prized road. The small bush reserves on the lowland portions of Tutira are now, during the breeding season, overrun by multitudes of sparrows.

In them the emancipated alien finds admirable accommodation for rearing his young. About such spots he thrives and multiplies, devouring during summer-time insect life, seeds and berries, formerly the exclusive property of native species. With the waning of the year, these fine-weather quarters are vacated; striking a road, sparrows follow it to the nearest homestead. There the company, or such of them as there is feeding for, remain till spring-time, when once again they move abroad.

Thus, according to the season of the year, sparrows spread abroad or closely congregate; when attention has been directed to the matter the double movement can hardly be missed. It is only, indeed, obscured by the rapid increase of fledglings in early summer, by the unostentatious plumage of the breed, and by the indifference with which so common a species is viewed. What I have myself seen occur on Tutira I believe took place when the birds debouched on to the plain of Hawke's Bay. They had become accustomed during their long trek to breed far from the dwellings of man, outside his pale of protection. As winter approached, however, their instinctive dependence on their human hosts reawakened: the birds flocked from the wilds into the few far-scattered homesteads. Arriving thus in swarms where few or none had been seen before, it is not surprising that settlers viewed "their enormous numbers" with "considerable alarm,"—the birds must have seemed to be appearing as if by magic.

Nowadays I find the winter numbers of the sparrow depend on the changing necessities of station management, on the amount of oats grown, and on the number of contract plough-camps, where teams are fed, where there is always grain spilt from nose-bags or unfinished in feeding-troughs. Probably a chart would show pretty accurately the relation of the sparrow population to the price of wool in London: with prices good more ploughing is done, more horse-feed grown; with prices bad, less. Like his fellow-mortals in New Zealand, the species is affected by events taking place at the other end of the world—events which he cannot control and for which he is in no degree responsible.

The sparrow, however, by no means has everything his own way on Tutira. Although indigenous to Britain, heavy rain does not suit him; no creature, indeed, can look more woe-begone than a wet sparrow, with unpreened, unoiled, and draggled plumage. Our local birds are from time to time decimated, I may almost say annihilated, by the

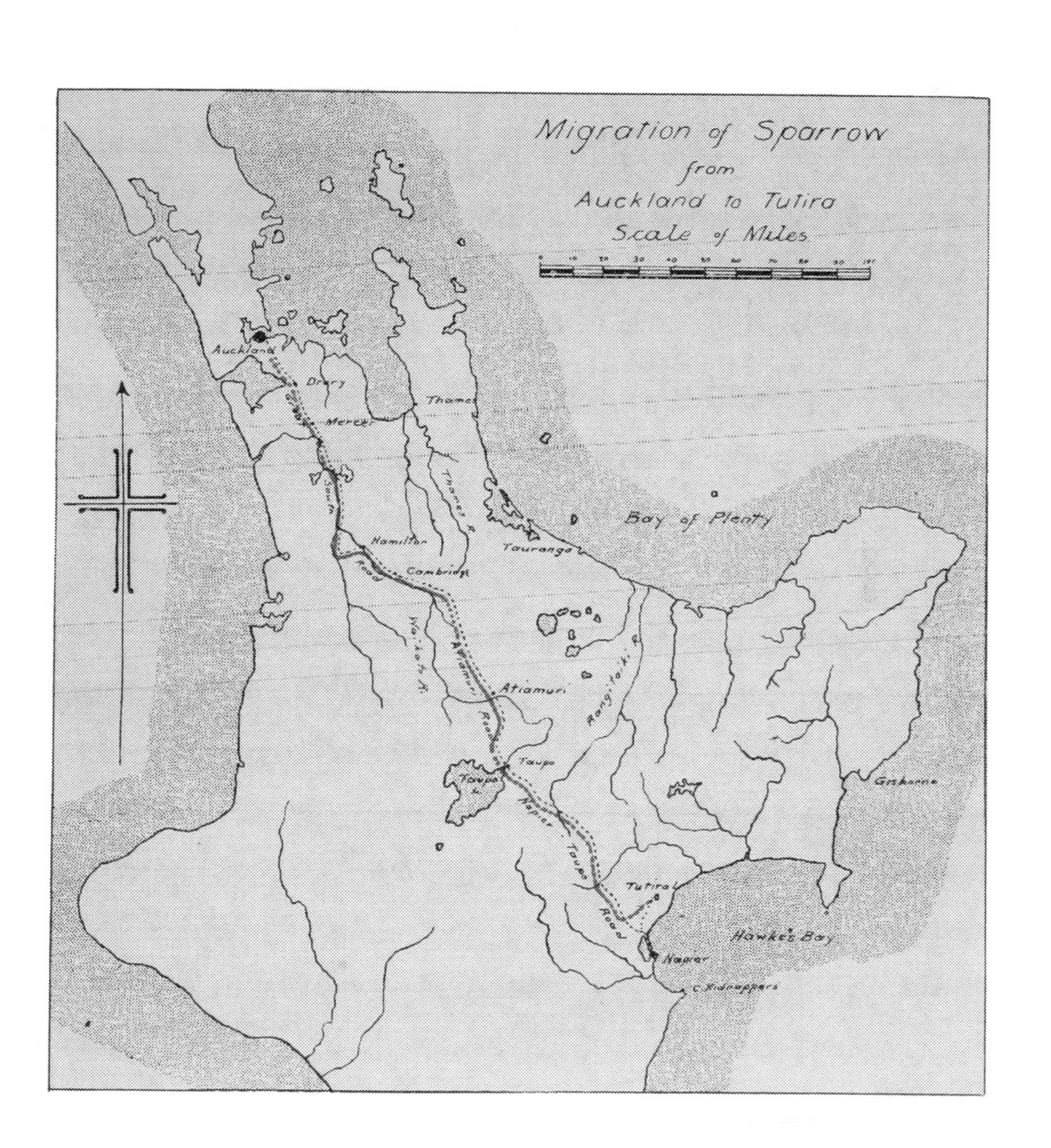
Migration of Sparrow
from
Auckland to Tutira
Scale of Miles
Auckland
Drury
Mercer
Hamilton
Cambridge
Thames R.
Tauranga
Bay of Plenty
Waikato R.
Atiamuri
Rangitaiki R.
Taupo
Napier
Taupo
Road
Tutira
Gisborne
Hawke's Bay
C. Kidnappers

great storms of three or four days' duration which, at intervals, past over the run. During one such gale, registering just under a foot and a half of rain in three sequent days, nearly every sparrow on the station perished; that any at all survived was owing to the ingenuity and adaptability of the race. On the afternoon of the third day of rain, a considerable number not only took refuge in the fowl-house, but actually ensconsed themselves amongst the feathers of the great silly Buff Orpingtons, broody on their nests or occupied in laying eggs.

The blackbird (*Turdus merula*) was liberated by the Auckland Acclimatisation Society in '67. Its naturalisation was at once successful, its increase in numbers immediate. Then there occurs a great blank in the history of the bird, a gap I have tried in vain to bridge. Times were very difficult, wars and rumours of wars, troubles of a hundred kinds pressed heavily on country settlers in particular. As can be imagined, there was little leisure for observation, for records, or for the amenities of life generally. At any rate, nothing is known of the blackbird until after many years it reappears scores of miles distant from its original site of liberation. We can but surmise the early stages of its long journey by elimination of lines obviously not pursued, and by the locality of its reappearance.

The blackbird, like the sparrow at an earlier date, shied off the poor lands immediately north of Auckland. The Great South Road, that lane of light cut through fern and forest which had allured the highway-loving sparrow, offered no inducement. Neither, apparently, did the species travel in a southerly direction down the west coast of New Zealand. It must have proceeded in an easterly direction to have been able eventually to re-emerge at Waiapu; we can be sure of one thing only, that the line taken was the line of least resistance. The species had the choice of the three natural routes of movement already named—river-bed, hill-top, and coast. The river-bed route certainly was not followed. From start at Auckland to finish at Waiapu throughout the whole of the way every stream and river flowed at right angles to it. The hill-top route offered as little encouragement. To begin with, there was no great natural backbone range such as had guided the red-deer and the rabbit northwards. Such chains of hills as did exist were broken and separate one from another, a mere jumble of rounded tops; one and all, moreover, were densely clothed with forest or with fern or scrub. There was no scrap of open ground for alighting, for exercise, for food. There were none of the great wind-blows, oases of open ground, naked summits and rockfalls, stepping-stones each of them to further progress, which existed on the mighty Ruahine Range; a warmer, wetter climate clothed every inch of the country in dense jungle. The line of least resistance was the coastal line. Mr J. N. Williams and his brother, the late Bishop of Waiapu, have never doubted but that this was the route followed, that the blackbird after leaving Auckland skirted the coast as far as the Bay of Plenty; they furthermore believed that then striking inland the blackbird topped the range and followed in a southerly direction the course of the Waiapu river.[1] As to the strike inland, it seems to me improbable that the birds should have crossed the Motu-Mangatu-Maungahamea-Arowhona-Aorangi range, with peaks reaching three, four, and five thousand feet; on the contrary, such a barrier, in my opinion, would be likely to pen them securely to a further stretch of coast. Personally I believe that the blackbird followed the coast line from Auckland to the Bay of Plenty, continued to follow it round East Cape, ultimately reaching the Waiapu river, where specimens were first seen, that river happening to flow for part of its course parallel to and at no great distance from the coast; there at any rate blackbirds were noticed "in the late 'seventies" by Mr J. N. Williams and the Bishop of Waiapu. "About" '87 they were seen by the late Mr John Hunter Brown between Poverty Bay and Wairoa, later at Waihua by Mr MacMahon, and at Tutira in '91 by myself.

The migratory current, at first a trickle, later increased to a stream. It was not, however, until 1912 that the progeny of the pairs breeding each season on the run seemed to stay. Then at once the increase was marked.

The history of the thrush (*Turdüs musicus*) is the history of the blackbird in duplicate. It was imported by the Auckland Acclimatisation Society at the same date as the blackbird, and like the blackbird, immediately became naturalised. The history of its migration is

[1] It may be that details at one time vivid had somewhat faded from Mr Williams' recollection when we first spoke of the matter, and that I may be unaware of some of the minor facts which originally led him to his conclusion. Details, after playing their part in a considered judgment, are apt to be relegated to the shade. In the case, for instance, of the sparrows' trek, I had many times, *apropos* of acclimatisation generally, heard of their discovery at Waipuna and of the specimens shot at Frimley without the additional item of information that birds "believed to have been sparrows" had been reported as having been seen at the old blockhouse at Opepe, at a period prior to their appearance at Waipuna. It may have happened, therefore, that I am not in possession of the minutiæ which led Mr Williams to the conclusion that the blackbird had struck inland.

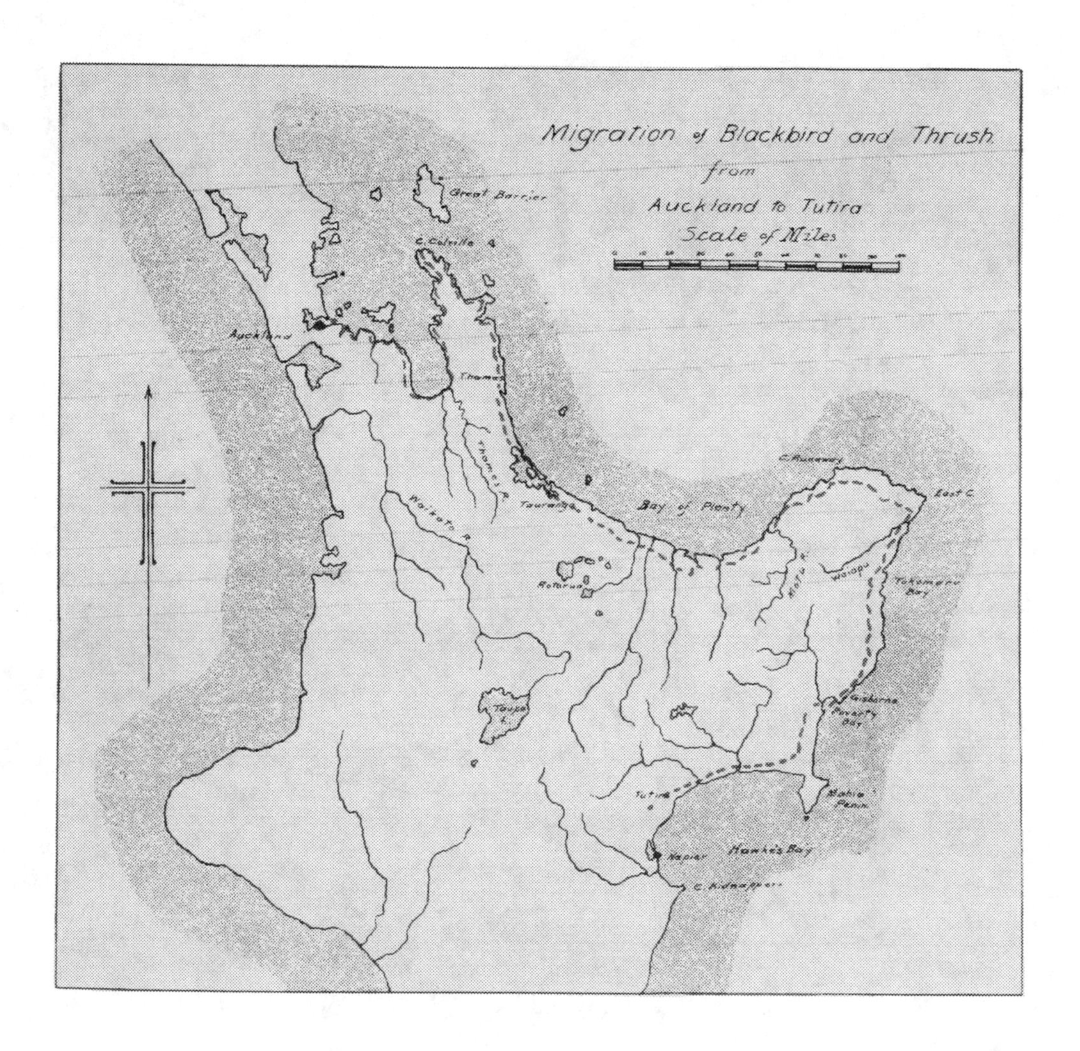

Migration of Blackbird and Thrush
from
Auckland to Tutira
Scale of Miles
Great Barrier
Auckland
Tauranga
Bay of Plenty
East C.
Tokomaru Bay
Rotorua
Taupo L.
Poverty Bay
Tutira
Napier
Hawke's Bay

likewise the history of the blackbird repeated. It was observed on approximately similar dates as the blackbird by Mr J. N. Williams and the Bishop of Waiapu in the Waiapu river-bed, by Mr MacMahon between Wairoa and Mohaka, and by myself at Tutira.

In the spring of 1902 I noticed near Petane a cock chaffinch (*Fringilla cœlebs*). A week later on Tutira I saw a hen chaffinch on the road between the wool-shed and homestead. Four days afterwards I marked a third chaffinch, and almost at the same instant a redpole (*Acanthus linaria*), the first of its breed seen by me in New Zealand. Both birds were hens: the redpole had her nest in the immediate vicinity, as I could tell by her angry protestations at my presence; that of the chaffinch lay on the ground, either pulled down by some accident or blown from its moorings by the gale on the previous day. The two nests must have been built within a few yards of one another.

Each of the species has an interesting history.

In '73 — four years, that is, after the liberation of the blackbird and thrush — a hundred pairs of redpole were imported by the Auckland Acclimatisation Society. The venture was successful, for next season the breed was pronounced to have become thoroughly established. After that, like not a few other aliens, redpoles seem to have vanished from the district where their naturalisation had been so speedily successful. Mr T. F. Cheeseman, for many years honorary Secretary of the Auckland Acclimatisation Society, tells me, indeed, that he has never himself seen a redpole wild in New Zealand; at any rate, after liberation and successful acclimatisation, this small species vanished, to reappear thirty years later at Tutira. Several months after observation of the first specimen, I saw another redpole, then a third, then a party of seven. A couple of years later the breed had become fairly numerous in the trough of the run, especially about belts of dry standing manuka over which fires had passed. There for several seasons considerable flocks maintained themselves. Nowadays, though still not uncommon, they are diminishing to normal numbers.

Chaffinches imported to Auckland, and freed by the Acclimatisation Society in '68, were pronounced an immediate success.

The appearance of the chaffinch at Tutira was no surprise. I had known for several years that the species was on the move towards the station. In '98, during inspection of forest-land thrown open for settlement in Poverty Bay, I had discovered chaffinches at the head-waters of the Mangatu stream — a mountain tributary of the great Waipaoa river. The birds were moving down-stream, for upon our return a few hours later they had proceeded coastwards a considerable distance. This original band of chaffinches I have always believed to have been one of many straggling across the watershed, following in a general way the course of the eastward-flowing stream.

Reasons have been given for thinking that the blackbird and thrush migratory movement followed the coast, without deviation, round East Cape. Settlement, however, had made vast strides since the 'seventies. Huge gaps had been cut out of the forest-lands of both coasts; woodlands had been fallen both east and west of the high watershed — Motu-Mangahamia-Arawhona; Opotiki on the one coast and Gisborne on the other were almost, though not quite, continuously linked by grassed lands and open country. The probabilities are that the chaffinch followed the blackbird and thrush coastal route until somewhere about the Opotiki region; there, tempted by the wealth of cultivated ground, it appears to have diverged from the sea, and, following the line of light, the open farm lands, moved inland up the Motu river, then up its tributary head-waters until the watershed was topped. Again utilising a river-bed route, the tributary streams of the Waipaoa, and later the channel of that great river itself, were followed to the opposite coast. The chaffinch, in fact, threaded one river from mouth to head-waters, another from head-waters to mouth. Reaching the east coast, the migration moved southwards — attracted, perhaps, by the greater quantity of low-growing scrub extending in that direction. Its vanguard was reported to me twice from the Wairoa and once from Waihua; it reached Tutira in the spring of 1902.

North of Gisborne, on the other hand, the chaffinch remained unknown, none having rounded the East Cape, none having diverged from the main body moving south. So small a species as the redpole — a bird, too, of the wilderness — might have been overlooked; the chaffinch, in a district where attention had been called to the matter of aliens by my inquiries, could hardly have been so passed over. During winter, moreover, the latter species draws into homesteads and farmyards; with the sparrow and yellow-

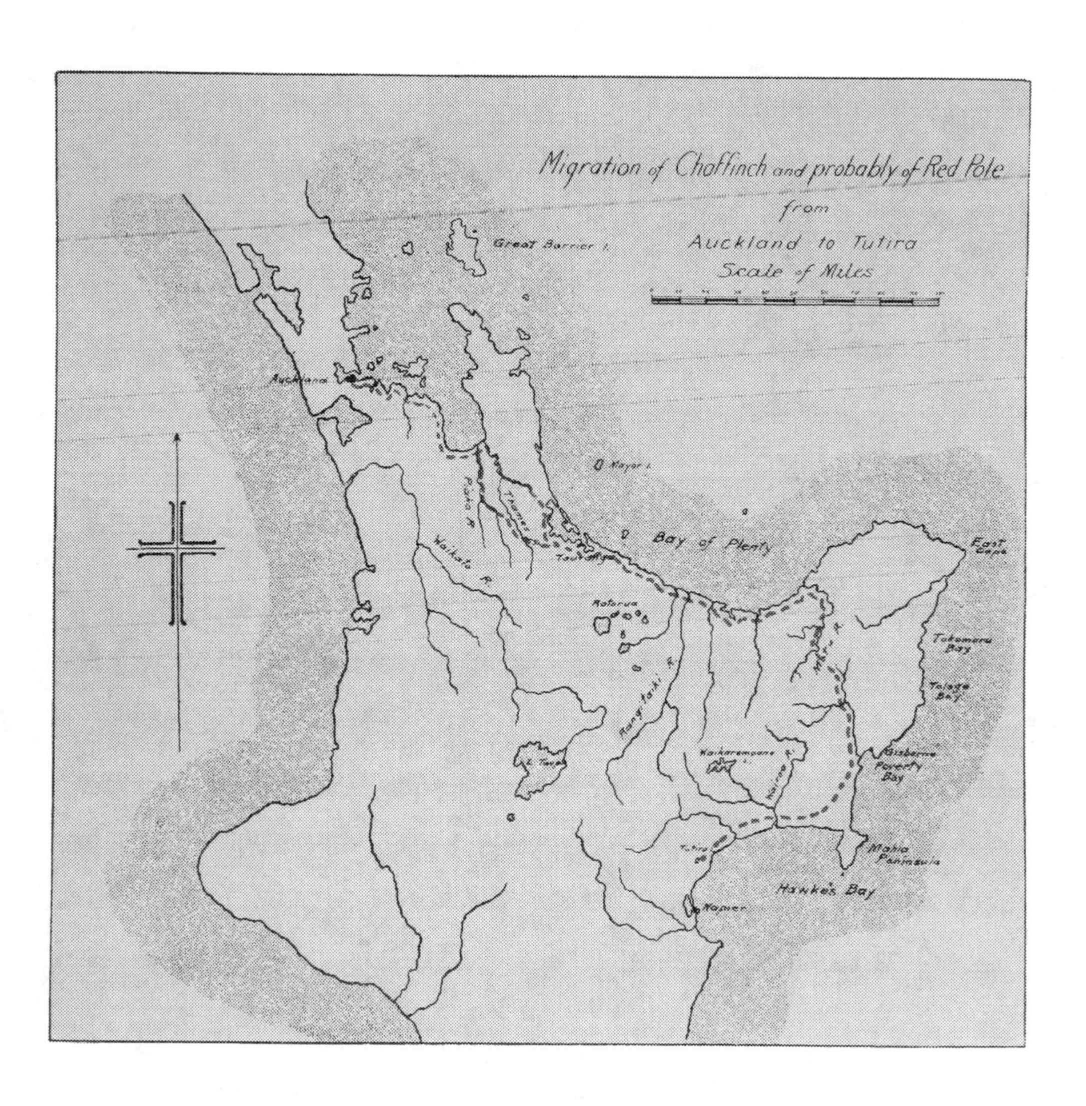
Migration of Chaffinch and probably of Red Pole
from
Auckland to Tutira
Scale of Miles
Great Barrier I.
Auckland
Piako R.
Thames
Waikato R.
Mayor I.
Bay of Plenty
Tauranga
Rotorua
East Cape
Rangitaiki R.
Tokomaru Bay
Tolaga Bay
Waikaremoana L.
Gisborne
Poverty Bay
Mahia Peninsula
Hawke's Bay
Napier
Tutira

hammer he claims his place in the sun, his share of the good things provided by man. It is highly improbable that the chaffinch could have escaped detection from one or another of my observers north of Gisborne.

Prominent amongst annuals fashionable in the early 'nineties were Shirley poppies. It was in a gorgeous bed of them that the first bumble-bee was noticed on Tutira in 1902. The following year I noticed a second specimen also in the flower garden; indeed, until red clover was sown as a fodder-plant, I never remember to have seen one on the station except in the garden.

The history of the bumble-bee (*Bombus terrestris*) in New Zealand is as follows: After several failures they were liberated on the Matamata estate, in the Thames district of the Auckland Province, in '84 — only two queens, however, surviving out of one hundred and forty-five. In '85 more successful shipments reached the South Island, and from there the North Island was again stocked. From Matamata stragglers reached Tutira five or six seasons later.

The experiment has been regarded as a success, and certainly since then large quantities of red-clover seed have been marketed. The bumble-bee has got the credit for this result; whether the alien insect altogether deserves it is, I think, more than doubtful. Red-clover blossom, to my certain knowledge, was fertilised long before the introduction of the bumble-bee. In 1880 I was a cadet on Peel Forest Station in South Canterbury. Even then I was on the watch for new plants and aliens in strange places; at any rate, I recollect scrambling up and gathering ripe clover-heads on the forest cutting between Peel Forest village and Holnicote, the beautiful homestead of the late Hon. John Barton Acland. Rubbed in my hand, these heads gave an excellent sample of plump seed; good seed was to be had on Tutira, too, long prior to the introduction of the bumble-bee. On a quarter-acre patch, ploughed and sown down in the 'seventies by the Stuart Brothers and Keirnan, it was always obtainable until the plants were eaten out. Lastly, in 1909, the Waterfall Paddock—300 acres of cow-grass—was thoroughly fertilised. There was so heavy a seeding that we thought of cutting and thrashing, and only did not do so because of difficulties in regard to the hire of machinery and traction over bad roads. The fertilisation of this clover-field was, I believe, accomplished by a small greyish moth, millions of which hid in the crop or rose in clouds if disturbed.

Experts can calculate the quantity of bumble-bees required to "set" 300 acres of tall cow-grass; in any case, it would be immensely greater than the district could produce. As a matter of fact, the number of bumble-bees was insignificant in 1909, nor are their numbers likely to increase. This alien, like others, cannot quite adapt itself to the peculiar climatic conditions; any large increase is checked by the deluges that from time to time pass over the run.

A second northern centre of dispersion from which aliens have reached Tutira has been Wairoa. From there have arrived the green frog, of Australian origin, and, I believe, the "opossum," also from Australia. Frogs reached Tutira in '94, being reported almost simultaneously on the east and west of the run. They are great climbers; I have got them not only on hill-tops up which they might have been tempted by a gradual rise, but on steep cones like the "Natural Hill" and the "Dome."[1]

They have never become plentiful on Tutira. About shallow lagoons and surface water-holes harrier hawks take them; in deeper waters, I believe, they are devoured by eels.

Opossum were turned out in the Waikaremoana forest reserve in 1900; we believe, though they were never actually seen on Tutira, that they passed southwards five years later—ribbon-wood saplings were peeled and succulent willow shoots barked bare, the hard cores of both showing the tooth-marks of a large rodent. Our willows and ribbon-wood had never before been thus peeled, and they have never been touched in that way again.

[1] Mr J. N. Williams tells me that frogs, liberated at a dam on Edenham, "almost at once" reached the top of the highest hill on the station.

CHAPTER XXXVI.

DOMESTIC ANIMALS "WILD."

BEFORE Tutira had been taken up pig were very plentiful; small numbers of horses, cattle, and probably sheep also were already running on the station in a "wild" state.

Of horses and cattle there is little to tell; of the former a score or so had strayed from the derelict *kainga* Waipopopo. During winter they were compelled, in search of food, to wade like cattle into the scrub; like cattle, too, they broke down and devoured the branches of shrubs and small trees. From time to time a few were run into extemporised stock-yards, roped and broken in, the last survivor of the herd being a grey ridden for many years by myself.

Concerning wild cattle there is hardly more to say. They were never numerous; cliffs, under-runners, and boggy creeks exacted too heavy a toll. There were never more than enough to provide us with meat and amusement; indeed, in early times most of our beef was shot on Kaiwaka.

Pig, descended from stock landed originally by Captain Cook, had been distributed over the colony as I have imagined pot-herbs to have been at a later date—passed, that is, as gifts from village to village. Conditions were favourable to their spread; as ample a food-supply existed without as within the bounds of the native villages. It is likely, indeed, that pigs, as their number increased, may have been purposely allowed to become half wild, that they may have been hunted out of the cultivation-grounds owing to their inveterate habits of trespass. At any rate, both in the north and south island swine increased and multiplied—in the early 'eighties, indeed, there were in the centre and west of Tutira more pig than sheep; there, beyond rooting and re-rooting the few sheep-camps scattered far apart, no damage could be done. In winter, however, hundreds left these wildernesses of fern and invaded the grassed portion of the station; a sort of migration set in then towards the sown lands, where considerable tracts of turf were ploughed up and turned over in search of roots and grubs. It was in spring-time that serious harm was done. Every old rusty boar in the vicinity seemed to be aware when lambing was in progress. Amongst the newly-dropped lambs, concentrated from dusk to dawn on camping-grounds, great havoc was wrought, the marauders, with returning light, retiring to their distant lairs. The harm was always done at night; only once have I detected a boar at work in full daylight. It was attempting to secure a lamb just old enough to stagger after its dam, barely beyond that stage of life when any moving object, animate or inanimate, will be followed. The sense of fear, however, develops very fast in young animals. Even as I watched, the little creature was beginning to comprehend something of the anxious bleating of its mother. At any rate, in spite of hesitancy, it continued to keep a few yards in front of the boar, sometimes lingering as if in doubt, and sometimes trotting up to the bleating, agitated ewe, just sufficiently ahead to lure her offspring onward. In the rear, dodging in and out of the flax-clumps, the boar maintained a stolid chase. To my surprise he never attempted a rush, either from brainlessness or perhaps because from former experience he knew the certain results of the wearing-down process. Whether he would or would not have secured this particular lamb I know not. He suddenly became aware of my presence and broke away in a clumsy gallop.

In colour "wild" pig are black, red and black, rusty red, red and white, and white, and bear a general resemblance to badly-fed, badly-bred swine of modern domesticated kinds. One incomprehensible trait in the wild sow is worth noting—her callous indifference to the fate of her suckers. I have killed at least hundreds, perhaps thousands, of pigs on Tutira, yet never have I known a sow evince the smallest concern for her progeny. Even a ewe will stand by her new-born lamb regardless of man and dog, but not the most heartrending squealings of a sucker worried by dogs will recall the cowardly, craven sow. The young are born into a warm comfortable nest of bracken or grass, and if taken in time make amusing and interesting pets. For that reason, and also doubtless because they eventually grew into pork, they were to be found in every Maori village, and in the old unsophisticated times used to accompany the shearing gangs on their rounds from station to station.[1]

Of the dog as a wild animal, Tutira has had but little experience. In my time there has been but one on the run; it has been fortunate in that respect, for a pack of wild dogs is one of the greatest curses a

[1] One brought up on the station was quite a character. "Tommy," as he was afterwards christened, when discovered some miles distant from the homestead, and selected out of five or six others, was a very baby but a day or two old. Taking thought of his frailty—I had yet a long day's work to do—he was swaddled in the waterproof strapped to my saddle, the reins drawn through a stirrup-leather, and my horse given his freedom. There were no fences in those days; the liberated nag trotted home without a halt to the homestead gate, where he was caught, the saddle removed, and the suckling exhumed safe and sound. "Tommy" had been taken—if I may say so, translated—so young that he grew up to consider himself half-man and half-dog, or rather half-man and half-puppy. He romped and ran with the station pups, who pretended to worry him, holding on to his long ears, growling, panting, yapping. He was very well able, however, to take care of himself, and could at any time terminate the play with a vigorous fling of his nose right and left. His more serious hours were spent in our company. He fed, if not with us, yet at our hands, at first drawing his milk from a teapot, later being promoted to a little trough of his own. The old dogs free and unchained condoned his presence as dogs do, accepting him as one of the eccentricities of masters whose whims are law. It was odd to see their greetings, to watch the nostrils of the two animals meet: on the dog's part, the cold curiosity of the salute, the instant disillusionment; on the pig's, the brusque discourteousness. His chief friend was the married man; him he followed everywhere, even into the water, for the story that pigs cut their throats swimming was apparently unknown to "Tommy." At any rate I have seen him swim after his friend in the boat and be lifted dripping and stiff out of the lake like a great black baby. With increasing age he became an adept at filching bones from the kennels of the chained-up dogs. It was then an advantage to him that he was conscience careless, that the higher standard of ethics reached by man and dog were unknown to him. There was no concealment of his thefts—he knew no better law. The dog, on the other hand, was handicapped by a deeper insight into the nature of things. After the first instinctive snarling rush to protect his property, conscience awoke in him; remembrance obtruded itself that the black marauder was in some way the property of his master, that he was *tapu*, that he himself as a dog had perhaps done wrong even in resenting the theft. Consideration like an angel came and whipped the offending Adam out of him; at any rate it was the dog who would bolt to kennel with his tail between his legs, whilst the pig brazenly enjoyed the stolen goods. It was impossible to watch "Tommy" and deny to him a genuine sense of humour. Often to our woolshed drafting-yards he would follow the shepherds with their dogs. There it was that his peculiar sense of fun found expression; a satisfaction that never palled was the stalking and rousing of a drowsy dog. The jest had probably its origin in chance, or in congenital rudeness, and only became with repetition established as a habit. After a long morning's work and not infrequently a gorge at the gallows, the satiated sheep-dogs are wont to lie half-asleep in a comatose or torpid condition in the shade of the rails. That was "Tommy's" chance: like a man attempting to outwit a horse hard to catch, his method was at first to contrive to be noticed moving away from his destined prey. With indeterminate movements he would further lull his victim, then advancing cautiously and quietly, would violently punch the dog in the paunch and listen to his howl of mingled anguish and surprise. I do not remember in "Tommy" any ostentatious sign of satisfaction: his line was that the dog's body had happened to meet his nose, that the whole affair was an accident. Watching from the yards, we used to credit him with the sly impudence of glancing up with head aslant after the manner of pigs, as if pretending to pause and listen, in doubt as to whether his ears had deceived him, whether he had really heard or not heard a howl of anguish. He would then continue his pretence of searching for something on the ground. Sometimes, with the predilection for wandering that characterises the pig race, "Tommy" strayed far afield. One morning, noticing on the hillside what I mistook for a wild pig, I put the hunting dogs on his trail, and arriving breathless on their heels found the three

station can endure. The anxiety consequent on its proximity is never-ending, nor can the damage done be estimated by the number of sheep actually found dead; in rough country a dozen or a hundred may be smothered or drowned for every one actually worried to death.

In early times on unstocked country, wild dogs lived on such ground birds as were procurable; but their mainstay was pig, and pig, rather remarkably, they continued to hunt long after their wilds had been stocked with sheep. The upland Patea country in Hawke's Bay was taken up and stocked in the 'sixties. Wild dogs were there unusually numerous, yet for several seasons sheep, I have been told by the late Mr W. Birch, remained untouched, the dogs for a time either desisting from fear of latent possibilities in the new animal, or from mere force of habit continuing to follow their original game.

When a mixed pack worries, the predominating trait of each breed asserts itself, the mongrels of bull-dog and mastiff extraction throttling the wretched sheep, the collie curs holding the huddled flock together and preventing it from scattering in a hundred directions.[1]

of them together, "Tommy" grunting out explanations, the mastiffs listening with friendly wagging tails; before my arrival they had settled it was all a mistake. "Tommy" was offered a few bits of fern-root by way of apology, and the four of us returned together. One of his great pleasures was to be scratched. Regardless of the sides of bacon dangerously exposed in the process, he would lie first on one side then upon the other, his eyes closed in ecstasy, enraptured, like a woman having her hair combed. In later life he became, like all pig pets, rough, brutal, even hunnish in his manners. He was finally presented, when he would no longer take "No" for an answer, to Joe Raniera and Hepe, his helpmate, then fencing on Tutira. With them for several seasons he shared bed and board.

[1] No shepherd can have owned a team of dogs without speculation as to the herding habit of the breed, the essence of which is to head and hold. Pups but a few days alive to light will on a hillside watch fowls or ducks or chickens, carefully keeping them together, running ahead of them, checking stragglers, "working" them carefully and correctly. This passion for watching and working stock is of so overmastering a nature that sometimes a young unbroken collie will instantly, when freed, bolt for the hills and there remain till dark, moving ahead of the little flock he has gathered, holding them together, shepherding them harmlessly and delightedly for hours without order or tuition, behaving as his forefathers did behave in the dim past, and as his descendants will behave in the remote future. In the pup there is no sign of a wish to drive, in the young unbroken dog there is no sign of a desire to heel stock—in short, herding and heading shows the collie instinct unmodified, driving shows it warped to the will of man. The origin of an instinct no wise man will attempt to fathom whilst the puzzle of priority in nature of the hen and of the hen's egg is unsolved, yet something may be ventured as to the use of the herding habit to its prehistoric possessors and of its exploitation by man. In the canine race, the stalk and momentary final pause ere leaping upon prey, the carriage of game dead and alive to den and earth, are, equally with the herding trait, primordial instinctive actions: basal, spontaneous, elemental, innate, they have no more been invented, superadded, or taught by man to beast than breathing, feeding, or perambulation. Each of these three instinctive actions has been originally wholly for the benefit of the animal itself. There exists, however, this vast difference betwixt two of these actions and the third, that whereas the stalk and momentary pause and the carrying of game are actions in each case concerning one animal only and its prey, the herding instinct could only have been useful to its possessor in combination with a partner. The ancestral collie has, I believe, herded and held together flocks of some sort in

Nothing good can be said of the wandering cat; the evil wrought by the roaming brutes outweighs by far the good. Though not rare on the run, they are seldom seen except during heavy rainstorms, when many half-drowned specimens crowd in for shelter. During one such period of heavy weather eleven unknown cats were shot in a day or two prowling about the men's quarters and cook-shop. Kittens must wander at an early age, for a splitter's camp, six miles from the homestead, and twice that distance from any other human habitation, was visited by a kitten which allowed itself to be tamed and was eventually carried back to the station.

In the 'eighties there were many patches of bush and scrub, chiefly on eastern Tutira, each of which maintained its little herd of wild sheep, rebels to station rule. These irreconcilables had never been yarded and wore their tails long as a visible sign of independence. They were a race apart, a peculiar people, maintaining their own customs like tribes driven to desert or mountain-top. Excepting after rain, when they emerged to dry themselves, they never mixed or fed with ear-marked, docked, domesticated stock. Then on the tops they were conspicuous by reason of the bright cleanness of the short rain-scoured wool on their backs. Their heads, necks, and the fore-end of their bodies were scraped almost bare by contact with timber; about their hind-quarters hung matted petticoats, the growth of years, often reaching to the ground. At the least alarm the brief connection of wild and tame was severed, the wild animals bolting downhill to their woodland fastnesses, the

conjunction with another beast less active though more powerful; the latter has struck down the huddled victims, their carcases have then been shared. There has existed, though now the connection is lost, one of those natural alliances or associations between two wild breeds, such as is to be found to this day in the relationship of the bee-bird to man, in the crocodile-bird to the crocodile, in the pilot-fish to the shark. The ousting of the dog's original comrade in this hunting partnership has been a matter of easy accomplishment. Pups have been captured and bred in domestication. Thus reared they have herded as perfectly for the savage ancestors of man as their descendants for the nineteenth century shepherds of Tutira. Often when mustering I have had occasion to remark the amazing combination of sense and senselessness in a collie's work, the instinctive portion of it so perfect, the rest so lacking in intelligence. It not infrequently happens on such ground as Tutira that a shepherd finds himself on one side of a long gorge or rift whilst sheep are on the other. To gather them, starting the collie from behind him, the dog is worked wide to head the gorge. The sheep are rounded up, the dog attempting to bring them in a straight line to his master. If left to his own devices he will continue to try to drive them across that gorge, a gorge which he knows to be impassable, for when whistled off he does not attempt it himself, but returns as he went. A collie in the same way will attempt to bring sheep in a straight line to his master through a strip of bracken so dense that when called off he will himself run round it. His work is perfect up to a point, but rigid, limited, incapable of modification. His instinct ordains that without movableness or shadow of turning all stock shall be brought in a straight line; so must it be. The gorge and the strip of bracken are barriers beyond his mental vision.

tame following ancestral habit and drawing on to the tops. The wild rams confined themselves to their own harems, or if, as very rarely happened, a more amative male emerged, he was detected and run down with dogs. Fom time to time these little companies or portions of them were raked in as chance favoured. Beyond leading the tame sheep astray at musters they did no harm; they were, indeed, rather an interest, the possibility of their capture giving a zest to the musters of early times. There were also larger mobs of wild sheep on several of the Waikoau cliff boundaries—cliffs not, as nowadays, densely overgrown with manuka, but then thick with anise and native grasses.

One of these little septs or clans deserves special mention, as doubtless its environment was the essential factor in a change of colour of fleece very rare or perhaps unique in wild nature. The flock in question ran on the wooded cliffs of a portion of the Opouahi block, a locality cut into irregularly sized sections by narrow waterless ravines beginning above as deep sink-holes and terminating below in the impassable gorge of the upper Waikoau. The Opouahi block, about three thousand acres in extent when first known to me in the 'eighties, formed a portion of Hindmarsh's Rakamoana station; as an outlying corner cut off from that run by intervening gorges and impenetrable belts of bush it was probably considered valueless to its real owner. At any rate he made no use of it; it was stocked by the then owner of Putorino, who wintered on it some three thousand hard-fed merino wethers. Now it always happens that in considerable mobs of dry sheep a proportion of ewes—two or three in a thousand, perhaps—are included by accident of ear-mark or by oversight in the drafting-yards. Thus there may have been ten or a dozen lambs born each season in Opouahi. In those days, however, the country was open and easy to muster clean. The likelihood is that any lambs born were swept up at the annual shearing muster, or if not, then at the secondary muster for stragglers.

In short, there were no more "wild" sheep in Opouahi than elsewhere. What makes me positive of this is that during the 'eighties one of my duties was to attend the Waikari draftings in order to pick out any "strangers" that might have boxed with those of that station. We were all shepherds in those days, and, as was natural, talk turned on events of camp and muster, the work of dogs, the good "turns" of particular collies, pig seen, feed, condition of stock—above all else, on the numbers of sheep brought in, for Putorino was at this date enduring the same pangs as Tutira: the number of sheep "short" after winter was a constant anxiety. Wild sheep were hardly mentioned; there was certainly no talk then of wild black sheep.

It was not until about 1892 that rumours began to circulate as to black sheep on Opouahi: a passing shepherd had noticed a mob of four or five; natives pig-hunting had seen black long-tailed rams. There could, however, have been but few as late as 1895, for during that excessively dry season I leased for six weeks the grass of the Opouahi block, still the property of Mr Macandrew. There during autumn we tupped our ewes, a step Stuart and myself would never have taken had

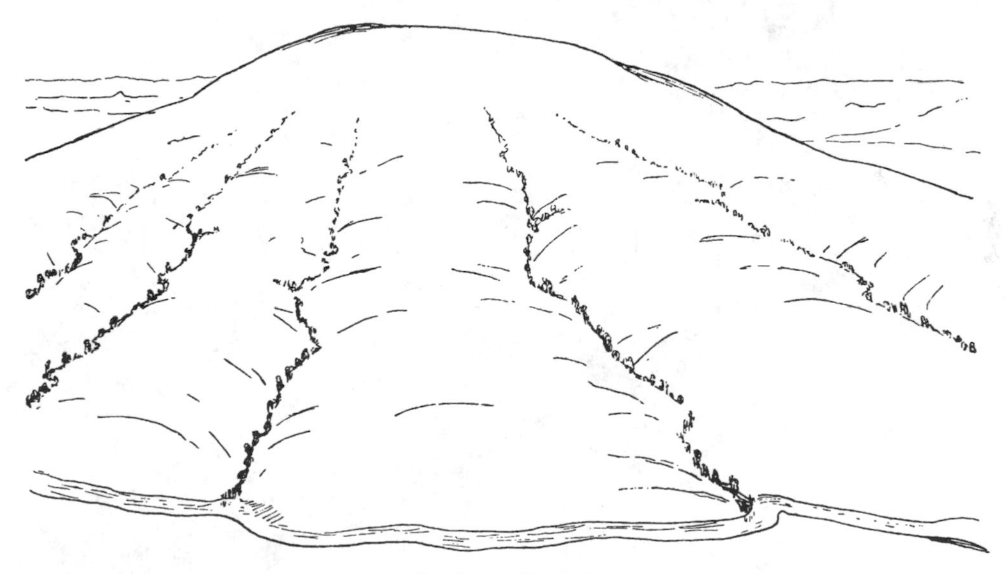

Ravines—Opouahi.

we considered that there might have arisen therefrom damage to our Lincoln-Romney ewe flock. It could not have been done, moreover, without the strongest possible protests from Harry Young, protests which would have been given full consideration. He, Stuart, and myself knew, as I have told, of a very few wild sheep, but judged that, as had happened on eastern Tutira in the 'eighties, their numbers were not sufficiently large to make them a menace, and that they would remain in their own haunts and not intermingle with the ear-marked stock. Our conjecture was justified, for when five months later the lambs were dropped, in none of them appeared any admixture of merino blood; the few wild sheep, black or white, on the Opouahi block had kept themselves to themselves.

We must now revert for a moment to those changes of vegetation which have been fully explained elsewhere. In the early 'eighties the Opouahi block was under native grass; gradually, however, on this block, as in other parts of the run, manuka began to take possession. From the 'nineties onward it spread at an increasingly rapid rate; each year the shrinkage of open country became more marked, until at length,

Ravine—Opouahi.

over hundreds of acres, it was impossible for a man to see more than a few yards in front of him. By 1900, except for certain tops, slopes, sheep-camps, and small oases in the scrub, the whole of the land in turf during the early 'eighties had reverted to manuka; only where native bush had grown it still grew thick and dense. The block, in fact, had become unworkable to shepherds; where over three thousand sheep had been fed less than six hundred were now carried, and carried moreover for but a few weeks each year. At last, except in winter, no sheep were run on the block at all. Feed, consequently, during spring, summer, and early autumn, was superabundant. By its luxuriance the wild sheep that had hitherto permanently remained in the scrub were now tempted on to the tops during twilight and early morn; passing at dawn *en route* for far Heru-o-Tureia I have again and again seen practically the whole Opouahi wild mob bolting downhill into the sheltering scrub. It was not a homogeneous flock, but consisted of seven or eight septs or clans, each of which returned to its own strip of territory, each of which moreover differed in numbers and proportion of blacks and whites.

During the last years of the expiring Opouahi lease, when no steps could be taken in regard to felling and clearing the land, the proportion of blacks increased fast. There was reason to believe that a pure race of black merino, self-evolved, was about to be established. Full accomplishment of this interesting natural development was frustrated by the obtainment of a proper tenure and the prosaic necessity of working country to the best advantage for which a large rent was being paid. Improvements, which have been the bane of Tutira from a field naturalist's point of view, now began also to desecrate Opouahi. Each year, starting from its east edge, a strip of the block was felled, fired, fenced, and sown. It was no longer possible for the wild sheep to escape their fate: their cliffs were bare, their hiding-places open to the light. Clan by clan they were rounded up and run into the drafting-yards.

The strips of territory thus successively reclaimed by axe and fire-stick gave the following results in black and white sheep. No. 4 lot is exact; the numbers of the other little clans, taken down from Harry Young's diary entries and counts, are approximately correct.

No. 1 lot—Eastermost strip, over 100 wild sheep; 20 to 25 per cent pure black, the larger number with white tips to their tails; 2 or 3 piebalds; the balance white.

No. 2 lot—40 in number; 18 black, mostly with white tail-tips; no piebalds; balance white.

No. 3 lot—About 50; rather more than half pure black, mostly with white tail-tips; no piebalds.

No. 4 lot—16 sheep, all black; all with white tail-tips.

No. 5 lot—About the same number of sheep as in No. 4; 2 or 3 piebalds or white; the rest black, with white tail-tips.

All of these wild sheep—black, piebald, or white—were pure merino; all of the rams carried magnificent heads. It will be noted that the proportion of piebalds is insignificant, that the percentage of black varies in different lots from 20 or 25 up to 100 per cent, and that oftenest pure black sheep showed white tail-tips.

In attempting to account for this instance of melanism, the idea that a return to a feral state had anything to do with the matter may be dismissed. No length of time, let alone a brief possible fifty years, would suffice thus to change the colour of sheep gone "wild." The sheep raided for mutton during the old starvation days of the 'eighties from the river faces of Land's End were white; the forty-three obtained by me in one haul from beneath the Razorback were white; the considerable proportion of wild sheep included in the nine hundred double-fleecers collected after weeks of work from the cliffs of the Mohaka by Mr George Bee contained only the normal proportion of blacks, two or three per thousand. In itself there is nothing remarkable in a black merino flock. There are, I believe, several such in Australia. It is the partial development of such a flock under conditions wholly natural that is noteworthy. Apparently there had happened on this wild corner of Tutira the very rare combination of suitable geological environment and happily fortuitous mating. Results were accomplished by chance which elsewhere have been obtained by a knowledge of the laws of breeding deliberately pursued.

Desire for high ground is still so marked among our modern breeds of sheep that it is difficult to believe that their ancestors did not live on mountain-tops. It is equally difficult to believe that animals feeding on ranges more or less covered with snow for long periods would have grown fleeces of any other colour than white. Be that as it may, certainly there is a strong tendency amongst domesticated sheep to break into black; it persists in the merino after hundreds of generations of elimination. White fleece will take dye, black will not; black rams have therefore been barred since the sheep has become a domesticated animal. In spite, however, of the selection of sires pursued through thousands of years, one lamb in three or four hundred is still born black.

My reading of the change of colour of the Opouahi wild flock assumes that in one of the ribbons of ground described, largely contained by natural boundaries, a black ram and black ewe mated and produced black progeny; that these again interbred until the black strain became fixed, until at length a sept became established reproducing only blacks. If this theory be accepted tentatively, the second step is to account for the remarkably rapid increase of blacks between the 'nineties and the early years of the present century. Configuration of the ground must again be taken into consideration—a series of narrow strips practically separate from one another except on the tops. On one of them we have supposed a minute flock of blacks to have evolved itself. In process of time this little sept would no longer be able to obtain feeding in its original strip. Other larger strips would be stocked, other little flocks would be invaded by members of our conjectural black flock. The smaller number of blacks found in the far selections may be in this way accounted for, the ewes of these strips having been for a lesser number of seasons in contact with No. 4 ribbon of ground, the strip assumed to be the black centre-spot. Interbreeding would have happened on a small scale, the rams of No. 4 strip only tupping the ewes of the sections into which they had spread immediately to right and left.

Increase of feed on the tops during the years when the block was unused throughout summer and autumn, tempting the wild flocks upwards from all the strips, would accelerate the change. From dusk to dawn on the tops there would occur interminglement of rams from every ribbon of land. Indiscriminate tupping of wild ewes would become the order of the day, black rams and white rams mating with black ewes and white ewes on the neutral ground of the tops. Furthermore, I think it is possible that all black ewes mating thus "held" to the black ram, and that it is likely that a large percentage of the white ewes also "held" to the black ram and produced black lambs, especially to males black for three or four or five or six or seven or eight generations; to rams, in fact, where the black strain may be supposed to have become more or less fixed.[1]

There may have been ancillary reasons, too, favouring the domination of the black—a greater virility and activity, a larger share, in fact, of the wildness of the feral ancestors of all sheep. Certain it is, at any rate, that in a bad hogget season the least well-woolled sheep survive, hoggets whose feeding has gone to the original qualities of frame and hair—stock, in short, that show the least effect of man's

[1] Passing through Canada on one occasion, I was taken to see a flock of Karakul on the foothills of the Rockies. Rams of this breed put to white "range" ewes throw about 99 per cent black lambs. Wensleydale ewes to this day, though black stock has been eliminated for generations, produce about 25 per cent of black lambs. The black drop seems in fact to be constantly attempting to reassert itself.

selection and mating. Pickers of stock for the frozen meat trade, too, are suspicious of blacks. I believe that subconsciously they are regarded as wild animals less likely to be really fat.[1]

To recapitulate: We have in Opouahi an area of some seven or eight hundred acres cut into six or seven ragged sections, each of them separated from the others by rents terminating at bottom in a river-bed impassable for sheep, and on top expanding into open feeding-grounds. There, on these narrow strips of precipitous land, sheep have possibly been in small numbers running wild since the date of the stocking of Heru-o-Tureia by Mr F. Bee half a century ago. Certainly in the 'eighties, when I first personally knew anything of Opouahi, no more wild sheep were in hiding there than elsewhere on Tutira.

By the middle 'nineties the lease of Opouahi was nearing its termination, the surface of the ground was fast deteriorating in carrying capacity because of scrub, the process continuing until at length it became impossible to work dogs. At last only a few hundred ear-marked sheep were wintered each year for a few weeks. In late spring, summer, and autumn the block lay vacant; feed consequently became rank; the wild sheep, hitherto subsisting on bark, shoots of trees and fallen leaves, now for the first time drew on to the open tops, each sept or clan ascending at dawn and at dusk again retiring to its particular strip in the scrub. Thus protected by favourable environment from shepherds, with extended areas of feeding-ground, kept in excellent condition because of unlimited feed, the wild flock or flocks of Opouahi increased and multiplied. Each season augmented the percentage of blacks until at last one small sept had become entirely black, whilst throughout the whole flock there were about as many blacks as whites. Then, alas! a fresh lease was obtained, the surface of the ground was bit by bit cleared of scrub and bush, the wild flock destroyed, and the natural evolution of a pure black race of merino terminated. To me the rapidity of the alteration in colour is not less remarkable than the change itself. Whatever intensification of black blood may have occurred in a single sept, to outward seeming the change over the whole little flock happened in ten, or at the utmost twelve seasons.

[1] I remember shortly after arrival in New Zealand watching the well-known dealer, Andrew Grant, picking "fats." When asked by me for his hesitancy and discrimination in regard to several apparently prime blacks, his answer was that they "killed" badly. So many statements, nevertheless, continue to be accepted because it is nobody's interest to disprove them, that this dealer's belief is given for what it was worth. On the other hand, what Sandy Grant did not know of sheep was not worth knowing.

CHAPTER XXXVII.

RECONSIDERATIONS.

BEFORE proceeding to the consideration of certain aspects of migration, it will be convenient to clear the air of misconceptions as to the effect on aliens of importation and acclimatisation. Not infrequently it has been assumed that in some measure capture, confinement, and the manipulation of man has altered, not the nature of the specimens captured—that is likely enough—but the nature of the offspring born of them; that, free in a new land, the trammels of civilisation still cling to and hamper their country-bred descendants. Such a belief would largely detract from the interest attaching to the journeyings of the many aliens that have passed through Tutira. It is not true; to the animals themselves, and indeed in the final result, the sails of a ship are no more than a prolonged gale, the deck of a steamer no more than a drifting timber mass, a floe, a fragment of sud; whether blown from their quarters by stress of weather, and borne abroad by currents of air, or whether trapped, railed to port, and shipped over thousands of miles, their descendants start level in the race of life.

The Wax-eye.

The wax-eye (*Zosterops cærulescens*), actually seen to have arrived on the Mahia by the Bishop of Waiapu, and the sparrow, known to have been liberated from cages at Auckland, have each arrived in New Zealand since its proclamation as a British possession. In a sense the arrival of the one was natural, the arrival of the other artificial, though the terms may give us pause, since every interference by man with the normal course of events may also be called natural. At any rate, once landed in New Zealand, the two breeds were essentially on the same plane. Each had reached its destination in its own way—the wax-eye by pinion and plume, the sparrow by sailcloth wings. Whatever the difference in manner of arrival may have been, once landed in New Zealand the breeds were on a par as to the future. Each was beyond the direct influence of man,—outside his pale, free to select the route of its wandering, its rate of increase, its climate. In truth, we may eliminate from our minds the long sea-voyage, the habitation of ships, the cramped confinement below deck. If we choose to do so, we may consider that the sparrow, blackbird, deer, and weasel arrived in New Zealand during the 'sixties, 'seventies, and 'eighties by a series of nearly connected islands, since submerged. At any rate their goal once attained, fullest liberty awaited them; they were free to pursue a future unshackled by the past.

The behaviour of aliens "wild" in a strange land can neither be compared to the seasonal migration of continental areas, nor on the other hand be passed over as mere incursions, mere irruptions, such as those of the sand-grouse, or the lemming, that ebb and leave no lasting mark. The treks of our aliens have a certain original place in the annals of migratory movement; they possess, moreover, the interest that attaches to an experiment which cannot be repeated. Conditions obtaining in New Zealand in the 'fifties, 'sixties, and 'seventies no longer exist. The world does not now contain a continent or a great island still virgin, still unmanned.

The particular history of certain aliens has been described. Two main facts stand out: the first, that amongst certain of the birds there still survive blind and broken traces of some sort of seasonal impulse, reminiscences, confused and indeterminate, of gatherings for flight. On Tutira I have again and again witnessed the assemblage in autumn of parties and congregations of birds, of petty local migrations. Blackbirds, greenfinch, and thrush gather in flights great or small; they travel somewhere for some purpose, though the whither and why will remain unfathomed until individual birds are marked and watched. The second fact standing forth pre-eminently is, that many of the aliens have moved at the prompting of a genuine migratory impulse.

New Zealand is in shape a long strip of territory pointing north and south. Its central ranges are rugged and high; the trend of settlement accordingly has been along the seaboard: settlement has moved north and south, not east and west. The line followed has been followed for the common-sense reason that it is easier to move parallel to a range than to traverse its heights. North and south has been the line of least resistance selected in turn by man and beast, first by Maori, later by European, and last by his introduced mammals and avifauna.

That has been the general direction followed; particular modifications have been the hill-top, the coastal and river-bed routes. The way of each species, moreover, human or brute, has been affected by special idiosyncrasies and predilections.

The wish for warmth has confined the Maoris, the earliest aliens of all, to the northern portion of the North Island, to coasts and estuaries elsewhere, and in the interior to the thermal region. To a later alien, the white man, ports and harbours have been essential. The Anglo-Saxon, a hardier breed from a colder climate, has spread not only along the coasts and over the northern parts of the North Island, but over the interior everywhere of both islands.

Aliens of a lower rank in the scale of nature have also followed diverse routes, influenced by diverse desires. Breeds sedentary in their original habits, as might have been expected, have shown similar characteristics in New Zealand. Others have felt in fuller degree the instinct that bids a race move forward. From treks of the latter sort some curious facts may be gathered. They exhibit a fire of restlessness, a passion of progress bred into the fibre of every member of the moving mass.

Rabbits turned down in the Wairarapa after a few seasons became a curse, increasing and multiplying until many of the local squatters were eaten out, their stations left desolate, and they themselves ruined. Between that period and the date of their invasion of Tutira the hand of man has lain heavy upon the rabbit. Everywhere also its advance has been retarded by the attacks of harriers, moreporks, and wekas, working not for eight hours a day like man, but for twenty-four. Rabbits nevertheless have forged ahead thinly on a vast front, reaching Tutira almost simultaneously by the hill-top, by the coastal route, and by the line of the human highway. There was no strip of open land

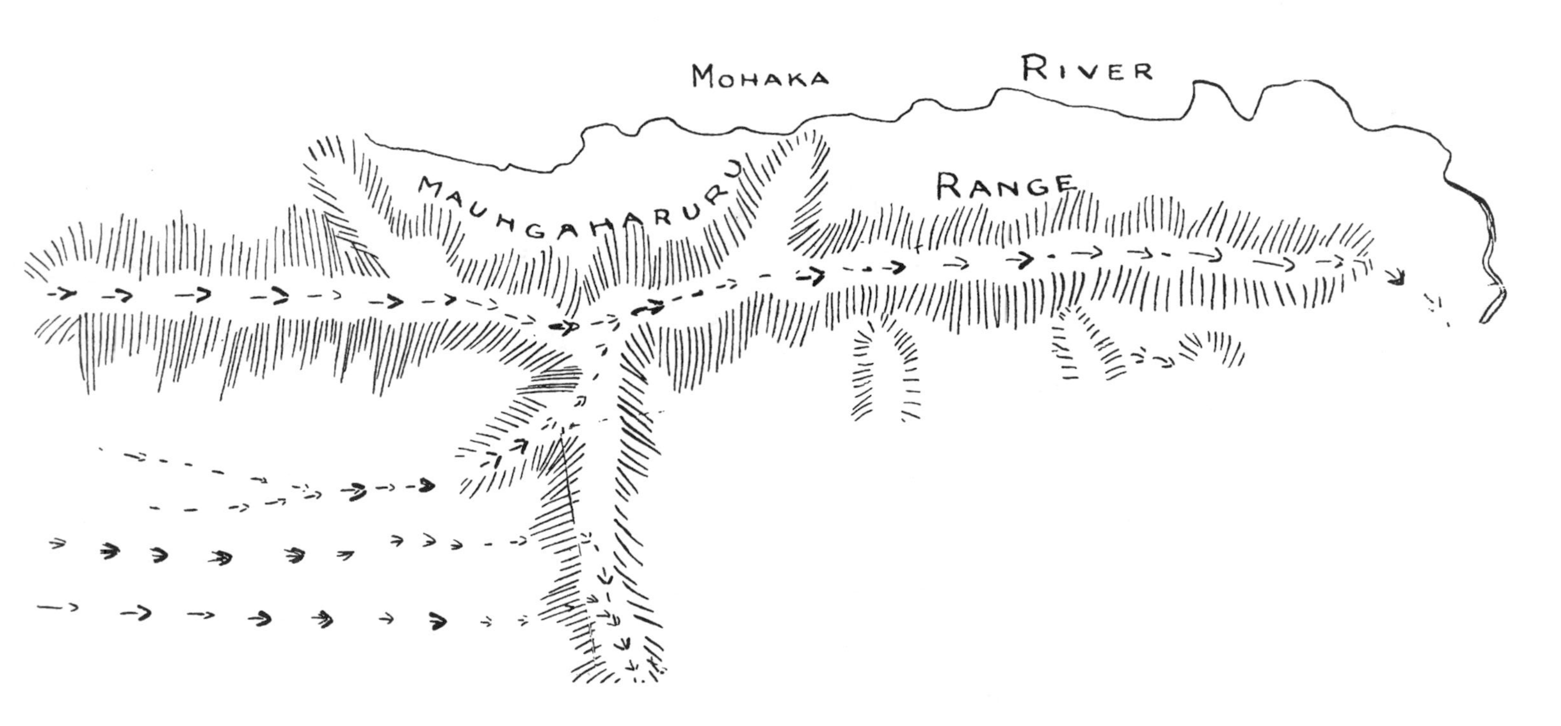

RABBIT ADVANCE CHECKED BY RIVER.

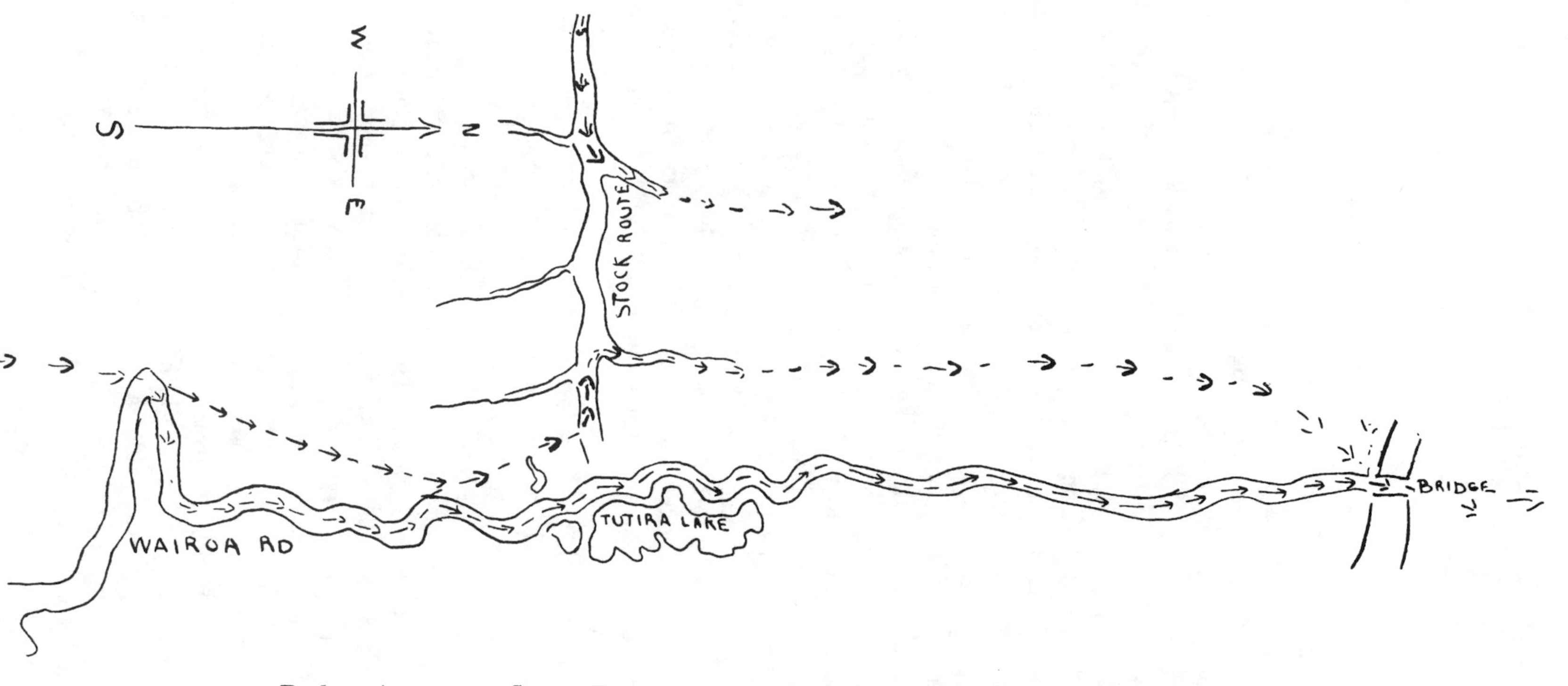

Rabbit Advance. Stock Route used only until Northern Exit discovered.

from coast to mountain ridge along which during the 'nineties rabbits were not filtering towards Hawke's Bay—towards Tutira. It was an invasion which, however disgusting to the sheep-farmer, was full of interest to the field naturalist.

The physiography of Tutira has been described,—a series of slopes precipitous to the west, tilted gently to the east, and fissured by ravines. It was a pattern well adapted to illustrate the rabbits' desire to follow a definite direction. The invasion moved from south to north, the ravines on the other hand ran east and west, at right angles, that is, to the line of trek. On the high ranges of the west these natural obstacles to progress were particularly formidable. On many the gorges began within a few chains, sometimes within a few yards, of the ridge-cap. Only rabbits, therefore, moving along the very summit could proceed. All others found their progress barred, discovered themselves on the brinks of precipices, many of which were fifty or sixty feet in height. It was impossible for them to remain for any length of time on barren cliff-edges. By natural enemies, necessity of food-supply, fear of the open, and lack of burrow accommodation, they were eventually forced either once more to the summit or downwards to more accessible country. Thus barred from their selected line of progress many were forced downhill, and reached after a time the great pumiceous trough of the run.

Through this, then, bleak forbidding country ran east and west the main stock-route of the station, the trail by which sheep are driven to the distant paddocks and mustered homewards for shearing. It varied in width from five to ten yards, and for miles passed through fern-lands bare of grass. Now it might have been anticipated that rabbits thus finding themselves on an open road, a smooth space hedged in by deep bracken, would have been content to accept it without question. Far from this happening, however, the east and west stock-route was used no longer than was necessary to find northward leading exits. No pig-trail however narrow, no sheep-track however overgrown, no fence-line however little trodden, that branched off northwards, was neglected. Rabbits caught by the contractor and his trappers were found in the extreme northern points of the blind spurs. On such spots the vermin stayed, blocked by impassable ravines, never attempting, as far as could be judged by signs, to retrace their path. Spots unexplored in former times, even by shepherds, even by myself whilst pig-hunting in early days, had now to be diligently searched for rabbits. They were always to be discovered on the northernmost extremities of northern-running spurs, never on the southernmost extremities of southern-running spurs.

On Tutira the plague has from the first been kept in check by contract, a most satisfactory system to all concerned. The duration of the original contract, which has been from time to time renewed, was for five years, a great proportion of each year's payment being withheld until the termination of the periods. Responsibility, moreover, for all fines and penalties that might be levied on the owner by the Hawke's Bay Rabbit Board was also accepted by the contractor. He had a personal interest, therefore, in the matter. One of the most practical proofs indeed of the rabbit's determined northern movement was the anxiety with which he watched the runs south of the countryside over which his agreement extended, and his indifference to lands north of Tutira. These were times when the rabbit was still an unknown quantity, when his advent was still viewed with little short of consternation. I knew, therefore, and sympathised with my contractor's uneasiness. Riding with him, I used to hear many aspersions as to the carelessness of the poisoning, trapping, and shooting work of southern neighbours. As to the conduct of stations north of Tutira he cared not a jot; their rabbits were not on his mind. As he phrased it, "the brutes are through us now and will never return." They were moving north.

Under the contract system described it may be supposed that no considerable number of rabbit congregations for long remained undiscovered. When, however, from time to time one such was first found, it did not seem to have increased beyond certain numbers. Whenever litters had become fit to move forward, apparently they had moved forward. There had never occurred multiplication to the limits of the local food-supply. These settlements, usually oases in manuka thickets, have been again and again trapped, shot, dogged, dug out, and poisoned; the rabbits inhabiting them annihilated. They are, in fact, caravansaries, favourite halting-places on the line of march, utilised by successive relays of rabbits, again and again restocked by fresh bands passing northwards.

Other illustrations of the resolute pursuit of a certain chosen line, and of the migratory fever possessing both old and young of the breed in motion, are afforded by the advance of several of the alien avifauna, by the behaviour of the blackbird, thrush, and chaffinch in particular. For years the blackbird and thrush passed through Tutira without perceptible increase of local birds, the two broods reared each season vacating the station, young and old alike merging themselves in the current of advance.

Self-interest is a trait in human nature ever alert and watchful; men's observations are whetted, their recollections sharpened, by events that harm or help. My contractor's jubilant exclamation in regard to the rabbits, "the brutes are through us now and will never return," was prompted by observation of certain facts in which he himself had a personal interest. The spread or rather settlement of the blackbird and thrush likewise became in a minor way a personal matter to every settler, shepherd, and station-hand on the east coast. Few, indeed, noticed their arrival or their habits and customs during the early years of their appearance. It was only when after a certain lapse of time that loss of ripe fruit from the wild cherry-groves, theft of raspberries and grapes from the garden, became but too obvious, that their presence was fully realised. Fruit which had ripened was no longer suffered to ripen.

Details of the arrival of the first blackbird and thrush have been given. They reached Tutira about the same time, in '91. Two years later considerable numbers were temporarily resident in spring, breeding with us. Now, had these station-born birds remained and bred, their descendants again remained and bred, the numbers both of blackbird and thrush would have increased so greatly that our cherry-groves and gardens would have been despoiled many seasons sooner than did actually happen. Had there been no forward movement, they would have locally multiplied like the price paid for the successive nails in the horse's shoe. As a matter of fact, no harm was done until ten or twelve years later. Since that time cherries have never been allowed to colour beyond a bright red; in olden days they remained for weeks, sweet, black, and wrinkled, on the laden trees.

Another fact worth noticing in the blackbird and thrush migration is its narrow width, the advance of young and old alike along a line within limits confined to the coast. For years later than the spoliation of groves on the route of march, the cherry plantations at Waikaremoana, inland from Wairoa, and at Maungaharuru, inland from Tutira, remained intact and continued to mature dead ripe fruit.

Other evidence of the non-stop character of the blackbird and thrush movement is the scant attention accorded to it. Had these birds, for instance, stocked Waiapu to the limit of winter food-supply—snails, worms, and insects—ere proceeding to Poverty Bay; had they stocked Poverty Bay to its limit of food-supply ere proceeding to the Wairoa, and so continued southwards down the coast, the striking contrast between districts swarming with birds and districts entirely void of them must have been appreciated by the most unobservant; local papers would have been full of letters from sentimental folk regarding the singing of feathered choristers from the homeland. There would have been bitter complaints from the resident growers of small fruit, for the blackbird and thrush by no means confine their depredations to cherries; every crank in the community would have been in favour of importing owls, hawks, eagles, and for aught I know, boa-constrictors, vampire-bats, and tigers to cope with the pest. The fact is, that our acclimatised birds have stolen upon us. Their advance has been so gradual, and at first so thin, that except for an observer here and there it has remained unnoticed, unchronicled. In a practical community little or no attention has been excited by the spread of imported aliens until, as has usually happened, they have become a nuisance.

About the chaffinch facts of a similar complexion are available. This species, it will be remembered, moving at first by the coast-line and afterwards by a river-bed route, had been first seen in Poverty Bay and later in the Wairoa before the earliest specimens reached Tutira in the beginning of the present century. Its arrival had been confidently anticipated; an especially keen watch had been kept there and elsewhere for the bird. Following its discovery at the apex of the Poverty Bay plain, I spent several weeks of two winters' quail and pheasant shooting in the Hangaroa district, five or seven miles inland from the coast as the crow flies. There were no chaffinches there when they had already reached Wairoa on the coast. There were still none to be seen at Hangaroa when the birds had penetrated Tutira, had passed through Petane, had moved onward as far as Havelock North. The trek, following the river-bed route, had struck the coast and then hung closely to it. It appears, indeed, to have been a mere lance-head thrust into the unknown.

The chaffinch, like the blackbird and thrush, passed through Tutira long before the limit of food-supply was reached. Small travelling parties, caught perhaps on the move by the impulse to build, seemed for several seasons, after the vanguard had reached the run, to breed, if not semi-gregariously, then at any rate thickly, in comparatively limited areas, thickets of tall manuka and the like. These areas in later years contained fewer pairs, sometimes, indeed, no birds at all, though covert, breeding accommodation, and food-supply were each and all, as before, superabundant.

Again, I think, there can be no doubt that old and young alike were drawn into the current of migration. There is no alternative by which the disappearance of broods reared on the line of march can be accounted for. It is, at any rate, as certain in regard to the chaffinch as in regard to the blackbird and thrush, that the advance was not solid, that there were not multitudes of birds north of a given line, whilst south of it not a specimen was to be found.

Consideration of the weasel's passage along the east coast shows more clearly than any other migrant movement that there exists in each unit of the mass, not only in its mature members, but also in those born during the journey, the instinct of adventure. The proof is complete; it lies in the fact that the pest passed through the district and disappeared. In this case we do not need to deduce and infer how the young have gone forward. We know they did go forward, for none of any age, young or old, remained. The three years' irruption of weasels through Tutira, and the district lying between Napier and Gisborne, resembled the progress of a comet across the heavens, the tail following the head, and at last leaving the sky once more clear. The invasion rolled itself up like a scroll; it came and went like a thunderstorm; ominous rumblings and mutterings, rumours of bitten babies, slaughtered fowls and lambs, lightning flashes from the Yellow Press, curses on the squatter class, heralded its approach. Then it rained weasels; they poured along the roads. There was a dissipation of the clouds; once again the sun shone bright through a blue sky—the weasel was a thing of the past, and remained so for very many years.

In the migratory movements of the starling and minah there is considerable proof of a general nature as to inclusion of young birds in the advance. In the march of each of them, however, there has been a parasitic clinging to man which places them in a different category from the weasel, rabbit, blackbird, thrush, chaffinch, and redpole. The trek of the sparrow through the very heart of the North Island is second to none in interest, but he followed the track of man—the king's highway.

Speaking in general, after a time the overpowering impulse to move onwards on a certain line, never to cease to follow a leader, begins to wane. The migratory fever dies down; a check occurs in the march forward, a check sometimes permanent, sometimes temporary. In either case every link in the long chain of migration is at once affected. The check may be compared in its action to the stopping of the engine of a train, when each carriage is in turn affected, or perhaps better, to the damming back of a stream. The sluice is closed, the stoppage of the head-waters stills the draw of the current behind until at last all seems quiescent, or—to continue the metaphor—until the dam breaks, giving way at last to irresistible pressure, until in one burst the waters once again rush forward.

The rabbit, for example, after only three or four seasons, began to settle on its tracks, its earnestness of endeavour to get forward became less keen. Its vanguard had reached the deep swift-flowing Mohaka. The presence of that barrier had been communicated backwards along the whole line, not in any occult mysterious fashion, but, I imagine, simply by a cessation of forward movement by the leading rabbits, by the leading files of rabbit—their halt reacting on the second file, that of the second on the third, and so on through the miles-long chain. It is for this reason that there is no huge immediate piling-up of rabbits against a newly-erected rabbit fence—the news of a check is automatically passed backwards along the line. There seem to be three phases in a migratory movement—the first, that of follow my leader, old and young alike moving onwards in a definite direction; the second period begins after the van has sustained a check, when the individual links of the living chain begin to breed circlewise, when they begin to stock the country in breadth as well as in length; the third, of which I have seen one notable example, is the bursting of the containing barrier and the sweeping bare of the breed from grounds previously overcrowded and overstocked.[1]

[1] At Peel Forest, South Canterbury, where in the early 'eighties I was cadeting, I was witness to what I have described as the third phase in a migratory movement. At the date mentioned hares swarmed about the westermost end of Peel Forest run. There were hundreds, there were thousands of them. During the breeding season I have counted seventeen and eighteen running amorously together. They were migrating westwards, and had reached a

The period between the first arrival on Tutira of the goldfinch and the time when a marked increase took place in the breed—when the station began to be stocked by the station-bred birds—was also noticeably brief. The goldfinch in its northern advance during the 'eighties

temporary *cul-de-sac*, hemmed in by mountains and forest on the one side, by a huge snow-fed river—the Rangitata—on the other, whilst directly across the westward exit lay the hamlet of Peel Forest. Two or three seasons later, upon my return, there was hardly a hare to be seen :

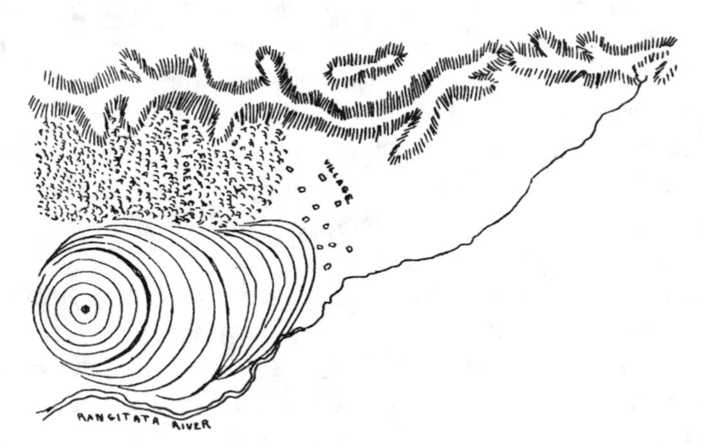

Hares east of village.

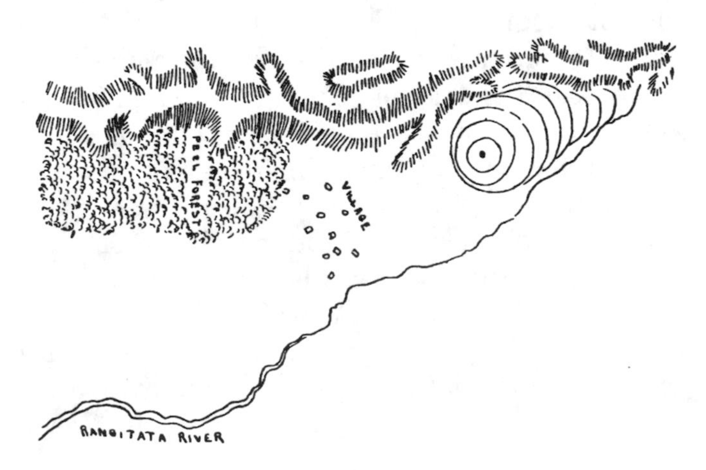

Hares west of village.

the impulse to move had swept the locality bare, the barrier had been burst, the hares had proceeded on their way.

Shooting-parties held in one decade east were in the next held west of the gorge.

had struck a barrier of another sort—a barrier of uncleared land, a barren countryside covered with scrub and bracken, where the thistle was unknown—a consequent insufficiency of food.

The period during which the migration of the redpole continued through Tutira was also brief. The barrier to the southern progress of the redpoles was of an exactly opposite sort—not bad land, but good land—unpropitious environment to a wild country breed,—vast stretches

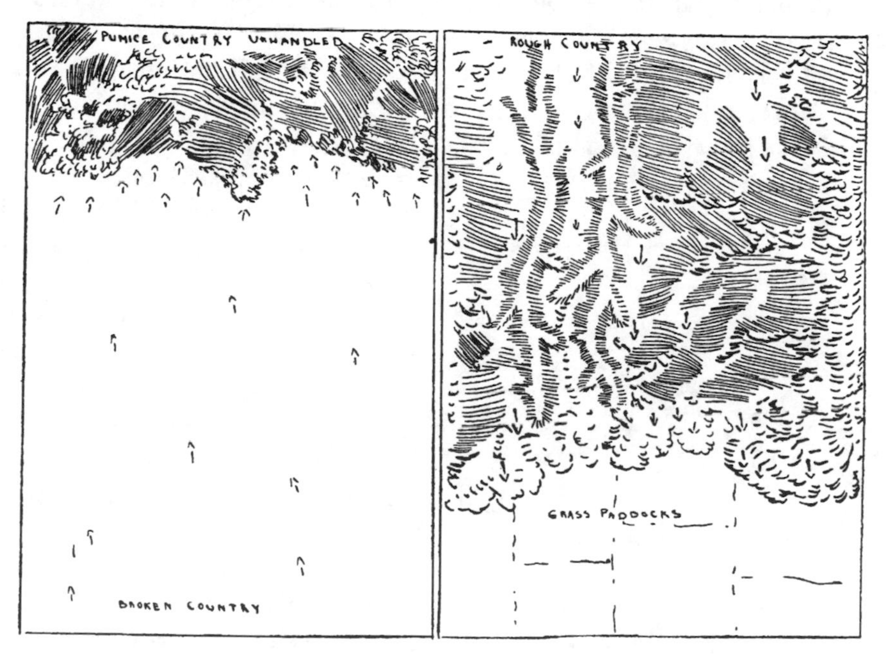

Goldfinch passing through "broken in" country—checked by unhandled bracken and scrub.

Redpole movement checked by grass-lands.

of grass-land completely cleared of scrub, lacking both food and nesting accommodation.

On the other hand, the passing of the blackbird and thrush through the station was prolonged for years. There was nothing to check the current of their flow for scores of miles beyond Tutira. There was every inducement to proceed through a warm, highly fertile coastal belt, amply supplied with shrubberies, plantations, and gardens.

The duration of the weasel trek through the run was more brief than that of any other aliens; rivers unbridged proved no barrier—he rushed through good and bad alike. Without a base, however, he perished from insufficient numbers; like the seed in the parable, when the sun was up he was scorched, and because he had no root he withered away. I could never trace the movement beyond the southern edge of the Poverty Bay plain. It passed along the east coast like a comet through the heavens, fading away in front and leaving no trace behind.

The origin of movements seems to be congestion. Where there is no great increase of numbers, as in the case of the pheasant, the quail, the rook, and other species, there is no accentuated movement. Spread is almost fortuitous, a merely negative pursuit of the *summum bonum* of animals—food, shelter, and breeding accommodation. On the other hand, species whose rate of increase is great multiply about the spot of liberation until the limit of food supply, clean quarters, and breeding accommodation is reached. The normal spread, perhaps, of a species is circlewise from a centre, perhaps also it would continue with equal speed in all directions for all time were exactly similar environments anywhere to be discovered. Sooner or later, however, weather conditions, physical and geographical barriers, limit such circular extension. Where a barrier is touched in any one quarter, pressure of growth is transferred to other segments of the circle,—its original shape is lost. At last, where resistance is least great, the living contents break forth. It may be that disinclination to move persists for a considerable period after the limits of comfort have been reached. There is a dread of new conditions, greater or less according to the nature of the breed, which prevents many species from severing widely the bond of one another's company. They prefer discomfort to disintegration; leadership is uncoveted — it is a post of danger and dread, — the desire of each is not to lead. As all, however, cannot be followers, the position of danger is distributed over a large number of individuals. A flock of sheep, not yet listless with dogging or careless on a well-known road, travels with a head formation rather blunt than broad,

Mob of travelling sheep.

such as water carries when poured on gently sloping even ground. Actual leadership is taken up moment by moment by different sheep, each temporary guide when a few inches ahead becoming scared, pausing, and letting his neighbours on either side precede and accept in their turn the responsibility and peril. By this method of advance no individual offers too prominent a mark, each feels himself able in a fraction of time to plunge sidelong into the mass, to lose himself in comfortable nonentity. If the sheep domesticated for hundreds of years cannot forget this sense of impending danger, wild creatures dare not for a moment cease to suspect the unknown. This dread it is, I think, which keeps large companies of animals voluntarily confined, penned close in limited areas for considerable periods. The absolute numbers of a congested horde are unimportant. It is the relation of numbers to environment that decides the genesis of the trek. The numbers of the rabbit in the Wairarapa, and of the goldfinch in lower Hawke's Bay before their advance began, must have been immensely greater than those of the sparrow or of the blackbird or thrush in Auckland. In the case of the rabbit and of the goldfinch there was room for the enlargement of the circle and a consequent postponement of the initiation of the trek. On the other hand, where the locality of liberation was limited in space or feeding-ground, as in the case of the sparrow, blackbird, and thrush, the time during which the circle could spread normally without striking barriers meteorological, physical, and geographical, was sooner reached, the movement of migration more quickly precipitated.

We have imagined a deep distrust and suspicion of the unknown sufficing to hold together a congregation of aliens in a strange land—sufficing to retard migration until further expansion has been blocked, this fear of what may lie beyond counteracting the disabilities of less clean feeding-grounds, less ample breeding accommodation; we have imagined the congestion becoming intolerable until the uneasy horde at length breaks forth. The begetting force of the actual moment of migration can be conjectured with a fair degree of likelihood; hardly a hint would be required. A premonition of that deep inherited kind we call instinctive would have permeated the inert and uneasy mass that movement was in the air; change of ground would have been long anticipated; the horde would be in a state of unstable equilibrium charged with an instinctive restlessness, expectant as duck after dusk for the signal to rise. Equality, however, amongst the lower creatures is no more existent than amongst men. There are no doubt degrees of fearfulness, of timidity, though perhaps as infinitesmal as the differences in height betwixt the lilliputian monarch and his subjects. Thus the actual incitement to move would, like the leadership of the travelling flock of sheep already cited, be shared by many simultaneously, or if that be impossible in time, then one of the migrants—a Moses about to lead his congregation into the wilderness—would exceed his fellows in celerity of rise by as little as the centre of a taut inch of thread differs in straightness from its extremities.

There are several lesser matters that may also be reconsidered,—sex of leaders, scouting, joint migration, climatic conditions, and general reasons for failure or success in the acclimatisation of aliens.

As to the first, the sex of leaders, I have little evidence to offer except in regard to rabbits. The pioneer of the advancing wave—the first rabbit taken on Tutira—was a male; of the first two or three dozen taken after the vanguard had appeared all were males. Amongst the first hundred or two, males still largely predominated.

In regard to scouting, there have occurred in my time at Tutira a couple of instances of single birds reappearing—reappearing, moreover, almost exactly twelve months later, and almost exactly on the same spots. The first case was that of the goldfinch seen near the wool-shed, and seen again on the same spot in the following season. The second instance was that of the minah, which, crouching close against the wire-netting of the newly completed hen-run, attempted to associate with the fowls. A year later the bird reappeared, again sitting close against the wire-netting exactly, as far as we could judge, on the site previously occupied. The very attitude was similar on the two occasions, an unwonted attitude for a species that under normal conditions never sits or crouches on the ground. It bespoke exhaustion, fear, and friendlessness. Like the rabbit which consorted with the Tangoio turkeys, this minah seemed glad to scrape acquaintance with any living thing. I feel as sure as a man can feel not dealing with marked birds that in each case it was the first seen individual revisiting us. The odds are enormous against another bird reaching by chance precisely, exactly, the same spot at the same time of the year on two sequent seasons. Were these birds scouts? Were they despatched as bees are said to be sent forth to discover quarters for the expected swarm? At any rate, both the goldfinch and minah appeared, presumably returned whence they came, and reappeared. After the lapse of a year they had remembered the route and repeated the experiment.

As to the joint migration of certain species, both Mr Williams and his brother have always maintained that the blackbird and thrush travelled together. They were first noted "at the same time" in the Waiapu and Poverty Bay; "within a short time of one another" in the Wairoa. At Tutira also the two breeds were heard and seen within a week or two of one another. If they had not travelled together, it is more than remarkable how evenly the species had kept step. The same may be said of the arrival of the chaffinch and the redpole many years later. If nothing more, it is a curious coincidence that two new aliens should twice have reached Tutira within a few days of one another. Personally I have no doubt they travelled in each other's company.

Of foreign species imported into New Zealand there have been more failures than successes. Many have perished immediately, like the nightingale, redbreast, and sand-grouse. In Hawke's Bay, black-game, partridge, and yellow wagtail seem never to have reached a second generation. The initial failure of the blackbird and thrush in Hawke's Bay is the more remarkable in view of their proved suitability in after years. Small numbers however only were freed. I am given to understand too that liberations were almost direct from the cages. Probably too the birds were cramped and gross from want of exercise; lastly, though liberated in localities cleared of cats, the destruction of rats and predatory native birds had been overlooked. Some species, again, after initial success, have disappeared, like the brown linnet; others, like the pheasant, the Australian and Californian quail, have for a considerable period increased, then begun to die away. The fate of game-birds has naturally attracted the largest share of public attention. Partridges have again and again failed, not only in Hawke's Bay, where their acclimatisation has been on too meagre a scale for much chance of success, but also in localities seeming to offer ideal conditions. The reason of the failure in such districts seems to be other than initial want of numbers or disturbance by vermin. They have perished, as in certain western islands of Scotland imported grouse have died off, and perhaps for the same reasons. For the decline of the pheasant many reasons have been assigned. Climate and vermin have been blamed, yet neither wholly account for the facts. Pheasants, we know, did in Hawke's Bay flourish and increase for a considerable period; neither climate nor predatory natives, such as the weka, hawk, or morepork at first prevented their multiplication. Search must be made for the lack of some formerly helpful factor; that factor was the absence of alien insect-eating birds. The decline of the pheasant synchronised with that advent of small birds whose rapid increase in the 'seventies "had been viewed with considerable alarm" by the Hawke's Bay Acclimatisation Society. The fact is, that owing to climatic conditions any great success with the pheasant was from the first foredoomed. It was only because of an unlimited supply of the most suitable and nutritive chick-food that the breed had been able to multiply even temporarily. Although Hawke's Bay "busters" must have always been detrimental, yet owing to the enormous supply of insect-food this disability had been more than counterbalanced for the time. It must be recollected, too, that about the date of introduction of the pheasant, the development of Hawke's Bay was proceeding apace. Each year on each sheep-farm large areas were being grassed; swamps were being drained; even a certain amount of ploughing done. Insect life during that period was increasing out of all proportion to natural checks.

Grasshoppers and caterpillars, native and alien, had multiplied on the succulent foreign grasses and fodder-plants by hundreds of millions. According to old residenters, they were plagues in the land. The pheasant had lived without competitors in the struggle of life; under more strenuous conditions the bird failed to hold its own. The breed is one hailing from a dry country; the chicks are peculiarly susceptible to cold and wet. With the deprivation of unlimited insect-food, for which formerly the hen had hardly to seek, the danger period to the chicks was extended. The brood was trailed over greater areas; the youngsters were not so fully fed; the number of eggs laid was less; in case of accident, perhaps, a second nest was not attempted. When account is taken of the care devoted by keepers at home to pheasant chicks, the special foods supplied to them, the short grass in the neighbourhood of the coops, and lastly, the absence of torrential rains, instead of wonder at the pheasant's decrease there should be marvel at the continuance of the breed at all in Hawke's Bay.

In this and previous chapters I have collected such evidence in regard to migratory movements as I have been able to gather from species that have crossed Tutira. The gist of it may be compressed into a few sentences. My beliefs are that where imported species, afterwards proved to have been adapted to their environment, have at first failed, it has been due to liberation in too scanty numbers; that aliens freed under these circumstances have scattered so widely and wandered so far apart from one another that in some cases they have lacked mates and in others have fallen victims to vermin; that individuals thus strayed associate with creatures in no way akin to them; that species successfully acclimatised increase circlewise until containing barriers are reached; that a migration once begun follows the line of least resistance, as may be determined by the leaders; that the line of trek having been decided upon, it is followed by all members of the horde, young or old; that broods or litters born by the way also pass forward; that rapidity of migration is determined by the speed of the leaders; that a check in front is communicated with remarkable celerity to the entire migrant chain; that species on the march retain the special habits and idiosyncrasies which mark them in quiescence; that the leaders are probably males; that certain alien species have travelled in company with one another; lastly, that possibly the single goldfinch and single minah cited may have been scouts—that, at any rate, they were not leaders of a trek already in motion.

CHAPTER XXXVIII.

VICISSITUDES.

COMING now to his last chapter, the writer would fain apologise in advance for what may at first appear its egotistical character. Really, however, as he hopes the reader will perceive, it is impossible altogether to dissociate the story of a bit of land from the story of the possessor or possessors of that bit of land. Without more words, then, he will proceed briefly to chronicle, not, alas! interesting changes of the earth's surface, the transformation of plant and animal life, but the doings only of man, the commonest species on the globe.

Yet even in this department of station life an evolution had occurred not without interest to the student of sociology. Readers will recollect how in the 'seventies and 'eighties, owners and employees had worked shoulder to shoulder as packmen, cooks, butchers, fencers, bullock-punchers, sawyers, and shepherds. In that arcadian life—for Romans were like brothers in the brave days of old—the lines of social demarcation had been unknown. All wore clothes—very few of them too—of the same cut, slept in the same hut, fed together on the spartan fare of those days—bread, mutton, potatoes, and duff.

Later, this primitive original relationship of master and man had given place to another slightly more complex, when management and clerical work had come necessarily to consume a greater portion of the owner's day, when he began to differentiate in outlook and responsibilities from his erstwhile fellow-labourers.

Later again, owing to a further extension of activities in station business, in fattening, in the multiplication of small paddocks, and in agriculture, the gap still further widened between the old life and the new.

Finally, the claims of personal labour were superseded by a more

Tutira Homestead.

leisurely life of oversight, when perhaps for the first time in the tiny microcosm Tutira, there might have been discovered latent the germs of those processes of cleavage which, developed, threaten disruption to more complex human organisations. Business the writer resigned to more efficient hands; perambulation of the run remained to him, a delight that can never fail or fade. It afforded a twofold interest, to the stockman and to the field naturalist, pleasant retrospection and pleasant anticipation. In these all-day-long rides, here we pass the spot where once a "placer" lived;[1] on this top are remembered wretched merinos where Lincoln-Romney sheep now thrive; there native grasses have supplanted

Foster-mother rock.

[1] Rock, log, nettle-clump, bush, or tree-stump may, as chance determines, become the foster-parent of the "placer." Like other small phenomena already noted on Tutira, it is the outcome of a combination of special conditions. To begin with, it must happen that the dam of the future "placer" shall perish within measurable distance of some such conspicuous object as one of those named; it must happen likewise that she shall perish when her lamb is young enough to miss greatly its former diet of milk, yet old enough to be able to support life on grass. The ewe, furthermore, must die in an out-of-the-way, thinly-stocked corner of a paddock, where the orphaned lamb cannot attach itself to another lamb of about similar age, cannot watch till that lamb is about to suck its dam, then rush to the unoccupied flank of the foolish ewe, seize the distended vacant teat, and kneeling opposite the true progeny, steal meals by stratagem. Lastly, the lamb must be of the female sex. The ewe dies then in the neighbourhood of one of the outstanding objects mentioned—say, a rock. At first the unfortunate lamb may be seen standing for long intervals by its dead dam, now and again bleating, cold, hungry, expectant still of its needed milk. When compelled by hunger to crop the turf, it never strays far; when disturbed by passing shepherds, it runs back bleating to the spot where its mother lies. With increase of age and appetite it feeds farther afield, but always when alarmed runs for protection and companionship to the patch of fast-disappearing fleece in the lee of the boulder; by the wool and bones it camps at night. Little by little the carcase flattens out, the wool, losing its brightness, becomes grey and stained, it sinks into the ground; but still at each disturbance the lamb rushes for protection to the now hardly-visible relics near the rock, still at night it sleeps close to them, close to the rock by which they lie. It forms no association with other sheep; at dusk, when they draw upward to the tops, it remains alone by the rock where its mother died; at dawn during musters, when shepherds shout and collies bark, it stands fast at the accustomed spot. In course of time wool and pelt alike become lost in the soil, the bleached bones become hidden by the autumn fall of foul rank grass—only, as always, the rock remains. Round it, as always, the lamb circles when feeding, to it as always she returns to sleep. At last feelings originally called forth by the dead ewe are entirely transferred to the rock, which becomes parent, protector, companion. I have never known a "placer" produce a lamb: I believe, when sought by the ram, it evinces the same sort of commingled terror and anger as is shown by a single bird for long a prisoner at the introduction of a companion into its cage. As it is impossible to part a placer from its foster-parent, there it remains always on the one spot separated from its kind, faithful unto death to the rock of its salvation. The particular sheep here shown was an unshorn six-tooth brought to the yards after two days' great trouble by a shepherd who wanted the animal for exhibition. Its foster-parent was a log near the Maungahinahina bush reserve. "Placer" is a term used to denote a gold digger who remains year after year on the one spot, on the one place.

manuka; here has been added to the local flora a new orchid, here a fresh alien has been discovered. Always there looms ahead the golden chance of such another find. Not one of a thousand rides on the station has been the duplicate of another; each has been for forty years a fresh page in the story—to be continued in our next—of the overthrow of the old world, and the slow re-establishment of a new equilibrium. Each ride too has supplied some hint for the remodelling of station management. None, moreover, but pleasant memories remain—even the disasters of the past retain not their sting, but the remembrance of the antidote applied. Addison wrote of his mistress, that to love her was a liberal education; to care truly for a bit of land anywhere the world over is a liberal education.[1] With this last change in the rôle of owner from workman to inspector of the work of others, station life from a sociological point of view passed into its final stage, the period of flux had ceased. One last great transformation, as yet however uncontemplated, lay dark in the womb of time,—a change destined to take place on every large sheep station in New Zealand.

For the last thirty-five or forty years the Governments of New Zealand—good, bad, indifferent—have agreed on one point, that the creation of a class of substantial yeoman farmers is beneficial to the colony—firstly, on the ground that a rural population is required;

[1] In youth's gay morn a man may possess land, in later life the land may possess him; the writer, circumstanced amongst friends virulently solicitudinous for the station's weal, barely evades his frankenstein. Though not yet devoured by this monster of his own creation, the station dominates his life, for always it comes first. His head shepherd, out of his thousands' increase every spring, grudges him a single prime lamb for the table. He is fobbed off with poor, pitiful, half-fat black brutes—"*blacks must be eaten, they spoil the look of the station lambs.*" For mutton he will be given tough old ewes which, if offered for sale, "*would ruin the appearance of a good station line.*" His wife—for a Scottish household too!—has to beg sheeps'-heads like a mendicant; his daughter—deliver my darling from the power of the dog—sobs for sweetbreads,—they are "*required for the station collies.*" He has been forced to pay off his mortgage "*to free the station.*" He cannot debit a door-scraper to the run but his book-keeper is up in arms cross-questioning him as to "*what interest the station can have in its purchase,*" as to whether it is not a "*luxury*" which, "*on a station run as a station pure and simple, could not be quite well dispensed with.*" Advice is lavished as to the amount "*the station can let him have*" each year. Posed by a balance-sheet over which he attempts to appear intelligent, he smothers a sigh for the good old-fashioned ledger of the 'eighties with its simple entries—

	National Mortgage and Agency Co. .	£9750
	Captain Russell	5

and all the nothings added. He understood *them* anyway. Then again, at the very suggestion, say, of some public-spirited scheme such as the utilisation of the ram paddock for an aviary, not one of his friends of twenty and thirty years' standing but would go back on him. Harry Young, omitting his customary benediction of "Well, ta-ta," would spark off in a rage. Jack Young, after a preliminary observation on the continued excellence of the weather, would swear it was impossible; Charlie Patterson would wag his red beard in negation; as for George Whatley, he would say nothing, one look would be enough—one of those frightful looks that have for years made him the real boss of the run.

The "Placer" Sheep

secondly, that the return from land held in small farms is relatively greater than from lands held in bulk; in the former case the ground is fully utilised, in the latter it is often not worked to its best advantage. In order to be able to satisfy the land hunger existing in the Dominion, legislation has been enacted which has necessarily affected the interests of those giants of the prime, the original holders, the squatters. Thus menaced, that class has not unnaturally, by procrastination and by the law's delay, striven to avert the sacrificial knife from its throat. Consequently, partly on this, and partly by reason of the natural antagonism between the house of "Have" and the house of "Want," a certain hostility has attached itself to all squatters, not as private citizens but as a class. As Meredith says of the sex, the individual has been rolled into the general, and a kick bestowed on the travelling bundle. A squatter—in the parlance of a certain set of the community, a bloated squatter—included any person who held land in a large way regardless as to whether he was a lease-holder or freeholder, whether the unearned increment accruing to the land occupied fell into his private pouch, into the coffers of the State, or into the bottomless pockets of the happy-go-lucky Maori. It has come about pretty regularly, therefore, that at election times the writer has found himself included in the general denunciation fulminated at the "wealthy squatters" north of Napier, wretches really, as the reader has seen, struggling against bankers, bad climate, bad land, and bad titles.

It was in an atmosphere thus charged with political electricity that with considerable alarm the writer heard of the appointment of a Royal Commission armed with powers to investigate the tenure of certain native leases—Tutira amongst them. As a matter of fact his concern was uncalled for, but at the time he could only recollect that he was a squatter, that his character was in no condition to endure the fierce light that beats upon a squatter's concerns in a court of judicial inquiry; that he was in possession of too many sheep and of too much land.

In due time the Commission arrived at Tangoio, and there heard the views of the native owners of Tutira; it arrived, as became a Royal Commission, in a coach and four. In front skirmished an irregular light horse of landlords. In the wake of the Royal vehicle flowed a stream of buggies, gigs, and dogcarts of every date and description: some brand new for the occasion, with shining buckles and clean leather; the harness of others patched up with rope, with wire, even with the humble necessary flax, but one and all containing landlords male and female, all agog and eager, like John Gilpin, for the treat. It was, in fact, high day and holiday with the owners of Tutira, for there is no event so exhilarating to a Maori as a *korero* over a piece of land; amongst the endless good qualities of the race, not the least pleasant is his affection for his patrimony. Firm that day in the mind of each owner was the resolve that it should have its rights. It was incumbent, therefore, that representatives of each "county family" should be in attendance to do honour to their property, to pay it due obeisance, to see that it was fitly honoured in speech and mythical allusion. Long and dignified, therefore, were the orations of the natives,—they are excellent speakers,—miserable was my counter-contribution to the general sum of speech. I have reason to believe that I appeared to the Commission rather in the light of fool than knave, and that my accumulation of 32,000 sheep seemed to their judicial minds rather the result of imbecility than of actual downright wickedness. I placed my affairs unreservedly in their hands, resigning my lease of the western half of Tutira, and asking in return that I should be given a fair deal.

Pakeha and natives alike, now that it was possible to do so, were eager to come to terms. Sincere were the felicitations that passed that day as to the reasonableness of all of us, I praising myself as a model tenant, the Maoris praising themselves as model landlords, and all uniting in praise of Sir Robert Stout and Mr Ngata. The natives were in high good humour that their land had been honoured by the approach of a Royal Commission, by my request for a fifty years' lease, by the greatly increased rental already fixed by assessment. Maoris, like other folk, appreciate the value of pounds, shillings, and pence: I believe, nevertheless, it was the aggregate sum of rental that was more pleasing to them than the calculation of each individual's share. Due respect, too, had been paid to the feelings of the land; its rights, its susceptibilities, had been fully recognised. To crown all, the Commission had decided with its own Royal eyes to inspect the honoured territory. Its coach and four accordingly was again put in motion, preceded by the station buggy, and followed as before by representative native owners. Tutira was inspected personally, and the Royal Commission, not snatched up to heaven in a chariot of fire as the assembled natives almost seemed to anticipate, proceeded in a cloud of dust on its way to Wairoa.

In dealing with the exceptional affairs of individual citizens, the State is always able to drive a hard bargain. Nevertheless, though the decision ultimately reached was good business from the point of view of the Commission, the tenant also was treated fairly. The Commission, if at all a beast in its terms, was, like our famous Rugby headmaster, a just beast. Political conditions, in fact, against all likelihood, made it possible that the matter should be treated with impartiality. Seddon was in the zenith of his power, a leader whom not his bitterest enemies could accuse of neglect of the interests of the masses. None but a Labour Government such as his could have—would have—dared to give fair play to a squatter.[1] Then, again, I was happy in the members of the Commission. Sir Robert Stout, amongst the innumerable activities of his great career, had played a prominent part during the 'eighties in the initiation of agrarian reform. He knew land at first hand from actual experience as he knew law. Mr Ngata, one of the four native members of Parliament, was an east coast sheep-farmer who had done excellent work in the organisation of native holdings in the Waiapu.

Well, making a long story short, and avoiding technical language, the recommendations of the Commission were that 18,000 acres should be leased to H. G.-S. for thirty years at a quadrupled rent, that certain lands should revert to the native owners, and that the rights of flax-cutting should be shared by tenant and landlords.

Much water was to flow beneath the Waikoau bridge before these recommendations were ratified. In the meantime, however, the satisfied tenant went home, possessed himself of shootings and fishings, and with thanks to an all-wise Providence who had ordained that he should remain on Tutira, heard by cable that in Section 45 of a certain Act of Parliament "the Board of the Ikaroa Maori Land District was hereby authorised to act for and on behalf of the native owners of the lands in the Hawke's Bay Provincial District, known as Tutira Block, and to give effect to certain recommendations of the Commissioners appointed by

[1] After the great snowstorm, which many years ago in the mountains of Canterbury destroyed whole flocks over a wide area, the Government of New Zealand reduced rents to its squatter tenants, extended their leases, and, when necessary, arranged that mortgages should be written off to reasonable amounts—actions not only wise in themselves, but proof that no class, however politically defenceless, was to be exempted from help when deserving of help.

the Governor." No prospects, in fact, could have been more pleasant than those obtaining when I left New Zealand. Wool was up, stock was up; as has happened before, however, and doubtless will happen again, these favourable—too favourable—conditions in the world markets culminated at last in a local land boom, one of those short-lived spurts of unwholesome prosperity that rage furiously until quenched by a fall in values. Wool, which had been at 11d., dropped to 5½d. Stock was unsaleable. I returned from the pleasant banks of the Dee to find my own prospects in particular unpleasantly dashed. My local agents had been unwise. They were hard hit; their position reacted on Tutira. It is extraordinary how on occasions of this sort gloom and despair seize persons who have passed through half a dozen crises and an equal number of recoveries. Temanites and Shuhites fill the land like frogs with their croaking, horrid rumours of failure vitiate the air. Banks and Mortgage Companies shut up like jack-knives, our very old acquaintance re-emerges,—that melancholy pessimist who has "offered his clip at fivepence for the next six years" and "cannot find a taker, mind you."[1]

Last, but not least, protests had been lodged against the signature of the lease recommended by the Royal Commission and sanctioned by Act of Parliament.

Glancing back now that the dust of battle has subsided, the writer can see that he himself was in part responsible for what had occurred, that his conduct had not been *tika*, not been correct. At the signing of former leases he and Stuart had always managed a haggard, anxious, careworn air when seen abroad. When in converse with groups, or followed by attendant trains of natives, or engaged with the all-important interpreter, gravity and a deep seriousness, as if overborne by weight of honour laid on these poor shoulders, had been "the thing." Now, too confident, with what countenance was he comporting himself, dallying after moor and river in Scotland whilst the run was enduring the birth-pangs of a new lease? His conduct could hardly have been

[1] Like conditions tend to breed, or at any rate to revive, like stories. The Timaru breakwater, now a success beyond all controversy, was, whilst still in its probationary stage, shaken severely by an exceptional gale. Amongst other yarns circulated by the faint-hearted as to the extent of damage done, was one to the effect that a Newfoundland dog had been washed through a crevice in the ruptured monoliths. Some fifteen years later the Napier breakwater was also damaged by a great gale. Will it be credited that the self-same dog-story was revived? It was rumoured abroad that a Newfoundland dog—*perhaps the same dog?*—had been forced clean through one of the breaches of the mole!!

construed otherwise than as a slight on Tutira,—a tacit denial of its full incorporation of him, body and soul. That sharp discipline was required to recall him to his former docility, that a wholesome reminder was necessary to let him understand that his reinstatement was no common transaction of everyday life, that the transference of such a run as Tutira for such a length of years was not to be consummated without the initiatory rites of fasting, scourge, and penance,—provided, not gratis, by the legal fraternity,—were all, I believe, ideas subconsciously at work in the native mind.

There is, I believe, in medicine a term—metastasis—signifying the transference of evil to which flesh is heir from one part of the human frame to another, suppression in one place leading to reappearance in another. There are racial analogues. Amongst the Maori tribes, the *pax Britannica* has stayed feuds which heretofore had found vent on the battlefield. Desire to war is still strong in the blood, but since it is no longer possible to slay each other decently in the open with *mere* and spear, the chiefs and elders of the tribes are constrained to bludgeon one another in the native Land Courts.[1]

The approaching litigation about the new lease of Tutira bore in its essentials no little resemblance to the famous cause between Jock o' Dawston Cleugh and Dandie Dinmont. The natives were apparently resolved that the land should have its rights, one of the most important of which was an appearance in court, vicariously represented by its owners. They were spoiling for a fight with some person or some Board. I am bound to confess, however, that they spared me to the last—that they took me on only when more legitimate game had failed.

The first attempt was made on the Conservation of Forests Board, who had, shortly after my departure, purchased from the native owners —all of whom were Tutira men—a scenic reserve around the picturesque Falls of Tangoio. This transaction, they now claimed, had been carried out without adequate explanation, the sum paid per acre had been insufficient, the lease had not been signed by every native entitled to sign,— any stick would do to beat the dog. They wished, in fact, to upset the deal; not a bit because they did not wish the amenities of the fall pre-

[1] R. L. Stevenson declares that the natives in certain Melanesian groups have died off chiefly from want of amusement—that the missionaries have killed them with decorum. If the true function of native Land Courts is to preserve the Maori race, to keep it alert, bright and happy, the writer can no longer regret his own involuntary contribution towards so good a cause.

served, quite the contrary; they would have been the first to remonstrate at its desecration. They wanted mental stimulus—amusement. A deputation accordingly arrived to interview my brother. They requested an advance of rent, funds to "fight" the Board—that was the crude barbaric term used by their spokesman. My brother, acting as I should have done myself, dismissed them with Pleydell's advice, "Go home, go home, take a pint and agree." They were not to be denied, however; a few months later a larger and more important deputation requested a still more considerable advance in rents. This time the object was more ambitious. It was neither more nor less than to upset the title of an extensive block of land in the South Island, a block in which several of my Tutira natives had, or imagined they had, an interest. This request seemed more preposterous than the other. Once more good advice, or what seemed to be such, was proffered; once more my brother advised them not to waste their money, as if indeed it was possible to waste money in litigation about land when the kudos of the conflict must revert to the land, whoever won! Twice did we interfere thus with their legitimate pastime; debarred from lesser game, they now determined to take on the writer himself.

Other influences, too, were at work utilising for their own purposes the litigious proclivities of the natives. The Royal Commission had, in dealing with another Hawke's Bay property, given mortal offence; to upset the Tutira lease would have been to discredit, perhaps to nullify, its labours.

Thus egged on, the safety-valves of other possible channels of litigation closed, balked of their desire to "fight" other Boards and other persons, the eyes of my landlords turned Tutira-wards. It was then, I imagine, that the idea of disputing the Tutira lease took form. The plan had many advantages over the Tangoio and South Island enterprises; it would raise the *mana* of Tutira to a height hitherto unattained. I feel sure that on the Maoris' part there was no malice. One of the Commission, indeed, had told me of the special wish of my landlords that I should remain on Tutira. They thought, not perhaps that I would actually enjoy a lawsuit, but that, as with themselves, it would dissipate ennui and boredom and provide me with a subject for thought and varied speculation. In the latter respect they were correct too; it did.

The old lease by which one-half of Tutira was still held was in the meantime steadily running out. With only seven years remaining, it was impossible to improve; without improvements the flock began to diminish in weight of wool and in general productivity. More than that, chance of success in the law courts militated for the time being against the run. It postponed work which would have been done in the ordinary course of events; fern and scrub, that otherwise would have been fired, were allowed to remain unburnt; the possibility of dealing properly with the large area of waste lands in central Tutira had to be taken into account. This postponement, although of future advantage in case of success, was in the meantime detrimental to the health of the flock; it was the sacrifice of a certain present to a problematic future. There occurred, on a greater scale than ever before, what has been fully explained in previous chapters. A huge contraction of the area over which sheep fed took place. Fern and scrub closed in everywhere, relegating stock on to the highly-manured tops, the sunny northern and western hill-slopes, the fertile alluvial flats.

Nothing, however good or evil, endures for ever. The Lower Court pronounced that, "unless notice of appeal be given within one month from this date, and proper security be found for the costs of the appeal," the Tutira leases were to be executed by the Board. That was the main matter.

I should have let it go at that. Not so the other parties. Why stop now and lose half the fun? Notice of appeal was launched, when again it was ordered that the leases must be executed.

As to the future, the Court very properly refused to stay execution, and also very properly refused to grant any interim prohibition; leave, however, on the other hand—quite unnecessarily, in my opinion,—was granted to appeal to the Privy Council; in short, the Court was disgustingly fair. Towards me, who deserved it, it was right that this should have been so, for, without a proper title, how in the world could I proceed with my improvements? Towards my opponents it was an open-mindedness absolutely chucked away; for there is no reason now to make a secret of the fact that, whilst the writer himself and his legal advisers were men of the highest moral standing, those acting against him showed a callous depravity sad to find in human nature. One great source of uneasiness to me at the time was, I remember, lest these wretches should contaminate my men. I could not but share in the amazement and horror of another litigant in a more celebrated case[1] when my adversaries' adviser from time to time had the audacity to wish good-morning to my people.

I was not destined, however, to perish in the supple-jack entanglements of the law. Unknown to me, out of Court, a life-long friend, who, by all the laws and rules of human nature should not have interfered, did so on my behalf. By his interposition and influence further interference with my affairs was barred. The appeal to the Privy Council never came off; the Ikaroa District Land Board signed the lease.

With the settlement of the trouble both the writer and the natives were satisfied. The former was pleased at any cost to have the matter definitely done with, to be free once again to improve; the latter had enjoyed a really excellent run for their money. Deputations had been again and again down to Wellington in connection with the matter. Members of the Lower House had been snowed under with pamphlets and counter-pamphlets. Native members had been consulted; sheets of oratory had been distributed broadcast. The *mana* of Tutira had been raised high above all other lands in the district; all knew of its name and fame; that beloved bit of land, its beautiful lake and ancient legends, had enjoyed its rights. So stimulating, indeed, and diverting had been the litigation to my native friends, that in all goodwill and gravity several of them were astonished at my demurring to pay legal costs for both sides. Why, I had won the case! Was it not natural, therefore, that I should pay? With childish frank-heartedness I was congratulated by my opponents on their defeat. I suppose what is inconceivable to one race of men may seem to another the baldest matter of fact. Almost without exception the native's house is built and his crops grown on communal land. Because there is ample room for each and all, trouble regarding these plots is practically unknown. It is impossible for the easy-going, simple-living Maori to comprehend the white man's need of finality, his mania for "improvements." Why could not the *pākeha* be content to live without a lease? They themselves possessed no leases. Why this extreme anxiety for the destruction of fern and manuka? It was excellent cover for pig, an animal favoured by the natives far above mutton. Perhaps the

[1] Bardell *v.* Pickwick.

native ideal, *taihod*—by-and-by, plenty of time, wait and see,—is right after all, and our British strenuousness a vain and mistaken waste of energy. Perhaps, after all, the old fellows who used in the 'eighties to threaten Cuningham and myself with prosecution for grass-seed sowing and drainage of marsh-land were the wise men and we the fools. At any rate the case was settled—a case only mentioned at all to show the financial scrape into which it nearly precipitated the station, and to emphasise the resilience of affairs in a country where everybody believes that in the past all has been done for the best, and that everything in the future is certain to succeed.

I had now to count the cost: one-quarter of the flock had been snipped off by the loss of Maungaharuru and Tutira lands reverting to the natives; my rents had been quadrupled. To make ends meet in station management was impossible with the low prices ruling for wool and meat. It was necessary, therefore, that the five thousand sheep gone should be at once replaced, that feed where there had been no feed should be created. It was my task as superpatriot to make five thousand blades of grass grow where none had grown before.

So huge a proportionate increase on soil such as that of the trough of the run would have been impossible, except that of late years the run had shrunk in feeding area, especially in its westermost portion. This had occurred partly because the lease of the Opouahi Educational Reserve, now also happily renewed, was about to expire, partly on account of wet years, partly because during the few dry spells available we had purposely abstained from burning the country. Further subdivision, the felling of manuka, and ploughing, were each expected to do their respective parts towards replenishment of the flock.

With wool phenomenally low, with bankers quite capable at any moment of quoting the man who had offered his clip at 5d. for the next six years, and "been refused, mind you," borrowing is never a cheerful job. I am a bad borrower, too; I know I do it with a countenance dismal enough to damn a Rothschild. Somehow or another, nevertheless, many thousand pounds had to be found for grass-seed sowing, ploughing, draining, bush-felling, scrub-cutting, and fencing. This book will have been read in vain if it has not been made clear that a return from "improvements" is not immediate. Threading miles upon miles of wire across deserts of bracken is but a means to an end; so also are the processes of draining, scrub-cutting, and stumping. In themselves they do not produce wool or fat stock. We did, however, rely on an immediate increase of stock through the firing of certain paddocks, especially of the Rocky Staircase, whose progress I have elsewhere described as typical of the whole trough of the run. We reckoned without our host. We found ourselves blocked by weather conditions. That summer was a remarkable one, not only in drizzle and windlessness, but in absence of sun. During each of the months of November and December there were ten days, and during the month of January eleven days, upon which the sun never shone—that is, during a third of the three hottest months of the year the sun registered no mark whatsoever on the sensitised papers of my sun-recorder. There was no weight of rainfall, but day after day the countryside was wrapped in a warm white dazzling drizzle. The growth of fern and scrub was prodigious; on permanently grassed lands the rush of feed was equally great. Where in ordinary seasons there grew a short sweet bite, now it lengthened into a fozy hay, which fell and lay in swathes on the saturated ground, and through which new green stuff forced itself. Growth of this sort sheep will not eat; they prefer to remain short of feed. During this abominable season, on the good country of Tutira as on the bad, the flock was pressed into the smallest compass, jammed into an area hardly quarter the size of the country over which they should have been feeding; they were concentrated on the foulest pasturage of the run—the tops, the camps, and the rich low-lying flats around the lake. There was a big shortage of lambs at weaning-time—always a bad beginning. In March a "buster" blew up from the south, a foot and a half of rain falling in three days.[1] The lambs, already in wretched order and full of disease, died as I had never before seen them die at Tutira. Even amongst the ewes, where 3 per cent is the normal rate of mortality, there was a considerable loss. The death-rate over the whole flock wintered was a fraction above 25 per cent. Nor is a set-back of this

[1] Reiteration of exceptional events may easily become misleading, the more so as it is the unusual that sticks most firmly in the mind. Thus references throughout this volume to heavy storms may give quite a wrong impression of normal Hawke's Bay weather. Certainly deluges do occur, certainly also there is a considerable annual precipitation, but because of its very vehemence whilst falling, the hours of actual rainfall are few; a splash of a couple of inches falls in an eighth of the time it would require elsewhere. It is an emotional climate—brief bursts of passionate tears, long spans of smiles and happy laughter, sad for an hour, serene for weeks.

magnitude adequately represented even by the actual loss of stock. Under such conditions the percentage of lambs is low; those born, moreover, are small in frame and meagre in condition. The clip is light in weight and poor in quality. The surviving young stock of such a season never become first-rate animals, the least well-woolled seeming to survive. Such years, in fact, leave a corporate mark in a flock in the same way as a severe illness can be noted in the nails and teeth. The survivors of a bad winter can be picked out as weedy, ill-grown beasts, not only as two-tooths, but until by process of time they pass out of the flock. Meteorological conditions affect a flock as vintages are affected; as connoisseurs in wine talk of Comet port and vintages of such and such dates, shepherds can tell from the general appearance of stock the conditions obtaining prior to its birth.

It is, however, an ill wind that blows no good. The very weather that had poisoned the flock with rank grass had suited perfectly the first crop of turnips, swedes, and red clover attempted on pumiceous ground. I had always hoped that something might be made to grow on the trough of the run, which had, at the revaluation of Tutira prior to the coming of the Royal Commission, been valued at 5s. an acre freehold. Instantly, therefore, upon signature of the lease, operations had been started. A patch of a hundred acres was fenced, the stunted bracken burnt off, the manuka fallen, the larger stumps grubbed, piled in heaps and burnt, the ground ploughed, disced, and tyne-harrowed. Turnip and swede seed was then drilled in with superphosphate, at the rate of one and a half cwt. per acre; attempts were made to consolidate the light porous land by means of Cambridge rollers; immediately prior to the rolling, and after the drilling in of the turnips, red-clover seed was scattered broadcast over the whole. The result of this heterodox farming was eminently satisfactory; there was a splendid take of both clover and turnip. The crop, certainly, was not a heavy one, yet it was a marvel to such as passed along the road—a vindication of my sanity to the travelling public who had scoffed to my ploughmen at the idea of any good thing coming out of a sahara of pumice grit. From the hill-tops miles away its bright verdure showed up an oasis of green in the desert of fern and manuka.

It was the one bright spot in this unfortunate year. With the station books showing a debit balance on the profit and loss account of several thousand pounds, prudence would have seemed to dictate a cessation of ploughing, grass-seed sowing, fencing, and stumping; inexorable necessity demanded, however, their full continuance,—a certain income was necessary to run the place, a certain-sized flock was necessary to provide that income. Improvements, therefore, in half a dozen different lines continued to be lavished on the run. Still, however, the station books showed a loss; it was a lesser loss—about half that of the previous year.[1]

Tutira was passing through a phase similar to that which had proved fatal to the pioneers of the 'seventies and 'eighties—the transition phase, when capital has been sunk and before returns have begun to pour in. Wool and stock were still low; although there was less to be heard of the man who had "offered his clip at fivepence," confidence was still far from having been restored.

It was then—the station books for a third sequent year continuing to show a loss, though a loss of but a few hundreds—that a letter arrived, one of those epistles "where more is meant than meets the ear," expressing my banker's opinion that it would be advisable to sell part of the run.

Another season passed away. The expenditure on improvements began to slacken: the most important operations had reached completion. Wool began to rise. Stock began to rise. Like Christian in his celestial journey, I was able to continue on my path regardless of the Bank and the Law, who could now only gnash their teeth at the

[1] Returns of the last few years are not available, but a sequence of nine years will show the range in number of sheep carried, average weight of wool, death-rate, and working expenses (including war taxation for last two years).

Number of sheep carried, weight of wool, and rate of mortality hinge chiefly on the rainfall.

Number of sheep wintered.	Average weight of wool per sheep.	Death-rate.	Working expenses per sheep.
	lb. oz.	Per cent.	*s.* *d.*
22,450	7 14	6	3 11
22,000	6 7	25	4 11
19,000	6 7	12½	5 5
17,800	8 7	5¼	5 6
19,100	7 1	5¼	5 4
19,024	7 14	5¼	5 8
19,771	8 15½	5·02	6 4
19,535	8 7	5·55	9 10
19,961	7 12	6·93	9 8

pilgrim whom they could no longer destroy. Under these circumstances what might have been reiterated as an imperious demand etherialised

"CHRISTIAN PASSES THE LIONS THAT GUARD THE PALACE BEAUTIFUL."

(*With acknowledgments to D. & B. Scott.*)

into a pious aspiration. There was no further request to sell any part of Tutira.

With that resilience which is so marked a feature in new countries, the position now began to improve. On the ploughed paddocks entirely open to the sun young stock throve splendidly. The miles of fencing transmuted themselves into sheep. On the principle that it never rains but it pours, the wet seasons passed away. They were succeeded by the kind of weather that best suits Tutira: fires were everywhere obtainable over a countryside tinder-dry; an expansion of feeding area occurred comparable only with the previous shrinkage. In one paddock alone the number of sheep carried rose in a single season from a few score to nearly two thousand. Native grasses spread enormously, suckling clover seed germinated in hundreds of millions on the burnt-out paddocks. The percentage of lambs was unprecedented, the lambs themselves of excellent quality; for the first time in the annals of the run fat sheep rolled off in thousands to the freezing-works.

Expenditure on improvements ceased, while the full effects of these same improvements proclaimed themselves alike in clip, condition, increase, and monetary return. The affairs of Tutira prospering progressively like those of another station whose possessor had also known bad times, for the third time the mortgage was paid off. If the writer did not, after escape from the hands of Satan, possess precisely "fourteen thousand sheep and six thousand camels, and a thousand yoke of oxen and a thousand she asses," he owned their equivalents; he had at least as much as was good for him. The Lord, in fact, had blessed the latter end of Job more than the beginning.

There is but little more to add. In August 1914, whilst the writer was at home, war broke out, the old world crashed and passed away. At first he was stranded, for possession, at the age of fifty odd, of a certain local knowledge of sheep-farming and of a certain field naturalist acquaintance with New Zealand bird-life, are difficult to fit into the plan of a European war. He was ashamed to be seen unemployed; for a couple of terms, therefore, in the scant company of British crocks and blackamoors, he hid himself as a sort of superannuated undergraduate at Cambridge. Later he had the good fortune, whilst soliciting employment as an orderly, to become acquainted with the famous and kindly physician then in command of No. 3 London General Hospital. There he took over and ran the grounds of that great hospital with his staff of artists, known locally by the bye as "the chain gang." It is not for him to say more; he believes he was of some use, and that were an

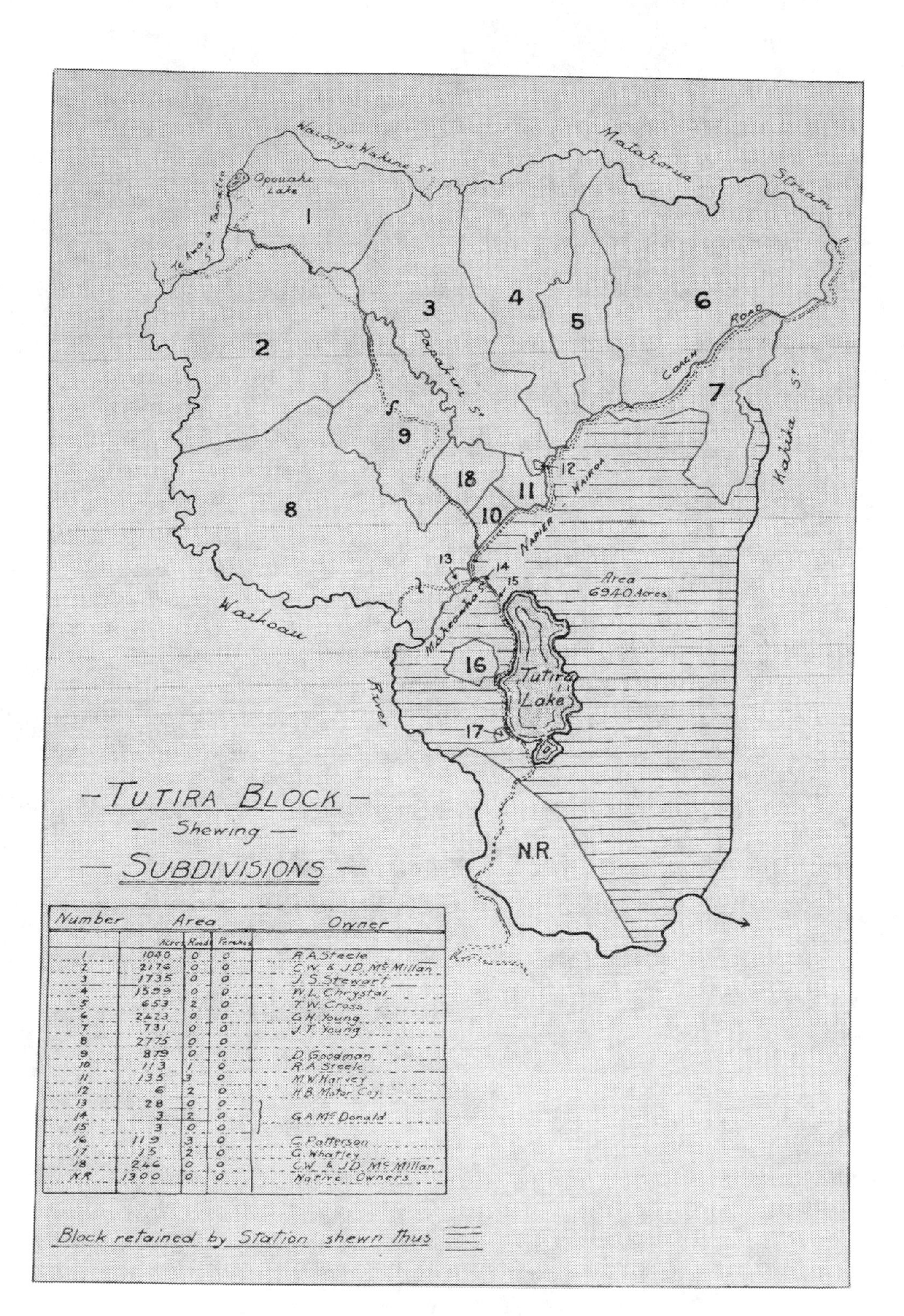

—TUTIRA BLOCK—
— Shewing —
— SUBDIVISIONS —

Number	Area			Owner
	Acres	Roods	Perches	
1	1040	0	0	R.A. Steele
2	2176	0	0	C.W. & J.D. Mc Millan
3	1735	0	0	J. S. Stewart
4	1599	0	0	W.L. Chrystal
5	653	2	0	T.W. Cross
6	2423	0	0	G.H. Young
7	731	0	0	J.T. Young
8	2775	0	0	
9	879	0	0	D. Goodman
10	113	1	0	R.A. Steele
11	135	3	0	M.W. Harvey
12	6	2	0	H.B. Motor Coy.
13	28	0	0	G.A. Mc Donald
14	3	2	0	
15	3	0	0	
16	119	3	0	C. Patterson
17	15	2	0	G. Whatley
18	246	0	0	C.W. & J.D. Mc Millan
N.R.	1300	0	0	Native Owners

Block retained by Station shewn thus

earthquake to swallow Tutira he could always get a billet from Sir Bruce Bruce Porter. Then it was he found salvation; then it was that he had leisure to meditate on his sins as a citizen, one of which was perhaps—only perhaps—occupancy of too large a tract of land. To this repentance, certainly somewhat of the leisurely eleventh hour or death-bed order, he was moved furthermore by many prosaic mundane reasons. He desired a greater portion of time for the pursuit of his own particular hobbies; lastly, he did not deem it wisdom to slave and save for the ravening wolves who determine the gradations of a New Zealand land and income tax; in short, like the lady in "Don Juan," who, swearing she would ne'er consent, consented, he decided to subdivide the larger remaining portion of the station into farms. This has been done; his interests in Maungaharuru had already lapsed, Putorino had been already sold; now, together with the Opouahi block, thirteen thousand acres of Tutira proper have passed out of his hands, and with the new era of settlement our history of the station may cease.

It but remains for the writer to advise any youthful readers who may have struggled through his book to go forth also into the wilds and to possess lands and flocks of their own, to become citizens of a country where content and moderate riches are within the reach of every man, where, without being quite aware of how the golden age has dawned, wealth has become something of a superfluity, and where, therefore, excessive toil in its pursuit is futile. Heredity and environment alike have conjoined to serve New Zealand; she has had no bad past painfully to live down. Her pioneers, gentle and simple, whether from north or south of Tweed, have come of the soundest stock; for eighty years, too, as like draws like, kinsfolk and friends of similar breeding, tastes, and aptitudes have been attracted to her shores. In the sowing of the nations other emigrants have sought homes in lands of easier attainment, the heaviest grain has been the furthest flung. New Zealand, if unlikely to produce a world poet or a world musician,—brains do not emigrate, no intellect of the highest order has yet arisen anywhere outside Europe,—can lay claim, in her founders, to courage and character; in her present population, to the saving virtue of simplicity. Her children arise and call their little country blessed in its absence of great cities, in its riches absorbed by none but shared by all, in its ideal of life measured in happiness rather than in wealth, in its climate of sunshine and rain though never of fog and gloom, in its sturdy children reared inland on green fields or along its coasts on clean sands washed by enormous breadths of sea, in its lowest death-rate in the world. The writer cannot but exclaim, when he thinks of what he could have found in his heart to say in praise of his dear adopted land throughout every page of 'Tutira,' that, like Clive before the wealth of India, he marvels at his moderation.

One last word: he hopes that his readers have played the game, that they have not indulged in the practice of skipping. If this has not been done, if every chapter has been read, they can rest assured that in examination, as it were under the microscope, of one station, they have discovered what is to be found in all. The writer cannot too strongly emphasise the fact that there is nothing exceptional in the little bit of land about which he has written. Taking chapter by chapter, every station in New Zealand has been moulded by a great rainfall, possesses legends and relics of a splendid aboriginal race; has been clothed with forest, flax, and fern; has been subdued by pioneers in desperate straits for credit and cash; has been overrun by an alien vegetation and alien beasts; has righted its equilibrium; has had its surface mapped by stock, its rivers affected by scour; has seen its original breeds of domesticated stock supplanted by others better fitted to meet changed conditions; and lastly, has been, or is in process of being, subdivided into smaller holdings. It has been the good fortune of the author to have witnessed these changes; they have been of enduring interest to himself; in bidding his readers farewell he hopes he may have been able to pass on that interest.

PRINTED BY WILLIAM BLACKWOOD AND SONS.

Printed in the United States
By Bookmasters